TOWARD ANTI-ADHESION THERAPY FOR MICROBIAL DISEASES

ADVANCES IN EXPERIMENTAL MEDICINE AND BIOLOGY

Recent Volumes in this Series

Volume 400A
EICOSANOIDS AND OTHER BIOACTIVE LIPIDS IN CANCER, INFLAMMATION, AND RADIATION INJURY, Part A
Edited by Kenneth V. Honn, Santosh Nigam, and Lawrence J. Marnett

Volume 400B
EICOSANOIDS AND OTHER BIOACTIVE LIPIDS IN CANCER, INFLAMMATION, AND RADIATION INJURY, Part B
Edited by Kenneth V. Honn, Santosh Nigam, and Lawrence J. Marnett

Volume 401
DIETARY PHYTOCHEMICALS IN CANCER PREVENTION AND TREATMENT
Edited under the auspices of the American Institute for Cancer Research

Volume 402
AIDS, DRUGS OF ABUSE, AND THE NEUROIMMUNE AXIS
Edited by Herman Friedman, Toby K. Eisenstein, John Madden, and Burt M. Sharp

Volume 403
TAURINE 2: Basic and Clinical Aspects
Edited by Ryan J. Huxtable, Junichi Azuma, Kinya Kuriyama, and Masao Nakagawa

Volume 404
SAPONINS USED IN TRADITIONAL AND MODERN MEDICINE
Edited by George R. Waller and Kazuo Yamasaki

Volume 405
SAPONINS USED IN FOOD AND AGRICULTURE
Edited by George R. Waller and Kazuo Yamasaki

Volume 406
MECHANISMS OF LYMPHOCYTE ACTIVATION AND IMMUNE REGULATION VI: Cell Cycle and Programmed Cell Death in the Immune System
Edited by Sudhir Gupta and J. John Cohen

Volume 407
EICOSANOIDS AND OTHER BIOACTIVE LIPIDS IN CANCER, INFLAMMATION, AND RADIATION INJURY, Volume 2
Edited by Kenneth V. Honn, Lawrence J. Marnett, Santosh Nigam, Robert L. Jones, and Patrick Y-K. Wong

Volume 408
TOWARD ANTI-ADHESION THERAPY FOR MICROBIAL DISEASES
Edited by Itzhak Kahane and Itzhak Ofek

A Continuation Order Plan is available for this series. A continuation order will bring delivery of each new volume immediately upon publication. Volumes are billed only upon actual shipment. For further information please contact the publisher.

TOWARD ANTI-ADHESION THERAPY FOR MICROBIAL DISEASES

Edited by

Itzhak Kahane

The Hebrew University – Hadassah Medical School
Jerusalem, Israel

and

Itzhak Ofek

Tel Aviv University – Sackler Medical School
Tel Aviv, Israel

PLENUM PRESS • NEW YORK AND LONDON

Library of Congress Cataloging-in-Publication Data

Toward anti-adhesion therapy for microbial diseases / edited by Itzhak Kahane and Itzhak Ofek.
p. cm. -- (Advances in experimental medicine and biology ; v. 408)
"Proceedings of the Bat-Sheva Seminar Toward Anti-Adhesion Therapy for Microbial Diseases, held February 25-March 1, 1996, in Zichron Yaakov, Israel"--T.p. verso.
Includes bibliographical references and index.
ISBN 0-306-45397-5
1. Bacteria--Adhesion--Congresses. 2. Anti-infective agents--Congresses. I. Kahane, Itzhak. II. Ofek, Itzhak. III. Bat-Sheva Seminar Toward Anti-Adhesion Therapy for Microbial Diseases (1996 : Zikhron Ya'akov, Israel) IV. Series.
[DNLM: 1. Bacterial Adhesion--congresses. 2. Adhesions--prevention & control--congresses. W1AD559 v.408 1996]
QR96.8.T69 1996
616'.014--dc20
DNLM/DLC
for Library of Congress 96-31493
CIP

Proceedings of the Bat-Sheva Seminar Toward Anti-Adhesion Therapy of Microbial Diseases, held February 25 – March 1, 1996, in Zichron Yaakov, Israel

ISBN 0-306-45397-5

A Division of Plenum Publishing Corporation
233 Spring Street, New York, N. Y. 10013

10 9 8 7 6 5 4 3 2 1

Printed in the United States of America

PREFACE

The emergence of pathogens resistant to conventional antimicrobial agents has forced us to intensify the efforts in search for new approaches to prevent infectious diseases. Such a direction was indicated in studies over the last two decades showing that adhesion of pathogens, primarily via glycoconjugate or protein receptors of the host tissue, is crucial for the infectious process. Moreover, it was found that infection can be prevented by blocking adhesion of the pathogen to mucosal surfaces of the host. The various aspects of interference with the process of microbial adhesion as a way of preventing diseases were the subject of the Bat-Sheva Seminar, "Towards Anti-Adhesion Therapy of Microbial Infectious Diseases," held in Zichron Yaakov, Israel, February 25 to March 1, 1996. A major aim of the Bat-Sheva de Rothschild Foundation for the Advancement of Science in Israel, which sponsors a series of seminars, ours among them, is to provide the necessary tools and settings for international forums and exposure of young scientists and promising students to the state of the art of the field. This goal has been achieved during the week's discussions, and its major aspects are presented in this compendium. The seminar's participants, as well as the readers of this book, thank the founder and Foundation for their support.

This book includes the major themes of this rapidly growing area. However, by no means do we intend to cover every bit and piece in it.

The book's first section deals with the lectin–sugar interactions and their inhibitors. This is followed by a section of studies on invasion and inflammation as consequences of adhesion. Because of its special importance, the adhesion process and its linkage to components of the extracellular matrix, including fibronectin, and also consequential effects of the adhesion process and the involvement on the eukaryotic cytoskeleton, is the focus of the next section. The fourth section deals with adhesion of specific microorganisms, including viruses, prokaryotes, and eukaryotes. A special section is devoted to studies of various inhibitors of adhesion with respect to their source and mode of action, such as synthetic, host-derived, or dietary inhibitors. The last section is devoted to aspects of oral microbiology, which also shed light on hydrophobic interactions, their importance in microbial adhesion, and the development of agents that interfere with the processes to reduce dental plaques.

In order to expand the scope of the volume even further, we have also included the abstracts of the posters presented during the seminar.

We thank the authors for their collaborative efforts, as well as those of the Plenum staff for publishing this compendium so relatively close to the end of the seminar. We also trust that highlights of this book will stimulate new ideas to develop practical designs for effective anti-adhesin agents for the treatment of infectious diseases. We would like to express our gratitude to all our colleagues and friends, especially to the rest of the members of the Organizing Committee (S. Ashkenazi, N. Gilboa-Garber, D. Mirelman, and N.

Sharon), who suggested, argued, and altogether helped a great deal and in many ways allowed the seminar to bloom.

We believe that a follow-up seminar should be held to present and discuss the results of new ideas on anti-adhesion therapy of infectious diseases that were illuminated here.

Itzhak Ofek and Itzhak Kahane, Cochairmen

CONTENTS

TOWARD ANTI-ADHESION THERAPY FOR MICROBIAL DISEASES

CARBOHYDRATE–LECTIN INTERACTIONS IN INFECTIOUS DISEASE

Nathan Sharon

Department of Membrane Research and Biophysics
The Weizmann Institute of Science
Rehovot 76100, Israel

1. INTRODUCTION

For a long time, carbohydrates were believed to serve solely as a source of energy and as structural elements, and to lack biological activity. They were therefore considered as dull molecules lacking the glamor of the other major cell constituents, the proteins and nucleic acids. This attitude started to change, albeit slowly, around 1970. An early indication for the change can be found in the book written by me some 20 years ago, where I stated that "*carbohydrates are ideally suited for the formation of specificity determinants that may be recognized by complementary structures, presumably proteins, on other cells or macromolecules*" (Sharon, 1975). This modern view of carbohydrates has now become widely accepted. Thus, the recently published 4th edition of the popular biochemistry textbook by Stryer, (1995) states that "*carbohydrate units on cell surfaces play key roles in cell-cell recognition processes*" and that "*carbohydrate-binding proteins called lectins mediate many biological recognition processes.*"

2. CARBOHYDRATES AND LECTINS IN BIOLOGICAL RECOGNITION

Theoretical considerations show that carbohydrates can serve as carriers of biological information even more effectively than polypeptides or polynucleotides. This is because, in contrast to amino acids or nucleotides that combine in one way only and form just linear oligomers or polymers, the monosaccharide units of oligosaccharides and polysaccharides can combine in several different ways, and form also branched structures (Table 1). Monosaccharides can thus be considered as letters in a vocabulary of biological specificity, in which the words are spelled out by variations in the constituent monosaccharides, differences in the linkages between them, and the presence or absence of branching. The messages encoded in the structures of complex carbohydrates are deciphered through interactions with complementary sites on carbohydrate-binding proteins, chiefly lectins

Toward Anti-Adhesion Therapy for Microbial Diseases, edited by Kahane and Ofek
Plenum Press, New York, 1996

Table 1. Complexity of oligosaccharides and polysaccharides

Sequence of the monosaccharides
GalNAc-Gal-Glc
Gal-GalNAc-Glc
Position of glycosidic linkage
Gal3GlcNAc
Gal4GlcNAc
Anomery of glycosidic linkage
Galα3Glc
Galβ3Glc
Branching
GlcNAcα 6 \ Galβ / GlcNAcβ 3
Ring size of monosaccharide
Furanose (5-membered ring)
Pyranose (6-membered ring)

(Sharon and Lis, 1993a, 1995). Lectins occur widely in nature, and are admirably suited to function in cell recognition because they are found on cell surfaces, exhibit fine specificity for mono- and oligosaccharides, and combine with them rapidly and reversibly.

The participation of carbohydrate-lectin interactions has by now been clearly demonstrated in a large number of biological processes. These include intracellular sorting of proteins, clearance of glycoproteins from the circulatory system as well as a variety of cell-cell interactions, ranging from the adhesion of infectious agents to host cells (Ofek and Doyle, 1994; Karlsson et al.,1994; Karlsson, 1995), to the control of leukocyte traffic and their recruitment to inflammatory sites(Table 2), (Etzioni et al., 1995; Nelson et al., 1995).

Table 2. Carbohydrates and lectins in cell-cell recognition

Process	Sugars on	Lectins on
Infection	Host cells	Microorganisms
Defense	Phagocytes	Microorganisms
	Microorganisms	Phagocytes
Fertilization	Eggs	(Sperm)[1]
Leukocyte traffic	Leukocytes	Endothelial cells
	Endothelial cells	Lymphocytes
Metastasis	Target organs	Malignant cells
	Malignant cells	Target organs

[1] Presumed, no experimental evidence available

3. MICROBIAL ADHESION TO CELL SURFACE SUGARS: SPECIFICITY AND ROLE IN INFECTION

The oligosaccharide repertoire on the host-cell surface, whether in the form of glycoproteins or glycolipids, is among the key genetic susceptibility factors in viral and microbial infection and in toxin action. Since the 1950's it has been known that to initiate infection, influenza virus must bind to *N*-acetylneuraminic acid (NeuAc) on the surface of the host cells. It took nearly a quarter of a century before it was demonstrated that attachment of bacteria via their surface lectins to cell surface sugars is a prerequisite for the initiation of bacterial infection (Ofek et al., 1978). Subsequently it was found that a large number of viral, mycoplasmal, bacterial and protozoan pathogens bind *in vitro* to carbohydrate structures of glycoproteins or glycolipids present on epithelial and other cells (Table 3) and that this binding can be readily inhibited by suitable mono- or oligosaccharides (Ofek and Doyle 1994; Karlsson et al., 1994; Karlsson, 1995). The bacterial and viral surface lectins are specific not only for terminal non-reducing sugars, but may recognize internal structures as well (Table 4).

Escherichia coli K99 provides an interesting illustration of the fine carbohydrate specificity of bacterial surface lectins, as well as of the animal tropism of bacteria. This organism binds to glycolipids containing *N*-glycoloylneuraminic acid (NeuGc), but not to

Table 3. Carbohydrates as attachment sites for infectious agents[a]

Organism	Target tissue	Carbohydrate structure	Form[b]
Viruses			
Influenza type A	Respiratory tract	NeuAc(α2-6)Gal	GP
B	Respiratory tract	NeuAc(α2-3)Gal	GP
C	Respiratory tract	9-O-AcNeuAc(α2-3)Gal	GP
Parvovirus B19	Erythroid cells	GalNAcβ3Galα4Galβ4	GSL
Polyoma virus	Epithelial cells	NeuAc(α2-3)Gal	GP
Bacteria			
E. coli type 1	Urinary tract	Manα3[Manα3(Manα6)	GP
P	Urinary tract	Galα4Gal	GSL
S	Neural	NeuAc(α2-3)Galβ3GalNAc	GSL
CFA/1	Intestine	NeuAc(α2-8)-	GP
K1	Endothelial cells	GlcNAcβ4GlcNAc	GP
K99	Intestine	NeuGc(α2-3)Galβ4Glc	GSL
Actinomyces naeslundi	Oral	GalNAcβ3Galβ	
Neisseria gonorrhoea	Genital	Galβ4Glcβ	GSL
		NeuAc(α2-3)Galβ4GlcNAc	GP
Streptococcus pneumoniae	Respiratory tract	GlcNAcβ3Gal	GP
Mycoplasma pneumoniae	Respiratory tract	NeuAc(α2-3)-	GP
Fungi			
Candida albicans	Skin and mucosa	Galβ4Glc	GSL
		Fucα2Gal	
Protozoa			
Entamoeba histolytica	Intestine	Galβ4GlcNAc	GP
Giardia lamblia	Intestine	Man-6-P	GP

[a] Based on Lis and Sharon, 1993b, and Sharon and Lis, 1996.
[b] Predominant form: GP, glycoproteins; GSL, glycolipids

Table 4. Binding of microorganisms to internal receptor sequences[a]

Bacteria	
Uropathogenic *E.coli*	
	GalNAcβ3Galα4Galβ4GlcβCer
	GalNAcα3GalNAcβ3Galα4Galβ4GlcβCer
	GalNAcα3(Fucα2)Galβ3GalNAcβ3Galα4Galβ4GlcβCer
Propionibacterium and others	
	GlcNAcβ3Galβ4GlcβCer
	Galβ3GlcNAcβGlcNAcβ3Galβ4GlcβCer
	GalNAcβGlcNAcβ3Galβ4GlcβCer
	Galβ3GalNAcβGlcNAcβ3Galβ4GlcβCer
Actinomyces naeslundi	
	Galβ3GalNAcβ4Galβ4GlcβCer
	Galβ3(NeuAcα6)GalNAcβ4Galβ4GlcβCer
Viruses	
Sendai virus and others	
	Galα4GalβCer
	Galβ4GlcβCer
	Galα3(Fucα2)Galβ4GlcβCer

[a] Modified from Karlsson et al., 1994. The recognized sequences are underlined.

those that contain *N*-acetylneuraminic acid. These two sugars differ in only a single hydroxyl group, present in the acyl substituent on the 4-NH group of the former compound and absent in the latter. *N*-Glycoloylneuraminic acid is found on intestinal cells of newborn piglets, but is disappears when the animals develop and grow. It is also not formed normally by humans. This explains why *E.coli* K99 can cause diarrhea (often lethal) in piglets, but not in adult pigs nor in humans.

In general the affinity of sugars to lectins is low, in the millimolar range. An increase of several orders of magnitude in the inhibitory potency of monovalent carbohydrates can be achieved by suitable chemical derivatization, as shown by us several years ago in studies of the mannose specific, type 1 fimbriated, *E. coli* (Firon et al., 1987) (Table 5). It can also be obtained by their attachment to polymeric carriers, to form multivalent ligands, as demonstrated with *Entamoeba histolytica* (Adler et al., 1995) (Table 6).

Recent analysis of the development of urinary tract infection in monkeys challenged with lectin positive versus lectin negative *E.coli* P strains (Galα4Gal-specific) has shown conclusively that the presence of a single lectin is necessary and sufficient to direct the

Table 5. Inhibition by aromatic α-mannosides of the adhesion of type1 fimbriated *E.coli* to yeasts and intestinal epithelial cells[a]

α-Mannoside	Relative. inhibitory activity (MeαMan=1.0) of:	
	Yeast agglutination (strain O25)	Adhesion to epithelial cells (strain O128)
p-Nitrophenyl	70	70
p-Ethylphenyl	77	140
p-Ethoxyphenyl	154	240
p-Nitro-*o*-chlorophenyl	717	470
Methylumbelliferyl	600	1075

[a]Data from Firon et al., 1987. Whenever tested, the corresponding α-glycosides were not inhibitory.

Table 6. Inhibition of *Entamoeba histolytica* induced hemagglutination by saccharides[a]

Saccharide	Min. inhibitory	R_{GalNAc}
	conc., μM	
GalNAc	710	1.0
Gal	2,500	0.3
LacNAc	480	1.5
NeuAc	(250,000)	
Fuc	(250,000)	
Man	(250,000)	
$Gal_{40}BSA$	0.15	4,700
$GalNAc_{39}BSA$	0.005	140,000

Data from Adler et al., 1995

pathogen to the kidney and to induce disease (Roberts et al.,1994). Similarly, it has been established that type 1 fimbriae increase the virulence of *E.coli* for the urinary tract by promoting bacterial persistence and by enhancing the inflammatory response to infection (Connell et al., these proceedings).

4. ANTI-ADHESION THERAPY OF MICROBIAL DISEASES: PROSPECTS AND PROBLEMS

Since lectin-mediated adhesion is crucial for the initiation of microbial infections, it should be possible to prevent such infections by blocking the lectins (Figure 1). This has

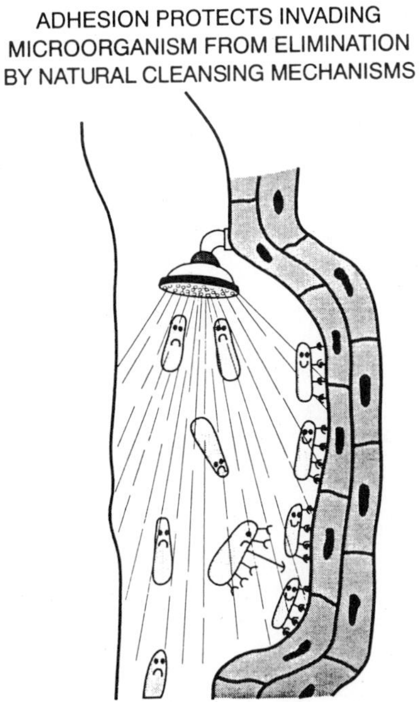

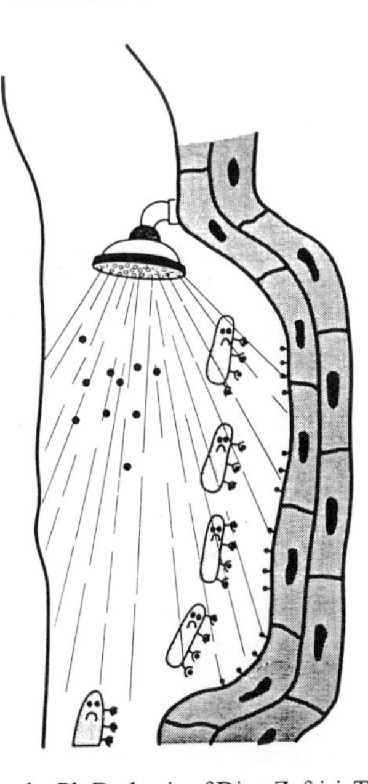

Figure 1. Microbial adhesion and antiadhesion therapy (adapted from the Ph.D, thesis of Dina Zafriri, Tel Aviv University, 1988).

Table 7. Inhibitors of sugar-specific adhesion prevent infection *in vivo*[a]

Organism	Animal, site	Inhibitor
Escherichia coli type 1	Mice, UT	MeαMan
	Mice, GIT	Mannose
	Mice, UT	Anti-Man antibody
Klebsiella pneumoniae type 1	Rats, UT	MeαMan
Shigella flexnerii type 1	Guinea pigs, eye	Mannose
Escherichia coli type P	Mice	Globotetraose
	Monkeys	Galα4GalβOMe
Escherichia coli K99	Calves. GIT	Glycopeptides of serum glycoproteins
Pseudomonas aeruginosa[b]	Human ear	Gal+Man+NeuAc
Helicobacter pylori[c]	Piglet stomach	Oligosaccharide NE0080
Streptococcus pneumoniae[c]	Rabbit lungs	Oligosaccharide NE1530
Streptococcus pneumoniae[b]	Mouse lungs	GlcNAc

UT, urinary tract; GIT, gastrointestinal tract

[a] Only references not included in Ofek and Sharon, 1990 are given.

[b] Beuth, these proceedings

[c] Zopf, these proceedings.

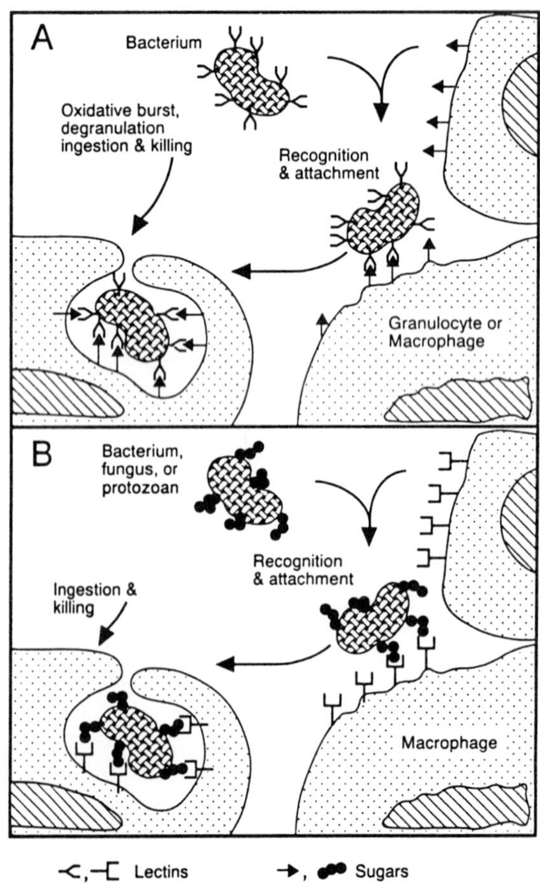

Figure 2. Two modes of lectinophagocytosis. A, mediated by microbial surface lectins that bind to carbohydrates on phagocytic cell, B, mediated by phagocyte surface lectins that recognize sugars on the microorganisms. (Modified from Ofek and Sharon, 1988).

indeed been originally demonstrated by our group in the late 1970s, when we showed that methyl α–mannoside can protect mice against urinary tract infection by type 1 fimbriated *E.coli*; methyl α -glucoside which is not recognized by the bacteria was not effective (Aronson et al., 1979). Further studies by many other groups have confirmed and extended our results, and have proven beyond any doubt the possibility to prevent bacterial infection by anti-adhesive agents (Table 7). These findings have provided an impetus for the development of carbohydrate-based anti-adhesion drugs to combat infections in humans.

The development of anti-adhesion therapy targeted at the microbial lectins has however been hampered be the great difficulty in large scale synthesis of the required inhibitors, usually oligosaccharides. Attempts are therefore being made to produce glycomimetics, compounds that structurally mimic the inhibitory carbohydrates but are more easily obtainable. Eventually, a cocktail of inhibitors, or a polyvalent one, will have to be used, since many infectious agents express multiple specificities.

Another process in which lectin-carbohydrate interactions play a role is lectinophagocytosis (Ofek and Sharon, 1988; Ofek et al., 1995) (Figure 2), well documented for the mannose specific *E.coli*. This mode of phagocytosis may result from binding of the bacteria to e.g. macrophages or neutrophils, which is followed by activation of the phagocytes and uptake and killing of the bacteria.

Viruses, bacteria, fungi and protozoa that express mannose containing polysaccharides or glycoconjugates on their surface, may also bind to the mannose specific lectin present on the surface of macrophages, with their resultant uptake and killing by the phagocytic cells. A particularly interesting example of such a microorganism is the pathogenic fungus, *Pneumocystis carinii*, a major cause of death among AIDS patients (Ezekowitz et al. 1991). Since lectinophagocytosis may provide protection against infection in non-immune hosts or in opsonin-poor sites, it should also be taken in consideration when trying to develop anti-adhesion drugs.

The design of such drugs will certainly benefit from more detailed knowledge of the specificity of microbial surface lectins, and the elucidation of the atomic structure of their combining sites, none of which is as yet known.

In conclusion, there is little doubt that in spite of these and other problems, anti-adhesion therapy of microbial diseases will soon move from the realm of dreams to reality.

REFERENCES

Adler, P., Wood, S.J., Lee, Y.C., et al., (1995) High affinity binding of the *Entamoeba histolytica* lectin to polyvalent *N*-acetylglucosaminides J. Biol. Chem. 270, 5164-71.

Aronson, M., Medalia, O., Schori, L, et al., (1979) Prevention of colonization of the urinary tract of mice with *Escherichia coli* by blocking of bacterial adherence with methyl α-D-mannopyranoside. J. Infect. Dis. 139, 329-332.

Etzioni, A., (1996) Adhesion molecules-their role in health and disease. Pediatr. Res. 39, 191-198.

Ezekowitz, R.A.B., Williams, D.J., Koziel, H., et al., (1991) Uptake of *Pneumocystis carinii* by the mannose macrophage receptor. Nature 351, 155-158.

Firon, N. Ashkenazi, S., Mirelman, D. et al. (1987) Aromatic alpha-glycosides of mannose are powerful inhibitors of the adherence of type 1 fimbriated *Escherichia coli* to yeast and intestinal epithelial cells. Infect. Immun. 55, 472-476.

Karlsson, K.A. (1995) Microbial recognition of target-cell glycoconjugates. Curr. Opin. Struct. Biol. 5, 622-635.

Karlsson, K.A., Abul Milh, M., Andersson, C., et al., (1994) Carbohydrate attachment sites for microbes on animal cells: Aspects on the possible use of analogs to treat infections. In Complex carbohydrates in drug research, Alfred Benzon Symp. 36, p.397-409, ed. Bock, K. and Clausen, H. Munksgaard, Copenhagen.

Nelson, R.M., Venot, A, Bevilaqua, M.P. et al., (1995) Carbohydrate-protein interactions in vascular biology. Annu. Rev. Dev. Cell Biol. 11, 601-631.

Ofek, I. and Doyle, R.J. (1994) Bacterial Adhesion to Cells and Tissues, Chapman and Hall, London.

Ofek, I. and Sharon, N. (1988) Lectinophagocytosis: a molecular mechanism of recognition between cell surface sugars and lectin in the phagocytosis of bacteria. Infect. Immun. 56,98-106.

Ofek, I. and Sharon, N. (1990) Adhesins as lectins: specificity and role in infection. Curr. Topics Microbiol. Immunol. 151, 91-113.

Ofek, I., Beachey, E.H.,and Sharon, N. (1978) Surface sugars of animal cells as determinants of recognition in bacterial adherence.Trends Biochem Sci. 3,159-160.

Ofek, I., Goldhar, J., Keisari, Y. and Sharon, N., (1995) Nonopsonic phagocytosis of microorganisms. Annu. Rev. Microbiol. 49, 239-276.

Roberts, J.A., Marklund, B.I., Ilver, D. et al., (1994) The Gal(α1-4)Gal- specific tip adhesin of *Escherichia coli* P-fimbriae is needed for pyelonephritis to occur in the normal urinary tract. Proc. Natl. Acad. Sci. 91, 11189-93.

Sharon, N. (1975) Complex carbohydrates, p 27. Addison Welsely, Reading, MA.

Sharon, N. and Lis, H. (1993a) Carbohydrates in cell recognition. Scientif. American 268 (1), 82-89.

Sharon, N. and Lis, H. (1993b) Protein glycosylation. Structural and functional aspects. Eur. J. Biochem. 218, 1-27

Sharon, N. and Lis, H. (1995) Lectins — proteins with a sweet tooth: function in cell recognition. Essays Biochem. 30, 59-75.

Sharon, N and Lis, H., (1996) Microbial lectins and their receptors. In Glycoproteins, ed. Montreuil, J., Vliegenthart, J.F.G. and Schachter, H. vol 2, Elsevier, in press.

Stryer, L. (1995) Carbohydrates. In Biochemistry, 4th ed, Freeman. Chapter 18, p 463-482.

2

ANTI-ADHESION AND DIAGNOSTIC STRATEGIES FOR ORO-INTESTINAL BACTERIAL PATHOGENS

Nicklas Strömberg, Stefan Ahlfors,[1] Thomas Borén, Per Bratt, Kristina Hallberg, Karl-Johan Hammarström, Charlotta Holm, Ingegerd Johansson, Magdalena Järvholm, Jan Kihlberg,[1] Tong Li, Mats Ryberg, and Golnar Zand

Department of Oral Biology
Faculty of Odontology
Umeå University
S-901 87 Umeå, Sweden
[1] Department of Organic Chemistry 2
Lund Institute of Technology
University of Lund
P.O. Box 124, S-221 00 Lund, Sweden

1. ADHESION–A RATIONAL TARGET FOR DIAGNOSTIC AND THERAPEUTIC STRATEGIES TO FIGHT INFECTIOUS DISEASES

Bacteria, viruses and eukaryotic cells bind to carbohydrate and protein receptors via cell adhesion molecules (Sharon and Lis, 1993). Adhesion is an essential step in microbial colonization and the development of infections, as well as in cell to cell communication, which affects leukocyte migration and tissue development (Paulson, 1993; Lasky, 1993). Anti-adhesion agents, such as carbohydrate and peptide receptor mimetics, thus represent potential therapeutic agents for the treatment of microbial infections, cancer and inflammatory diseases.

A large number of carbohydrate and peptide receptor sites and microbial adhesion molecules (adhesins) have been described (Karlsson, 1989; Hultgren et al., 1991). Saccharide inhibition and crystal structure analyses have shown that hydrogen bonds and hydrophobic interactions are responsible for the binding of the receptor to its adhesin (Kihlberg et al., 1989; Weiss et al., 1988). The crystal structure of the influenza virus hemagglutinin complexed with its sialic acid receptor points to the design of powerful drugs (Weiss et al., 1988). Similarly, powerful receptor mimetics have been discovered for mannose binding bacteria (Firon et al., 1987). In addition, the organization and assembly of adhesive organelles (pili) and the selectivity (tropism) of bacterial infections is being elucidated. Peptide domains of periplasmic chaperones, which assemble pili in various pathogenic bacteria, have

Toward Anti-Adhesion Therapy for Microbial Diseases, edited by Kahane and Ofek
Plenum Press, New York, 1996

been used to generate inhibitors of pili assembly (Hultgren at al., 1993). It has been shown that small differences in the architecture or conformation of cell surface receptors determine the host range of bacterial adhesins (Strömberg et al., 1990; Strömberg et al., 1991). Thus, the broad range of host susceptibility of infectious diseases may be explained by individual differences in receptor architectures.

Adhesion of bacteria to receptors on eukaryotic cells induces cellular responses, such as cytokine production from epithelial cells, bacterial uptake into cells and phagocytosis (Cossart et al., 1996). Phagocytosis and inflammatory responses are thus modulated by microbial adhesins. To perturb host defense mechanisms, bacteria also mimic the action of growth factors and cell adhesion molecules, inducing cellular events including membrane ruffling and active actin polymerization. Thus, adhesion of bacteria or bacterial toxins to cell surfaces induces intracellular signaling and protein kinase activities. Microbial adhesins are thus suitable bioligands for studying cell biology and inflammatory processes. Research in this area will lead to the development of novel therapeutics for infectious diseases and other pathological conditions.

A therapeutically important, but poorly understood, issue is the link between adhesion and growth, the two principal components of bacterial colonization and persistence. Regulatory systems, for example stimulons, modulons, and regulons, sense different conditions including oxidative stress, acid stress and starvation. In response, they alter the pattern of gene expression. It may also be that the adhesin itself acts as a two component system sensing changes in the environment and responding to it. Therapeutics regulating the expression of bacterial adhesins represent another strategy to modulate bacterial colonization and persistence.

2. HUMAN DENTAL PLAQUE–Pro-Gln AND GalNAcβ BASED DIAGNOSTICS AND THERAPEUTICS

Dental caries and periodontal disease are global diseases caused by dental plaque, the bacterial communities on teeth. In western countries, both diseases show a dramatic decline due to lifestyle directed efforts. In spite of these efforts, a minor portion of the population (about 10% in Sweden) show high disease activity, suggesting that genetically determined host factors play an important role. Among such factors, the host-microorganism interactions which underlie bacterial adhesion and colonization are anticipated to provide novel diagnostic and therapeutic strategies to fight dental diseases.

The human oral ecosystem involves more than 325 species of bacteria. These bacteria repopulate a cleaned tooth first with primary colonizers (e. g. *Actinomyces*, *Haemophilus*, *Streptococcus* and *Veillonella*) then secondary colonizers (e. g. *Actinomyces*, *Actinobacillus*, *Capnocytophaga*, *Eubacterium*, *Fusobacterium* and *Porphyromonas*). Many host regulatory systems determine this colonization pattern and combat bacteria associated with caries (*S. mutans*) and periodontitis (*P. gingivalis*). To study this complex system, we have established a simplified model in which two general receptor sites, Pro-Gln and GalNAcβ, account for a large portion of human dental plaque formation.

The Pro-Gln site serves in attaching primary colonizers, such as streptococci and *Actinomyces*, to acidic proline-rich proteins (APRPs) and statherin adsorbed on teeth (Gibbons, 1989). Primary colonizers (e. g. *S. gordonii*, *S. sanguis*, *S. mitis*, and *S. oralis*) are associated with good dental health. Thus, APRPs play a role in the attachment of a tooth protective bacterial flora. APRPs are members of a polymorphic protein family encoded by the PRH1 and PRH2 gene loci on chromosome 12p13.2 (Azen, 1993). Tooth protective roles, such as regulation of calcium phosphate precipitation and hydroxyapatite crystal growth, are also associated with APRPs (Hay and Moreno, 1989). Like many multifunctional salivary proteins which release antimicrobial peptides upon proteolytic cleavage (e.g. histatins) (Lamkin and Oppenheim, 1993), the adhesion promoting 150 residue APRPs are subject to proteolytic cleavage into inactive 106 residue and Pro-Gln-containing 44 residue fragments. In addition, APRPs serve adhesive function on mucosal surfaces (Hammarström et al., 1996).

The GalNAcβ site, present in streptococcal capsular polysaccharides, attach secondary *Actinomyces* colonizers onto the streptococcal layer. The partner specificity of these intergeneric interactions are described by the coaggregation patterns of six streptococcal (1-6) and six *Actinomyces* coaggregation groups (Kolenbrander, 1988). Other than for streptococci, intrageneric interactions are uncommon. As a prominent primary and secondary colonizer, *Actinomyces* have been considered important in transforming early plaque into more mature plaque associated with gingivitis and periodontal disease. A wide variety of gram-negative secondary colonizers, including *Fusobacterium nucleatum* with multiple bacterial partners, form the extensive multigeneric network of mature human dental plaque. Because of its many partners and numerical prominence in subgingival plaque, *Fusobacterium nucleatum* is thought to play an important role in the multigeneric communities associated with gingivitis and periodontal diseases. Lactose (a structural analogue of GalNAcβ) blocks most bacterial interactions. The latter are divided into lactose dependent and lactose independent interactions (Kolenbrander, 1988). On host molecules, GalNAcβ promotes adherence of bacteria to salivary proteins, including secretory IgA, oral epithelial cells and to polymorphonuclear leukocytes (Strömberg et al., 1992; Strömberg and Borén , 1992, Bratt et al., 1996a).

2.1. Mapping of the General GalNAcβ and Pro-Gln Receptor Sites

In studying GalNAcβ- and APRP-receptors, several observations supported the use of *Actinomyces* as model bacteria. First, among fresh isolates of *Actinomyces*, GalNAcβ and APRP binding properties play a role in colonizing teeth and mucosal surfaces, while GalNAcβ-noninhibitable binding properties play a role on tongue surfaces (Hammarström et al., 1996). Thus, *Actinomyces* commonly express the three major specificities of oral bacteria. Secondly, isolates of *Actinomyces* express a variety of APRP and GalNAcβ adhesin biotypes of different biological specificities (Strömberg et al., 1992; Strömberg and Borén , 1992, Strömberg et al., 1996a; Hallberg et al., 1996; Järvholm et al., 1996).

In probing glycosphingolipids, the lactose sensitive adhesin of *A. naeslundii* strain 12104 recognizes a GalNAcβ residue in which the β-configuration at carbon 1, the acetamido group at C2 and the axial hydroxyl at C4 are crucial for binding (Strömberg and Karlsson, 1990). Substituents at C1, C3 and C6 do not block receptor function. Because many lactose sensitive bacteria lack binding to glycolipids, saccharide inhibition was used to delineate additional GalNAcβ receptor sites. Despite their lack of binding to glycolipids and distinctly different partner specificities, saccharide inhibition was used to map the receptor site(s) for many lactose sensitive bacterial adhesins to the originally delineated GalNAcβ residue (fig. 1). The lactose sensitive *Actinomyces* type-2:1 and type-2:2 adhesins, which possess distinct streptococcal partner specificities, are most efficiently inhibited by GalNAcβ1-3GalαOethyl and GalNAcβOethyl (Strömberg et al., 1996b). Similarly, these GalNAcβ-structures are the most potent inhibitors of lactose sensitive multigeneric bacterial interactions, irrespective of partner specificity or bacterial species (Bratt et al., 1996b). Lactose noninhibitable coaggregations are not inhibited by GalNAcβ-saccharides. To explain these findings, a working model where the bacterial capsular polysaccharides exposes different GalNAcβ configurations or conformations may be proposed (fig. 1). First, strikingly similar GalNAcβ-saccharide inhibition patterns characterize lactose sensitive bacterial interactions of diverse biological specificities. Second, the rotational freedom of the saccharide-ceramide linkage

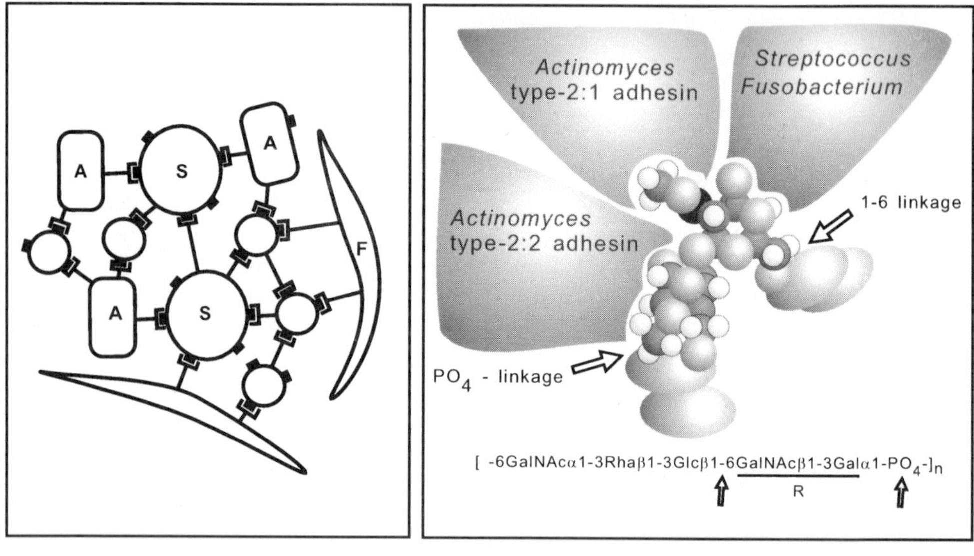

Figure 1. Model of different GalNAcβ-conformations in generating the selectivity and specificity of multigeneric bacterial interactions of human dental plaque. The complexity of GalNAcβ-dependent multigeneric bacterial interactions of dental plaque is illustrated to the left. Different receptor epitopes on GalNAcβ1-3Gal for the *Actinomyces* (A) type-2:1 and type-2:2 adhesins and for *Fusobacterium* (F) and streptococci (S) are illustrated to the right. Different GalNAcβ-conformations resulting from 1-6 and phosphate linkages of high rotational freedom are thought to expose the different epitopes on the repetitive hexasaccharide of the bacterial capsular polysaccharides. The model is in part based on the ability of GalNAcβ1-3GalαOethyl to inhibit multigeneric bacterial interactions *in vitro* and *in vivo*.

of glycolipids is known to generate different glycolipid saccharide chain orientations recognized by G-adhesins of uropathogenic *E. coli* (paragraph 4. 2). Third, the capsular polysaccharide of streptococcal partners is composed of a repetitive hexasaccharide containing a conserved domain, a GalNAcβ-disaccharide surrounded by 1-6 and phosphate linkages, and an immunogenic and antigenic domain (Abeygunawardana et al., 1990; Abeygunawardana et al., 1991). Fourth, 1-6 and phosphate linkages are recognized for their high degree of rotational freedom (Landin and Pascher, 1995; Homans et al., 1986).

The Pro-Gln receptor site was delineated by a series of peptides covering the carboxy terminal portion of APRPs (Hay et al., 1996; Gibbons et al., 1991). The terminal glutaminyl and the penultimate prolyl residues account for a major portion of the adhesion promoting activity for *A. viscosus* strain LY7 and *S. gordonii* strain NCTC 10231. In probing various Pro-Gln-containing proteins, variant APRP adhesin biotypes were observed (Järvholm et al., 1996) (fig. 2). Although their receptor sites and functional properties remain to be elucidated, they are tools to study the role of the receptor conformation in generating biological specificity. Recombinant fusion proteins and site directed mutagenesis may be used to interchange adhesin specificities.

2.2. Mapping of GalNAcβ and APRP Adhesin Binding Domains

Adhesion of *Actinomyces* to APRPs and GalNAcβ is mediated by type-1 and type-2 fimbrial adhesins, respectively. The type-1 gene of strain T14V (type-1:1 adhesin) encodes a 503 amino acid protein, and the type-2 gene of strain 12104 (type-2:2 adhesin) a 502 amino

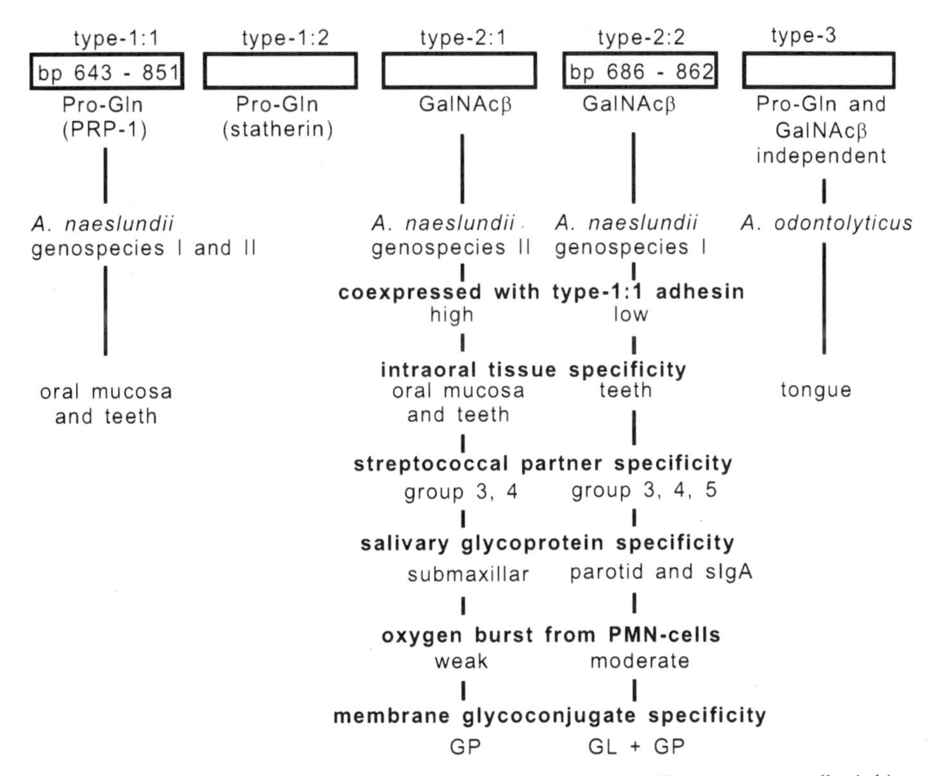

Figure 2. *Actinomyces* adhesin biotypes and their biological characteristics. The *Actinomyces* adhesin biotypes have been deduced from genetic analyses and by examining a large collection of fresh *Actinomyces* isolates for biological characteristics. The biological specificity of the adhesin biotypes is thought to reside in a distinct portion of the adhesin genes (outlined for the type-1:1 and type-2:2 adhesin genes within boxes). The type-1:1 and type-1:2 adhesins binds preferentially to PRP-1 and statherin, respectively. The type-2:1 and type-2: adhesins are thought to recognize different GalNAcβ-conformations, while the typ-3 adhesin exhibit GalNAcβ and APRP independent interactions with streptococci.

acid protein (Yeung and Cisar, 1990). The type-1 and type-2 adhesin genes, which share 34% amino acid sequence identity, were used to design DNA probes specific for the GalNAcβ and APRP adhesins.

Digoxigenin-labeled DNA probes, generated by the polymerase chain reaction, were used in Southern and slot blot hybridization analyses with genomic DNA from fresh *Actinomyces* isolates of defined GalNAcβ and APRP specificities (Hallberg et al., 1996). DNA probes generated from a non-homologous 200 bp region hybridized with fresh isolates in agreement with their APRP and GalNAcβ specificities (fig. 2). Thus, the type-1:1 probe was specific for isolates with a type-1:1 adhesin, while those with a type-1:2 adhesin or nonbinding isolates were negative. Similarily, the type-2:2 probe was specific for isolates with a type-2:2 adhesin, while those with a type-2:1 adhesin or nonbinding isolates were negative. Sequencing of the DNA region encoding the type-1:1 and type-2:2 adhesins revealed highly conserved regions. The finding that DNA probes spanning the entire type-2 gene are positive for both the type-2:1 and type-2:2 adhesins confirms their relatedness. The association of a type-2 DNA region with receptor specificity could mean that it encodes the adhesin binding domain. Furthermore, DNA probes of entire type-1 and type-2 genes also hybridized with *Actinomyces* tongue isolates showing GalNAcβ-noninhibitable interactions with streptococci (type-3 adhesin) (fig. 2). In contrast, nonbinding isolates of *A. israelii* and *A. meyeri* did not hybridize with any of the DNA probes. These findings suggest the existence of a family among *Actinomyces* of related adhesins with different biological properties.

None of the type-1 or type-2 DNA probes reacted with APRP or GalNAcβ binding strains of other bacterial species (Järvholm et al., 1996). Although probes corresponding to the entire type-1 gene were positive for the type-1:1 and type-1:2 adhesins, they were negative for *S. gordonii* strain NCTC 10231 with a receptor specificity identical to that of the type-1:1 adhesin (Hay et al., 1996; Gibbons et al., 1991). APRP and GalNAcβ binding adhesins may reflect highly species-specific structures with relatively little structural homology, possibly due to a long vertical evolution.

2.3. Pro-Gln and GaLNAcβ Adhesin Biotypes and Their Biological Characteristics

Among the many APRP and GalNAcβ binding bacterial species, multiple adhesins seem to be at work. The biological properties of different adhesins have been most extensively investigated for *Actinomyces* (fig. 2). While the type-1 (APRP) and type-2 (GalNAcβ) adhesins are expressed on mucosal and plaque isolates of *A. naeslundii* genospecies I and II, the type-3 adhesin (APRP and GalNAcβ independent) is expressed on *A. odontolyticus* tongue isolates. The type-2:1 and type-2:2 adhesin biotypes have been thoroughly investigated, and differ in the following ways (fig. 2);

a. The type-2:1 adhesin is associated with isolates of *A. naeslundii* genospecies II, while the type-2:2 adhesin is associated with *A. naeslundii* genospecies I (Hammarström et al., 1996; Hallberg et al., 1996).

b. The type-2:1 adhesin is highly coexpressed with the APRP-binding type-1:1 adhesin, while the type-2:2 adhesin is not (Hammarström et al., 1996).

c. The type-2:1 adhesin is found on isolates from mucosal and teeth surfaces of most individuals. In contrast, the type-2:2 adhesin is common only on isolates from tooth surfaces of certain individuals (Hammarström et al., 1996; Hallberg et al., 1996; Strömberg and Borén, 1992).

d. The type-2:1 adhesin interacts with the streptococcal coaggregation groups 3 and 4, while the typ-2:2 adhesin interacts also with group 5 (Strömberg et al., 1996a; Hallberg et al., 1996).

e. The type-2:1 adhesin exhibits GalNAcβ-inhibitable binding to submaxillar glycoproteins, while the type-2:2 adhesin preferentially recognizes parotid glycoproteins (Strömberg at al., 1992). The type-2:2 adhesin also differs in its recognition of O-glycosidic carbohydrates on the heavy chain of secretory IgA (Bratt et al., 1996a).

f. The adhesins differ in their interactions with polymorphonuclear leukocytes, such that the type-2:1 adhesin induces a weak, and the type-2:2 adhesin a moderate, oxygen burst upon adherence to leukocyte cell surface glycoconjugates (Karlsson et al., 1996).

g. The type-2:1 adhesin recognizes only cell surface glycoproteins, while the type-2:2 adhesin recognizes both glycolipid and glycoprotein on cell surfaces (Strömberg et al., 1996b; Karlsson et al., 1996).

Thus, the type-2:1 and type-2:2 adhesin biotypes possess distinct biological specificities and interact with different glycosylation patterns on different molecules, tissues and individuals. Because they are associated with different *A. naeslundii* genospecies and have different partner specificity and type-1:1 expression, they may well participate in different microbial communities and have a different temporal appearance in plaque formation. Because of their different membrane glycoconjugate specificities and oxygen burst responses from leukocytes, they are suitable bioligands to study cell biological events associated with glycoconjugates.

2.4. Pro-Gln and GalNAcβ Diagnostics and Therapeutics

Because of their prominence in saliva and their tooth protective roles, including modulation of the commensal bacterial flora, the polymorphic APRPs are a rational target for studies on diagnostic markers of host susceptibility to dental caries. Clearance factors, such as *S. mutans* binding agglutinin glycoproteins, represent other diagnostic markers for risk assessment in preventive strategies.

Screening of parotid salivas from different subjects showed a high frequency of salivas with low or moderate *S. mutans* binding, while only a few mediated high binding (Carlén et al., 1996). These three salivary phenotypes differ also in their APRP-mediated binding of *Actinomyces*. The *S. mutans* binding activity, which was blocked by an anti-agglutinin monoclonal antibody, correlated with the level of agglutinin. The *A. viscosus* binding activity of the three salivas correlated with the degree of cleavage of the 150 residue APRPs into 106 residue fragments, i. e. low *A. viscosus* binding coincided with high levels of 106 residue fragments. The subject with high *S. mutans* binding saliva had a high rate of dental plaque formation, high counts of *S. mutans* and low counts of other streptococci and *Actinomyces*. Thus, the relative level of receptor active 150 residue and inactive 106 residue APRPs and agglutinins may determine the level of commensal primary colonizers and cariogenic bacteria on teeth. In a recent multivariate statistical analysis of numerous lifestyle associated and biological parameters and dental caries, the level of agglutinin was correlated with dental caries activity (Nordlund et al., 1996).

Because of the high prevalence of the GalNAcβ-binding among *Actinomyces* plaque isolates and the ability of GalNAcβ1-3GalαOethyl to inhibit multigeneric interactions *in vitro* (Strömberg et al., 1996b), we investigated the ability of GalNAcβ1-3GalαOethyl to prevent plaque formation *in vivo* (Strömberg et al., 1996b). The effect of GalNAcβ1-3GalαOethyl on *in vivo* plaque formation was assessed in five human subjects who received professional tooth cleaning and then abstained from oral hygien procedures for 48 hours (fig. 3). Although the disaccharide was applied only on one tooth quadrant, an overall reduction in plaque formation was observed for all individuals compared to a placebo regimen. In individual 5, who had the most dramatic plaque reduction, the disaccharide dose (80 μl; 2 mg/ml) was applied 32 times in 48 hours rather than the usual sixteen times. The GalNAcβ1-3GalαOethyl treatment tended to reduce the relative proportion of *Actinomyces* (*A. naeslundii*), while increasing that of *Streptococci* (mainly *S. sanguis*, *S. oralis* and *S. mitis*). Thus, GalNAcβ-saccharides may both arrest and modulate human dental plaque formation *in vivo*.

3. THE GASTRIC PATHOGEN *HELICOBACTER PYLORI* – THE FUCOSYLATED Le^b RECEPTOR

The gastric pathogen *H. pylori* is the agent responsible for the development of chronic active gastritis, gastric and peptic ulcer disease, and gastric adenocarcinoma (Marshall, 1983;

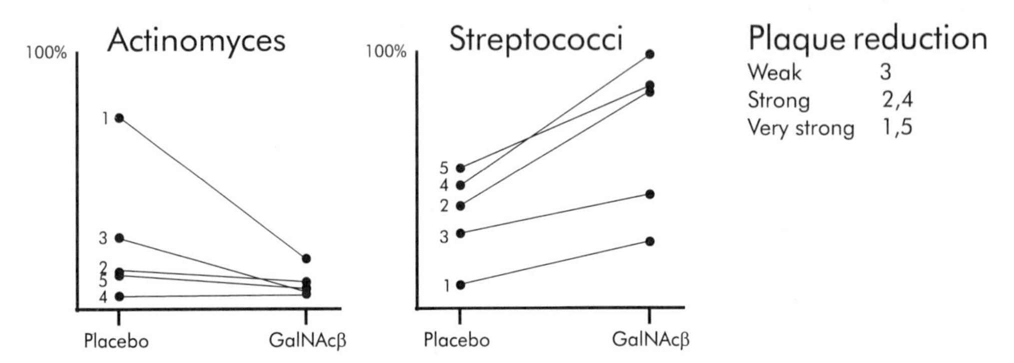

Figure 3. Effect of GalNAcβ1-3GalαOethyl on plaque formation *in vivo*. The disaccharide administrated 16 times (2 mg/ml) over 48 hours reduces the total plaque amount and results in a plaque containing fewer *Actinomyces* and more streptococci.

Forman et al., 1993). *H. pylori* is preferentially found in the less acidic, lower part of the stomach. *H. pylori* is actively motile and can move through the highly glycosylated and protective mucus layer. Here, the bacteria may attach to sialic acid and sulphate receptors in mucus. *H. pylori* also express a Le^b binding adhesin which attaches it to epithelial cells, allowing it to occupy a new habitat. On the epithelial cells, *H. pylori* release inflammation triggering substances including the vacuolating cytotoxin. *H. pylori* is probably transmitted by an oral-fecal route. The bacterium was recently detected in saliva and dental plaque, suggesting possible transmission by oral secretions (Li et al., 1995). Receptor mimetics and vaccine strategies making use of conserved adhesins or the vacuolating cytotoxin are possible therapeutic strategies.

3.1. Identification of the Le^b-Receptor and the Blood Group Antigen Binding (BAB)-Adhesin

In order to localize the cell surface receptors mediating adhesion of *H. pylori*, we overlaid human gastric mucosa with fluorescently labeled *H. pylori*. Binding of the bacteria to tissue sections was specific for surface mucous cells (Falk et al., 1993). Periodate oxidation of gastric mucosa sections identified the *H. pylori* receptor mediating cell adhesion as a glycoconjugate. Various glycosidase treatments identified fucose residues required for *H. pylori* binding. Furthermore, immunohistochemical analyses demonstrated that there is a perfect correlation between the cellular distribution of *H. pylori* receptors and fucosylated blood group antigens. The fucosylated blood group antigens, typically found on erythrocytes, are also expressed on epithelial cell surfaces such as the pyloric and duodenal mucosa in humans. The adhesin carbohydrate specificity was further elucidated by pretreatment of *H. pylori* with structurally defined glycoconjugates prior to adhesion to tissue sections *in situ*. Human colostrum Le^b glycoproteins are potent inhibitors, whereas Le^a glycoproteins are not (Borén et al., 1994). Fucosylated neoglycoproteins are also potent inhibitors, and were used in solid phase binding experiments to confirm the *H. pylori* specificity for the Le^b structure (Borén et al., 1993).

Epidemiological studies show that individuals with the blood group O phenotype have an approximately 1.5 fold increased risk of developing ulcers. Thus, the receptor specificity for Le^b and the H-antigen suggest less available *H. pylori* receptors in individuals of blood group A and B phenotypes, compared to blood group O individuals (Borén and

Falk, 1994). In fact, addition of the blood group A GalNAcα-determinant on Le^b blocks binding. Blood group antigens (e.g. ABO, Lewis, P) and sialylated carbohydrate structures show a high degree of variability among individuals and populations. Such differences affect the glycosylation patterns of all epithelial cell surfaces and, in addition, of the soluble glycoconjugates present in e.g. breast milk and saliva. The latter may act as our natural anti-microbial clearance factors, and partly govern our predisposition to bacterial infections. In this respect, the presence of high affinity salivary glycoprotein receptors for *H. pylori* is noteworthy.

In experiments where Le^b neoglycoproteins were crosslinked to *H. pylori* cell surface molecules, an adhesin candidate was identified on Le^b binding isolates. The amino terminal sequence of the purified adhesin is currently being used for cloning and sequencing of the *H. pylori* BAB adhesin (Ilver et al., 1996).

4. P-PILI OF UROPATHOGENIC *ESCHERICHIA COLI*–Galα1-4GaL AND CHAPERONE THERAPEUTICS

Expression of P-pili by *E. coli* is important for its ability to colonize the human gut and urinary tract, and therefore represents a virulence determinant in recurrent urinary tract infections (Leffler et al., 1986). P-fimbriae consists of a rigid stalk composed of several hundred helically arranged PapA subunits to which the Galα1-4Gal-binding PapG-adhesin is linked via a flexible tip (Hultgren et al., 1993). The linear tip fibrilla is composed of the PapE, Pap F, PapK and PapG proteins. The molecular chaperone PapD is required for proper assembly of P-pili (Kuhen et al., 1993). In the bacterial periplasm, PapD forms bimolecular complexes with the pilus subunits and escorts them to the outer cell membrane, where they attach to the base of the growing pilus. The P-pilus architecture serves as a platform to present three known classes of G-adhesin variants: G-I, G-II and G-III (Kihlberg et al., 1989a).

We have investigated the recognition of cell surface glycolipids by G-adhesins and have delineated the Galα1-4Gal receptor epitopes. Deoxy- and deoxyfluoro derivatives of oligosaccharides revealed the hydrogen bonding patterns between the G-adhesins and their Galα1-4Gal-receptor. Bacterial binding to Galα1-4Gal-containing glycolipids in cellular membranes (membrane directed orientation) and affixed to artificial surfaces (random orientation), in conjunction with molecular modeling, demonstrated that the G-adhesins recognize different glycolipid saccharide conformations. Synthetic peptides have also been used to elucidate the details of the binding of the PapD chaperone to pilus subunits.

4.1. Mapping of the Galα1-4Gal Receptor Structure for G-Adhesins

The Galα1-4Gal receptor determinant was originally identified by binding *E. coli* to globo series glycolipids sharing a Galα1-4Gal sequence (Bock et al., 1985; Leffler et al., 1986). The receptor epitope on Galα1-4Gal was delineated by generating all monodeoxy analogs and some deoxyfluoro analogs of methyl β-D-galabiose (Galα1-4Gal) (Kihlberg et al., 1986; Kihlberg et al., 1988; Kihlberg et al., 1989b). The analogs were used for mapping of the binding epitopes of the G-adhesins of *E. coli* strains $PapG_{J96}$ (G-I adhesin) and $PapG_{AD110}$ (G-II adhesin). Both strains are of human origin, although the G-II adhesin is more closely associated with human urinary tract infection. Inhibition of hemagglutination of human red blood cells carrying globoseries glycolipids by the saccharide derivatives was used to map the binding epitopes (Kihlberg et al., 1986).

The inhibitory efficiencies of the deoxy- and deoxyfluoro galabiosides (Striker et al., 1995; Kihlberg et al., 1989a) is depicted in fig. 4. Both G-adhesins (figs 4A and B) require

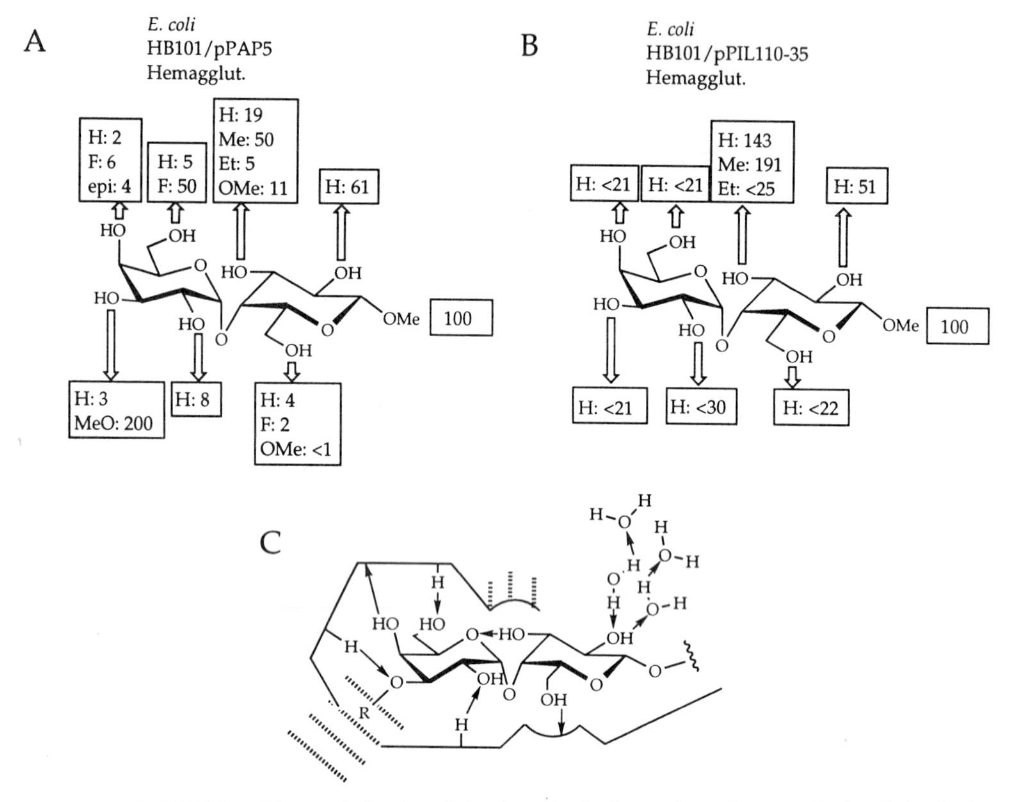

Figure 4. A-B: Inhibition of hemagglutination of *E. coli* strains by deoxy, deoxyfluoro, and other derivatives of methyl β-D-galabioside. The hydroxyl group substitution is shown in the squares together with the inhibitory power of the analog relative to that of the parent methyl galabioside, which was set to 100%. C: Receptor model for the binding of methyl β-D-galabioside by the two *E. coli* adhesins (arrows indicate intermolecular hydrogen bonds).

the same five oxygens of the Galα1-4Gal unit for efficient recognition, as demonstrated by the low relative inhibitory powers of the 6-, 2′-, 3′-, 4′-, and 6′-deoxy compounds. The 6-, 4′-, and 6′-deoxyfluoro compounds revealed the directionality of the saccharide-protein intermolecular hydrogen bonds. The analog carrying a MeO group at carbon 3 in the Galα moiety is twice as efficient as the parent compound, while replacement of the methyl glycoside by a TMSEt glycoside increased the inhibitory power four-fold (Kihlberg et al., 1989a). In the globo series glycolipids, the Galα1-4Gal moiety is extended at both these positions. The inhibition data (figs 4A and B) leads to the saccharide-protein recognition model for *E. coli* shown in fig. 4C. Interestingly, both adhesins recognize roughly the same saccharide epitope despite the fact that they have only 46% sequence homology (Striker et al., 1995; Hultgren et al., 1993).

4.2. Glycolipid Conformation and Host Tropism

In binding studies of diverse bacteria to lactosylceramide, a single hydroxyl group, irrespective of its position in the lipophilic ceramide, modulated glycolipid receptor function (Strömberg and Karlsson, 1990; Strömberg et al., 1988)). The possibility of the saccharide conformation affecting glycolipid receptor function was subsequently confirmed in studies on the G-adhesins. The G-II adhesin, represented by $PapG_{AD110}$, is frequent among isolates from the human gut and human upper urinary tract. The G-III adhesin, on the other hand, is quite common in human acute cystitis, and is the only adhesin present among isolates from the canine intestine and canine urinary tract (Hultgren et al., 1993).

The G-adhesins were compared with respect to their binding to globo series glycolipids affixed onto artificial surfaces or present in natural cellular membranes (Strömberg et al., 1991). The G-I and G-II adhesins both bound strongly to globotriaosylceramide (GbO_3), and the Forssman antigen (GbO_5) affixed onto either microtiter wells or onto thin-layer plates. In contrast, the G-III adhesin mediated strong binding to GbO_5, weak binding to globoside (GbO_4) and no binding to GbO_3. Although all three G-adhesins bound strongly to GbO_5 affixed on artificial surfaces, only the G-III adhesin bound this glycolipid embedded in a biological membrane. Only the G-III adhesin bound efficiently to dog-derived MDCK cells and to goat and sheep erythrocytes which all contained GbO_5 as the dominant isoreceptor. Furthermore, although both the G-I and G-II adhesins bound strongly to GbO_3 affixed onto artificial surfaces, only the G-I adhesin agglutinated rabbit erythrocytes with GbO_3 as the dominating isoreceptor. Finally, the G-II adhesin promotes strong adherence to human bladder adenocarcinoma T24 cells with GbO_4 as the dominating isoreceptor, while the G-I adhesin binds considerably more weakly to these cells. These binding characteristics are in good agreement with the glycolipid composition of target tissues for the G-adhesins. Thus, the G-adhesins recognize different globo series glycolipids in natural membranes, GbO_3 for G-I, GbO_4 for G-II, GbO_5 for G-III.

To understand how the different glycolipid isoreceptors bind G-adhesins in membranes, the membrane conformations of the three globo series glycolipids were elucidated (fig. 5) (Strömberg et al., 1991). Minimum energy calculations using the GESA program and molecular modeling (MM3) show well defined and rather rigid conformations of the sugar-sugar linkages and similar basic saccharide chain conformations of the three glycolipids. The calculations however indicate a fairly high rotational freedom for the O-C(1)-C(2) bonds (torsion angles ψ and θ) of the saccharide-ceramide linkage. Changes of the ψ and θ torsion angles lead to drastic changes in the orientation of the saccharide in relation to the lipid part of the glycolipid, as shown by the 9 different conformations of GbO_5 (fig. 5). Energy diagrams generated as a function of ψ and θ show three favoured conformers (nos 2, 5 and 6). Two of these, nos 2 and 6, are in accordance with crystal structures of monoglycosylceramides. To account for steric interferences from the membrane surface a restriction plane was placed 4 Å above the sphingosine atom C1. In the case of GbO_5, conformers 2 and 5 are allowed, while conformer 6 interferes with the membrane surface. Only GbO_3 can adopt conformation 6. In conformer 6, the Galα1-4Gal moiety is oriented parallel to the membrane surface. A slightly strained version of conformer 6 (conformer 6′, not shown) is slightly lifted up in relation to the membrane. Conformers 2 and 5 ,which can be adopted by all three glycolipids, are the only allowed orientations for GbO_5. In these conformers, the Galα1-4Gal unit is oriented almost vertical to the membrane surface. It is believed that the three classes of G-adhesins recognize different conformers of the globo GSLs. Thus, only the G-I adhesin is believed to recognize GbO_3 in conformer 6. The G-II adhesin binding epitope is thought to be slightly displaced toward the GalNAcβ unit of GbO_4 (conformer 6′), while the G-III adhesin has a binding epitope displaced toward the trisaccharide terminus of GbO_5 (conformers 2 and 5). The five hydroxyl groups on Galα1-4Gal that were shown to be involved in hydrogen bonding with the G-I and G-II adhesins are optimally exposed in conformers 6 and 6′.

In recent experiments, these data were used to design highly active and stable globoside structures recognized by G-II adhesin. Ethanol based large scale preparation procedures allowed the evaluation of these globoside structures as therapeutic and prophy-

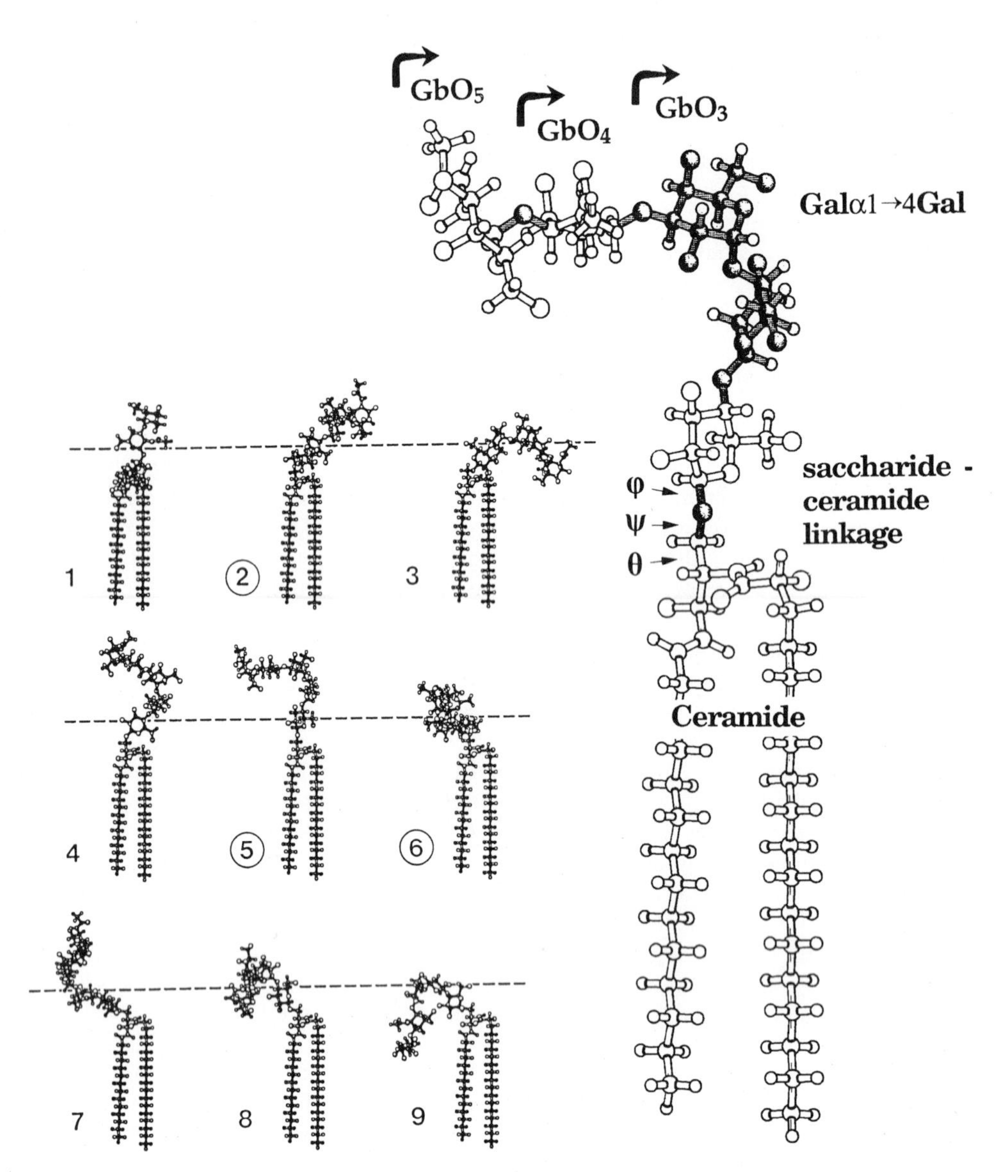

Figure 5. Molecular models of globo series glycolipids, illustrated with GbO_5 (conformer 5). The curved arrows denote the terminal saccharides in GbO_3, GbO_4 and GbO_5. The torsion angles (Φ, ψ and θ) of the saccharide-ceramide linkage are indicated. Different orientations of the saccharide chain of GbO_5, due to changes in the torsion angles ψ and θ of the saccharide-ceramide linkage, are shown (nos 1-9). Molecular modeling (MM3) of GbO_3, GbO_4 and GbO_5 shows that conformers 2, 5 and 6 are energetically favored. Arranged in a membrane (dotted line), most conformations (nos 1, 3, 6-9) are disfavored because of steric interference of the saccharide chain with the membrane surface. However, conformer 6 is possible in the case of GbO_3 due to its short saccharide chain. While the G-III adhesin must recognize conformers 2 or 5, the G-I adhesin is thought to recognize conformer 6 and the G-II adhesin a sligthly strained version (conformer 6′, not shown). [GbO_3 = Galα1-4LacCer, GbO_4 = GalNAcβ1-3Galα1-4LacCer and GbO_5 = GalNAcα1-3GalNAcβ1-3Galα1-4LacCer].

lactic agents against urinary tract infections in man. Systems for delivery of globoside to the large intestine of human subjects after per-oral administration were designed, and are currently being evaluated clinically.

4.3. Chaperone Inhibitors of Pilus Assembly

The assembly of adhesive pili requires periplasmic chaperones. The PapD chaperone has been found to have significant homology to more than 20 other chaperones which assemble pili in various pathogenic bacteria (Kuehn et al., 1993; Knight and Hultgren, 1996). The subunit binding site, proposed from the PapD crystal structure, is situated in the cleft between the two PapD domains and contains several amino acids which are conserved in the family of periplasmic chaperones, including the invariant Arg^{8} and Lys^{112} (Hultgren et al., 1993). The carboxy termini of pilus subunits show several conserved features (Kuehn et al., 1993), and appear essential for the binding of the PapG-adhesin to the PapD chaperone (Hultgren et al., 1989).

Synthetic peptides corresponding to the 19 carboxy terminal residues of the pilus subunits were prepared to investigate complex formation of the subunits with the PapD chaperone. In an enzyme-linked immunosorbent assay (ELISA), PapD bound well to the 19 residue peptide from the PapG-adhesin (PapG296-314: Gly^{296}-Lys-Arg-Lys-Pro-Gly-Glu-Leu-Ser-Gly-Ser-Met-Thr-Met-Val-Leu-Ser-Phe-Pro^{314}), and moderately or not at all to the carboxy terminal peptides of other pilus subunits (Kuehn et al., 1993). The potent 19-mer PapG296-314 was also found to inhibit PapD-mediated folding of denatured PapG (Kuehn et al., 1993), and binding of PapD to a carboxy terminal domain of PapG (Flemmer et al., 1995), suggesting that the peptides indeed bound to the subunit binding site of PapD. Truncation of the 19-mer peptide revealed that even a peptide as small as the octamer PapG307-314 was able to inhibit complex formation between PapD and PapG (Flemmer et al., 1995).

Futher insight into the structural basis of the recognition of pilus subunits and related peptides by PapD was obtained from the crystal structure of PapD complexed with the peptide PapG296-314 at 3.0 Å resolution (Kuhen et al., 1993). The peptide was bound in an extended conformation with the carboxylate group of the carboxy terminal Pro^{314} anchored by hydrogen bonds to the side chains of the invariant Arg^{8} and Lys^{112} in PapD. The peptide forms a parallel β-strand interaction with one of the β-sheets in PapD, mediated by main chain intermolecular hydrogen bonds. In the complex, the side chains of the peptide residues Met^{307}, Met^{309}, Leu^{311} and Phe^{313}, which belong to a conserved region in pilus subunits, make significant contacts with PapD and are most likely at least partly responsible for the specificity of PapD for pilus related peptides and subunits (Kuhen et al., 1993). It can be expected that other parts of PapG also contribute to the interactive surface in the PapD-PapG complex and we have recently shown that the peptide PapG175-190 represents one such epitope on PapG (Xu et al., 1995).

CONCLUDING REMARKS

With the rapidly increasing development of antibiotic resistant strains, alternative therapeutics against infectious diseases are urgently needed. Indeed, anti-adhesion therapeutics, such as carbohydrate and peptide receptor mimetics, are currently being evaluated clinically for various pathologic conditions, including inflammatory processes.

The generation of high affinity receptors with suitable pharmaceutical properties will require drug design by such approaches as computer aided drug design and screening of combinatorial chemical libraries. Knowledge of the structure and action of naturally occurring clearance and colonization factors in secretions such as milk and saliva, may help in designing such anti-adhesion therapeutics. In addition to receptor mimetics, resolving the molecular mechanisms of microbial adhesion should generate additional anti-adhesion strategies, such as inhibitors of pili assembly and use of adhesins in vaccine development. Through our studies we have indicated the importance of receptor conformation in generating biological specificity, a feature which has to be considered in the design of anti-adhesion therapeutics. The possible involvement of linkages with high degrees of rotational freedom in generating this biological specificity also pertain to cell-cell interactions, including those involving different RGD-conformations. Different individual receptor conformations or architectures could also be involved in host susceptibility to infectious diseases, and thus be used to screen for particularly susceptible individuals in the context of preventive health programs.

Finally, greater understanding of the degree of turnover of bacterial clones and involved mechanisms is needed. Studies on the link between bacterial growth and expression of bacterial adhesins would also improve our understanding of bacterial adhesion, colonization and persistence.

ACKNOWLEDGMENT

This work was funded by grants from The Swedish Medical Research Council, The Swedish Natural Science Research Council and The Swedish National Board for Industrial and Technical Development.

REFERENCES

Sharon, N. and Lis, H. (1993) Carbohydrates in cell recognition. Scientific American. 268 (1), pp. 82-89.

Paulson, J.C. (1993) Selectin/carbohydrate-mediated adhesion of leucocytes. In: Adhesion. Its role in inflammatory disease. Harlan, J.M. and Liu, D.Y. eds. pp 19-42, W.H. Freeman and Co., New York.

Lasky, L.A. (1993) The homing receptor (LECAM1/L selectin): A carbohydrate-binding mediator of adhesion in the immune system. In: Adhesion. Its role in inflammatory disease. Harlan, J.M. and Liu, D.Y. eds. pp 19-42, W.H, Freeman and Co., New York.

Karlsson K.-A. (1989) Animal glycosphingolipids as membrane attachment sites for bacteria. Annu. Rev. Biochem. (58) pp 309-350.

Hultgren, S.J., Normark, S. and Abraham, S.N. (1991) Chaperone-assisted assembly and molecular architecture of adhesive pili. Annu. Rev. Microbiol. (45) pp. 383-415.

Kihlberg, J., Hultgren, S.J., Normark, S. and Magnusson, G. (1989a) Probing of the combining site of the PapG adhesin of uropathogenic *Escherichia coli* bacteria by synthetic analogues of galabiose. J. Am. Chem. Soc. 111, pp. 6364-6368.

Weiss, W., Brown, J.H., Cusack, S., Paulsson, J.C., Skehel, J.J.and Wiley, D.C.(1988) Structure of the influenza virus haemagglutinin complexed with its receptor, sialic acid. Nature (333) pp. 426-431.

Firon, N., Ashkenazi, S., Mirelman, D., Ofek, I. and Sharon, N. (1987) Aromatic alpha glycosides of mannose are powerful inhibitors of the adherence of type 1 fimbriated *Escherichia coli* to yeast and intestinal epithelial cells. Infect. Immun. (55) pp.472-476.

Hultgren, S.J., Abraham, S., Caparon, M., St. Geme III, J.W. and Normark, S. (1993) Pilus and nonpilus bacterial adhesins: Assembly and function in cell recognition. Cell, 73 pp. 887-901.

Strömberg, N., Marklund, B.-I., Lund, B., Ilver, D., Hamers, A., Gaastra, W., Karlsson, K.-A. and Normark, S. (1990) Host-specificity of uropathogenic *Escherichia coli* depends on differences in binding specificity to Galα1-4Gal-containing isoreceptors. EMBO J. (9) pp. 2001-2010.

Strömberg, N., Nyholm, P.-G., Pascher, I. and Normark, S. (1991) Saccharide orientation at the cell surface affects glycolipid receptor function. Proc. Natl. Acad. Sci. (88) pp. 9340-9344.

Cossart, P., Boquet, P., Normark, S. and Rappuoli, R. (1996) Cellular microbiology emerging. Science. 271 pp. 315-316.

Gibbons, R.J. (1989) Bacterial adhesion to oral tissues: A model for infectious diseases J. Dent. Res. 68 (5) pp. 750-760.

Azen, E.A. (1993) Genetics of salivary protein polymorphisms. Crit. Rev. Oral Biol. Med. (4) pp. 479-485.

Hay, D.I. and Moreno, E.C. (1989) Statherin and the acidic proline-rich proteins. In: Human saliva: Clinical chemistry and microbiology. Vol I. pp. 132-150. Tenovuo, J.O., ed. CRC Press, inc., Boca Raton, Florida.

Lamkin, M.S. and Oppenheim, F.G. (1993) Structural features of salivary function. Crit. Rev. Oral Biol. Med. 4(3/4) pp. 251-259.

Kolenbrander, P.E. (1988) Intergeneric coaggregation among human oral bacteria and ecology of dental plaque. Ann. Rev. Microbiol. (42) pp. 627-656.

Strömberg, N., Borén, T., Carlén, A. and Olsson, J. (1992) Salivary receptors for GalNAcβ-sensitive adherence of *Actinomyces* spp.: Evidence for heterogenous GalNAcβ and proline-rich protein receptor properties. Infect. Immun. (60) pp. 3278-328.

Strömberg, N. and Borén, T. (1992) *Actinomyces* tissue specificity may depend on differences in receptor specificity for GalNAcβ-containing glycoconjugates. Infect. Immun. (60) pp. 3268-3277.

Bratt, P., Landys, D., Borén, T. and Strömberg, N. (1996a) Secretory immunoglobulin A heavy chains present Galβ/GalNAcβ-saccharide receptors for *Actinomyces naeslundi*. Manuscript in preparation.

Hammarström, K.-J., Hallberg, K., Dahlén, G., Gibbons, R.J., Hay, D.I. and Strömberg, N. (1996) The GalNAcβ- and PRP-binding specificities are common among *Actinomyces* spp. from plaque and buccal surfaces but absent among tongue isolates. Manuscript in preparation.

Strömberg, N., Nyholm, P.-G. and Kolenbrander, P. (1996a) *Actinomyces* specificity for *Streptococci* depends on different GalNAcβ receptor epitopes on streptococcal capsular polysaccharides. Manuscript in preparation.

Hallberg, K., Öhman, U., Hammarström, K.-J. and Strömberg, N. (1996) Association of type-1 and type-2 fimbrial genes with APRP and GalNAcβ specificity and *Actinomyces naeslundii* genospecies. Manuscript in preparation.

Järvholm, M., Bratt, P., Hallberg, K., Johansson, I., Hay, D.I. and Strömberg, N. (1996) Different acidic proline-rich protein and statherin binding-modes among bacteria colonizing tooth surfaces. Manuscript in preparation.

Strömberg, N. and Karlsson, K.-A. (1990) Characterization of the binding of *Actinomyces naeslundi* (ATCC 12104) and *Actinomyces viscosus* (ATCC 19246) to glycosphingolipids, using a solid-phase overlay approach. (265) pp. 11251-11258.

Bratt, P., Kolenbrander, P. and Strömberg, N. (1996b) GalNAcβ-receptors mediates multiple intra- and intergeneric interactions among human dental plaque bacteria. Manuscript in preparation.

Abeygunawardana, C., Bush, C.A. and Cisar J.O. (1991) Complete structure of the cell surface polysaccharide of *Streptococcus oralis* ATCC 10557: A receptor for lectin-mediated interbacterial adherence. Biochemistry 30 pp. 6528-6540.

Abeygunawardana, C., Bush, C.A. and Cisar J.O. (1990) Complete structure of the polysaccharide from *Streptococcus sanguis* J22. Biochemistry (29) pp. 234-248.

Landin, J. and Pascher, I. (1995) *Ab initio* and semiempirical conformation potentials for phospholipid head groups. J. Phys. Chem. (99) pp. 4471-4485.

Homans, S.W., Dwek, R.A., Boyd, J., Mahmoudian, M., Richards, W.G. and Rademacher, T.W. (1986) Conformational transitions in N-linked oligosaccharides. Biochemistry (25) pp. 6342-6350.

Hay, D.I., Gibbons, R.J., Schlessinger, D.H. and Schluckebier. (1996) Adhesion of *Actinomyces viscosus* LY7 to human salivary acidic proline-rich proteins: Effects of substitutions at residues 149 and 150. Manuscript in press.

Gibbons, R.J., Hay, D.I. and Schlessinger, D.H. (1991) Delineation of a segment of adsorbed salivary acidic proline-rich proteins which promotes adhesion of *Streptococcus gordonii* to apatitic surfaces. Infect. Immun (59) pp. 2948-2954.

Yeung, M.K. and Cisar J.O. (1990) Sequence homology between the subunits of two immunologically and functionally distinct types of fimbriae of *Actinomyces* spp. J. Bacteriol. (172) pp. 2462-2468.

Karlsson, A., Hammarström, K.-J., Markfjäll, M., Dahlgren, C. and Strömberg, N. (1996) The capacity of clinical isolates of *Actinomyces* to activate the neutrophil NADH-oxidase is related to the exposure of bacterial structures reacting with mammalian cell surface glycoconjugates containing GalNAcβ and sialic acid. Manuscript in preparation.

Strömberg, N., Dahlén, G., Markfjäll, M., Kolenbrander, P. and Sandros, J. (1996) Modulation of human dental plaque formation by an anti-adherence carbohydrate. Manuscript in preparation.

Carlén, A., Bratt, P., Olsson, J. and Strömberg, N. (1996) Agglutinin and acidic proline-rich protein receptor patterns modulate bacterial adherence and colonization patterns on tooth surfaces. Submitted manuscript.

Nordlund, Å., Carlsson, J., Ericson, T., Johansson, I., Sjöström, M. and Strömberg, N. (1996) Evaluation of factors related to dental caries incidence: A mulivariate approach. Manuscript in preparation.

Marshall, B. (1983) Unidentified curved bacilli on gastric epithelium in active chronic gastritis. Lancet, pp. 1273-1275.

Forman, D.,Webb, P., Newell, D.G., Coleman, M., Palli, D., Møller, H., Hengels, K., Elder, K. and DeBacker, G. (1993) An international association between *Helicobacter pylori* infection and gastric cancer. Lancet (341) pp. 1359-1362.

Li, C., Musich, P.R., Ha, T., Ferguson Jr, D.A., Patel, N.R., Chi, D.S.and Thomas, E. (1995) High prevalence of *Helicobacter pylori* in saliva demonstrated by a novel PCR assay. J. Clin. Pathol, 48 (7). pp. 662-665.

Falk, P., Roth, K.A., Borén, T., Westblom, T.U., Gordon, J.I. and Normark, S. (1993) An *in vitro* adherence assay reveals that *Helicobacter pylori* exhibits cell lineage-specific tropism in the human gastric epithelium. Proc. Natl. Acad. Sci. USA. (90) pp. 2035-2039.

Borén, T., Normark, S. and Falk. P. (1994) Molecular basis for host recognition in *Helicobacter pylori*: Trends in Microbiology. (2) pp. 221-228.

Borén, T., Falk, P., Roth, K.A., Larson, G. and Normark, S. (1993) Attachment of *Helicobacter pylori* to gastric epithelium mediated by blood group antigens. Science. (262) pp. 1892-1895.

Borén, T. and Falk, P. (1994) Blood type and the risk of gastric disease: Response letter. Science.(264) pp. 1387-1388.

Ilver, D., Arnquist, A. and Borén, T. (1996) Purification and characterization of the *Helicobacter pylori* blood group antigen binding (BAB)-adhesin. Manuscript in preparation.

Leffler, H. and Svanborg Eden, C. (1986) In Microbial Lectins and Agglutinins, D. Mirleman ed., John Wiley and Sons, New York, pp. 82.

Kuehn, J., Ogg, D.J., Kihlberg, J., Slonim, L.N., Flemmer, K., Bergfors, T. and Hultgren, S.J. (1993) Structural basis of pilus subunit recognition by the PapD chaperone. Science (262) pp. 1234-1241.

Bock, K., Breimer, M.E., Brignole, A., Hansson, G.C. Karlsson, K.-A., Larsson, G. Leffler, H., Samuelsson, B.E., Strömberg, N., Svanborg-Eden, C. and Thurin, J. (1985) Specificity of binding of a strain of uropathogenic *Escherichia coli* to Galα1-4Gal-containing glycosphingolipids. J. Biol. Chem. (260) pp. 8545-8551.

Kihlberg, J., Frejd, T., Jansson, K. and Magnusson, G. (1986) Synthetic receptor analogues: Preparation of the 3-O-methyl, 3-C-methyl, and 3-deoxyderivatives of methyl 4-O-α-D-galactopyranosyl-β-D-galactopyraniside (methyl β-D-galabiose). Carbohydr. Res. (152) 113-130.

Kihlberg, J., Frejd, T., Jansson, K., Sundin, A. and Magnusson, G. (1988) Synthetic receptor analogues: Preparation and calculated conformations of the 2-deoxy, 6-O-methyl, 6-deoxy, and 6-deoxy-6-fluoro derivatives of 4-O-α-D-galactopyranoside (methyl β-D-galabiose). Carbohydr. Res. (176) pp. 271-286.

Kihlberg, J., Frejd, T., Jansson, K., Kitzing, S. and Magnusson, G. (1989b) Preparation and calculated conformations of the 2′-, 3′-, 4′-, and 6′-deoxy, 3′-O-methyl, 4′-epi, and 4′- and 6′-deoxy-fluoro derivatives of methyl 4-O-α-D-galactopyranosyl-β-D-galctopyranoside (methyl β-D-galabiose). Carbohydr. Res. (185) 171-190 .

Striker, R., Nilsson, U., Stoneciper, A., Magnusson, G. and Hultgren, S.J. (1995) Structural requirements for the glycolipid receptor of human uropathogenic *Escherichia coli*. Mol. Microbiol. (16) pp. 1021.

Strömberg, N., Ryd, M., Lindberg, A.A. and Karlsson K.-A. (1988) Studies on the binding of bacteria to glycolipids. Febs Letters. (232) pp. 193-198.

Knight, S. and Hultgren, S. (1996) Personal communication.

Hultgren, S.J., Lindberg, F., Magnusson, G., Kihlberg, J., Tennent, J.M. and Normark, S. (1989) The PapG adhesin of uropathogenic *Escherichia coli* contains separate regions for receptor binding and for the incorporation into the pilus. Proc. Natl. Acad. Sci. USA (86) pp.4357-4361.

Flemmer, K., Xu, Z., Pinkner, J.S., Hultgren, S.J. and Kihlberg, J. (1995) Peptides inhibit complexation of the bacterial chaperone PapD and reveal potential to block assembly of virulence associated pili. BioMed. Chem. Lett. (5) pp. 927 .

Xu, Z., Jones, C.H., Haslam, D., Pinkner, J.S., Dodson, K., Kihlberg, J. and Hultgren, S.J. (1995) Molecular defection of PapD interaction with PapG reveals two chaperone binding sites. Mol. Microbiol. (16) pp 1011.

3

THE GALα1–4GAL-BINDING ADHESIN OF *STREPTOCOCCUS SUIS*, A GRAM-POSITIVE MENINGITIS-ASSOCIATED BACTERIUM

Sauli Haataja,[1] Kaarina Tikkanen,[2] Jukka Hytönen,[1] and Jukka Finne[1,2]

[1] Department of Medical Biochemistry
University of Turku
FIN-20520 Turku, Finland
[2] Department of Biochemistry and Biotechnology
A. I. Virtanen Institute
University of Kuopio
FIN-70211 Kuopio
Finland

SUMMARY

Streptococcus suis causes septicaemia and meningitis in pigs and occasionally in humans. A major galactose-inhibitable adhesin recognizing the blood group P -related disaccharide Galα1–4Galβ1– present in the GbO_3 glycolipid was identified in *S. suis*. Two variant adhesins, inhibitable by galactose and N-acetylgalactosamine (type P_N) or galactose only (type P_O) both preferred the disaccharide in terminal position. The hydrogen bonding patterns were determined using deoxy and other derivatives of the receptor disaccharide, and were compared to that of *E. coli* $PapG_{396}$ adhesin. The essential hydroxyls were the HO-4′, HO-6′, HO-2 and HO-3 hydroxyls; type P_O adhesin also weakly interacted with HO-6 and HO-3′. The mechanism differed from that of *E. coli* which binds to a cluster of five hydroxyls, HO-6, HO-2′, HO-3′, HO-4′ and HO-6′. The purified adhesin had a molecular weight of 18 kDa and an isoelectric point of 6.4. The agglutination of latex-bound purified adhesin was inhibited by the same inhibitors as agglutination with whole bacteria. The adhesin was detected by immunoblot analysis in all 23 *S. suis* strains examined representing different serotypes, was highly immunogenic and showed opsonizing activity. This represents the first example of the comparison of the saccharide receptor hydrogen bondings of two bacteria of different origin and shows that the same saccharide may be recognised by two different mechanisms. As a potential virulence factor present in different serotypes the adhesin represents a potential vaccine against *S. suis* infections.

Toward Anti-Adhesion Therapy for Microbial Diseases, edited by Kahane and Ofek
Plenum Press, New York, 1996

INTRODUCTION

Bacterial meningitis continues to be an important health problem with high mortality and high rate of neurologic sequelae even in developed countries (Tunkel and Scheld, 1993). Cell surface carbohydrate of bacteria and the host are involved in the pathogenesis of meningitis in at least two ways. Most pathogenic bacteria are surrounded by a negatively charged polysaccharide layer which gives advantages in the pathogenic process, and sometimes—as in the case of polysialic acid—mimics host cells surface carbohydrates in a highly specific manner (Finne, 1985). On the other hand host surface carbohydrates are often involved as the binding targets of bacteria in the initial infection of the body by the pathogen (Duguid et al., 1979; Beachey, 1981), but subsequent interactions with other molecules and cells occur during the invasion, spread and the establishment of secondary sites of infections in other parts of the body like the meninges. Dissection of the molecular basis of the interaction of the bacteria with body components is a prerequisite for the development of efficient and safe vaccines or novel antibacterial drugs based on interference with adhesion.

Most of the knowledge on bacterial adhesion mechanisms has been gained from studies of gram-negative bacteria (Sharon, 1987; Karlsson, 1989). Although some gram-positive bacteria are also known to recognize cell surface carbohydrates, there have been essentially no systematic studies, and the molecular basis of these interactions is poorly understood. *Streptococcus suis* is an important gram-positive pathogen associated with porcine meningitis, and occasionally also causes meningitis in humans (Windsor, 1977; Twort, 1981; Arends and Zanen, 1988). The increase in the frequency and severity of invasive infections and bacterial resistance to traditionally effective antibiotics (Cantin et al., 1992; Reams et al., 1993) have increased the need for novel therapies and vaccines. To date more than 30 serotypes have been reported (Higgins et al., 1995), and no effective vaccine is available. Thus characterization of the infectious mechanisms of *S. suis* is both an important practical goal and a useful animal model for streptococcal neonatal meningitis.

CARBOHYDRATE-RECOGNIZING ADHESION ACTIVITIES IN *STREPTOCOCCUS SUIS*

In order to study the presence of carbohydrate-based adhesion activities, *S. suis* strains were screened for hemagglutination activities using the human erythrocyte model (Korhonen and Finne, 1985). The following types of erythrocytes were used: untreated erythrocytes capable of reacting with multiple agglutinins including sialic acid- and poly-N-acetyllactosamine-specific adhesins, sialidase-treated erythrocytes to detect galactose-specific adhesins, and endo-β-galactosidase-treated erythrocytes to detect N-acetyl-glucosamine-specific agglutinins. Sialidase cleaves sialyl linkages of cell surface glycoproteins and glycolipids (Corfield et al., 1983) exposing terminal galactosyl-residues, and endo–β–galactosidase cleaves specifically the β-galactosidic linkages in –GlcNAcβ1–3Galβ1–4GlcNAc/Glc– oligosaccharides (Fukuda et al., 1979), exposing terminal GlcNAc–residues (Väisänen-Rhen et al., 1983).

From a total of 121 *S. suis* strains isolated from the brain, lungs and other organs of the diseased pigs, and from the tonsils and the nose of healthy carriers 13 (11 %) were hemagglutinating (as found out later, the low percentage of hemagglutinating strains is a reflection of the low sensitivity of the hemagglutination method in detecting the presence of the adhesins; see below). In most cases the streptococci agglutinated all three kinds of erythrocytes, and therefore no indication of the specificity of the agglutination was found (Table 1). No differences were found with endo-β-galactosidase-treated cells compared to

Table 1. Detection of the presence of carbohydrate-specific adhesion activities in strains of *S. suis* using the hemagglutination method with differentially modified erythrocytes

	Hemagglutination of erythrocytes treated with		
Number of strains	Untreated	Sialidase	Endo-β-galactosidase
108	–	–	–
2	–	+	–
1	+	++	+
1	+	+	+
7	+++	+++	+++
2	+++	–	+++

untreated cells, which suggested that the hemagglutination activities were not based on N-acetylglucosamine recognition.

Two *S. suis* strains were positive for untreated erythrocytes but lost their agglutination after the erythrocytes were treated with sialidase. This indicated that these strains recognized sialic acid. Studies involving mild periodate treatment, use of erythrocyte variants, specific enzyme treatments, resialylation of desialylated erythrocytes with linkage-specific sialyl-transferases, hemagglutination inhibition experiments with sialyl oligosaccharides and purified poly-N-acetyllactosaminyl glycopeptides, and binding of radiolabelled bacteria to blots of erythrocyte membrane proteins indicated that these strains recognized the structure

NeuNAcα2–3Galβ1–4GlcNAcβ1–3Galβ1–

present in the poly-N-acetyllactosaminyl chains of the Band 3 and Band 4.5 glycoproteins and polyglycosyl ceramides (Liukkonen et al., 1992).

Of the 13 hemagglutinating *S. suis* strains 9 were inhibited with neutral monosaccharides (Table 2). Of these strains 4 were inhibited with both galactose and N-acetylgalactosamine, and 5 with galactose only. Only 4 of the *S. suis* strains were not inhibited with monosaccharides, and two of them recognized sialic acid.

IDENTIFICATION OF THE RECEPTORS OF GALACTOSE-BINDING *S. SUIS*

Radiolabelled galactose-binding *S. suis* bacteria did not bind to human erythrocyte membrane glycoproteins on a nitrocellulose filter, and trypsin and pronase treatment of erythrocytes did not abolish hemagglutination. This suggested that glycolipids rather than glycoproteins are receptors of *S. suis*. In a thin-layer chromatography overlay experiment, the bacteria bound to two glycolipid bands of the total glycolipid extract of human erythro-

Table 2. Inhibition of the hemagglutinating activities of *S. suis* with monosaccharides

	Hemagglutination inhibition by 100 mM		
Number of strains	Galactose	N-acetylgalactosamine	Other sugars
4	+	+	–
5	+	–	–
2	–	–	–
2 (sialic acid specific)	–	–	–

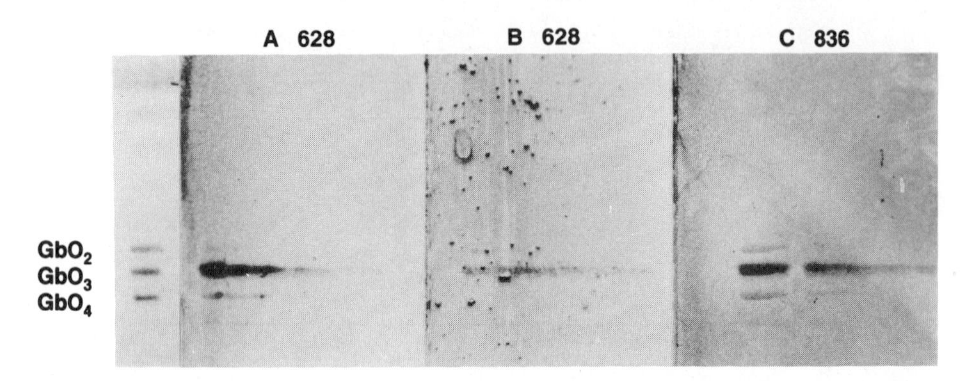

Figure 1. Binding of *S. suis* bacteria to glycolipids on TLC plates. Radioactively labelled bacteria of type P_N (strain 628) and P_O (strain 836) were incubated with polyisobutylmethacrylate-coated thin-layer chromatography plates containing equal amounts of two-fold dilutions (from left to right 5–0.04 nmol) of the glycolipids galabiosylceramide (GbO_2), globotriosylceramide (GbO_3), and globoside (GbO_4). Panels *A* and *C*, plates coated with 0.3 % polyisobutylmethacrylate (Plexigum P28) in ether–hexane (3:1); Panel *B*, plate coated with 0.5 % polyisobutylmethacrylate (M_r 300 000) in ether. The left panel shows the glycolipids (0.5 nmol) stained with anisaldehyde.

cytes. These corresponded to the globo-series glycolipids GbO_3 and GbO_4, which indicated that Galα1–4Gal-containing glycolipids are receptors of *S. suis* in human erythrocytes. This was also confirmed by the strong binding to purified GbO_3 (Fig. 1, A) and P_1 glycolipid and weaker binding to GbO_2 and GbO_4 (Haataja et al., 1994).

Agglutination studies with erythrocytes differing in their glycolipid composition indicated that bacteria recognize globo-series glycolipids, because human blood group P erythrocytes, which lack Galα1–4Gal- containing glycolipids, were not agglutinated. The bacteria most strongly agglutinated human erythrocytes of type P_1^k and P_2^k, and rabbit erythrocytes, which have as their predominant receptor GbO_3, indicating that this glycolipid is the primary receptor of *S. suis*.

At the monosaccharide level *S. suis* prefers galactose with aglycons in an α configuration, as indicated by the hemagglutination inhibition studies (Haataja et al., 1993). Hydrophilic aglycons (monosaccharides) were better accepted than hydrophobic (Me or pNO$_2$Phe). Of the galactose-containing α1–3, α1–4 or α1–6 disaccharides, Galα1–4Gal was preferred and was a 2000-fold better inhibitor than the other disaccharides. At the oligosaccharide level *S. suis* most strongly recognizes the trisaccharide Galα1–4Galβ1–4Glc. Further extensions to the non-reducing end reduced its affinity. Interestingly, lactose (Galβ1–4Glc), which constitutes a significant part of the inhibitory receptor trisaccharide, was inactive as an inhibitor, which suggests that the terminal α-galactose moiety has a dominating role in the binding. The inhibitory activity of free galactose can thus most likely be ascribed to inhibition of the binding of the terminal α-galactose.

THE ADHESIN VARIANTS P_N AND P_O

The *S. suis* variants, inhibited with galactose and N-acetylgalactosamine or galactose only were designated type P_N and P_O respectively. Both types were found to recognize Galα1–4Gal (Haataja et al., 1994). In addition type P_O differed from type P_N in that it was better inhibited with oligosaccharides containing terminal α-galactosyl residues, Galα–OMe and Galα1–6Glc.

At the oligosaccharide level the variant adhesion activities differ in respect of the influence of further extensions to the terminal α-galactose moiety. In hemagglutination inhibition assays type P_N recognized globotetraose better than type P_O. Further differences were observed in solid phase inhibition assays using receptor derivatives which did not have glucose at their reducing end. Type P_N was inhibited by Galβ1–3Galα1–4Galβ–*O*–$(CH_2)_2SiMe_3$ at nanomolar concentrations, whereas type P_O was not inhibited even at millimolar concentrations. It seems that the N-acetyl group in the terminal GalNAc moiety does not fit in the binding site because it abolishes the inhibitory activity of the oligosaccharide. Also in hemagglutination assays the removal of the N-acetyl group increased the activity of the globoside tetrasaccharide with type P_N but not with type P_O. Thus the combining site of the type P_N adhesin tolerates further extension to the non-reducing end by galactose whereas type P_O is more specific for unsubstituted α-galactose.

In binding of type P_N bacteria to Galα1–4Gal-containing glycoproteins and glycolipids immobilized on microtiter plate wells, bacteria bound equally to neoglycoproteins containing either Galα1–4Galβ1–4Glcβ1–O–CETE (the oligosaccharide part of GbO_3 and the P^k epitope) and Galα1–4Galβ1–4GlcNAcβ1–O–CETE (the P_1 epitope), which indicates that the bacteria primarily recognize terminal Galα1–4Gal and that the N-acetyl-moiety in the glucose has little effect on binding. The same was observed with type P_O. In microtiter wells *S. suis* bacteria bound more strongly to glycoproteins than to glycolipids, whereas a control strain of *E. coli* had a higher capacity to bind glycolipids.

S. suis type P_N specifically recognized GbO_3 in TLC overlay studies. The bacteria did not bind to globoside even at high concentrations. Because the plastic coating of the TLC plate can alter the binding specificity, assays were also performed using lower concentrations and different preparations of polyisobutylmethacrylate. Both variant types of bacteria bound most strongly to GbO_3, but also showed some binding to GbO_2 and GbO_4 (Fig. 1, B and C).

HYDROGEN BONDING PATTERN OF *S. SUIS* ADHESIN

Using deoxy- and deoxy-fluoro-D-galactose derivatives and epimers of D-galactose the contribution of the hydroxyl groups of terminal galactose were determined by hemagglutination inhibition assay (Haataja et al., 1993; Haataja et al., 1994). The complete hydrogen bonding pattern required for the binding of the adhesin to the receptor disaccharide was determined with synthetic Galα1–4Gal oligosaccharide derivatives.

The contributions of the individual hydroxyls of the receptor disaccharide Galα1–4Gal were indicated to be the following (Fig. 2):

HO-6′: the hydrogen is indicated to be bonded to the adhesin since 6-deoxy-D-galactose (D-fucose), 6-deoxy-fluoro-D-galactose, 6-deoxy-6-sulfate-D-galactose and 6′-deoxy-Galα1–4Galβ-O-Me were inactive as inhibitors. The inactivity of 6-deoxy-fluoro-D-galactose suggests that HO-6′ is hydrogen bond donor.

HO-4′: the hydrogen is indicated to be bonded to the adhesin since 4-deoxy-D-galactose, 4-deoxy-fluoro-D-galactose, D-glucose (4 epimer of galactose) and the 4′-deoxy- and 4′-epimer derivatives of Galα1–4Galβ-O-Me were not inhibitors. Inactivity of 4-deoxy-fluoro-D-galactose indicates that HO-4′ is hydrogen bond donor.

HO-3′/type P_N: the 3-epimer D-gulose was twice as good an inhibitor as galactose, which suggests that the adhesin may have non-polar interactions on the β-side of

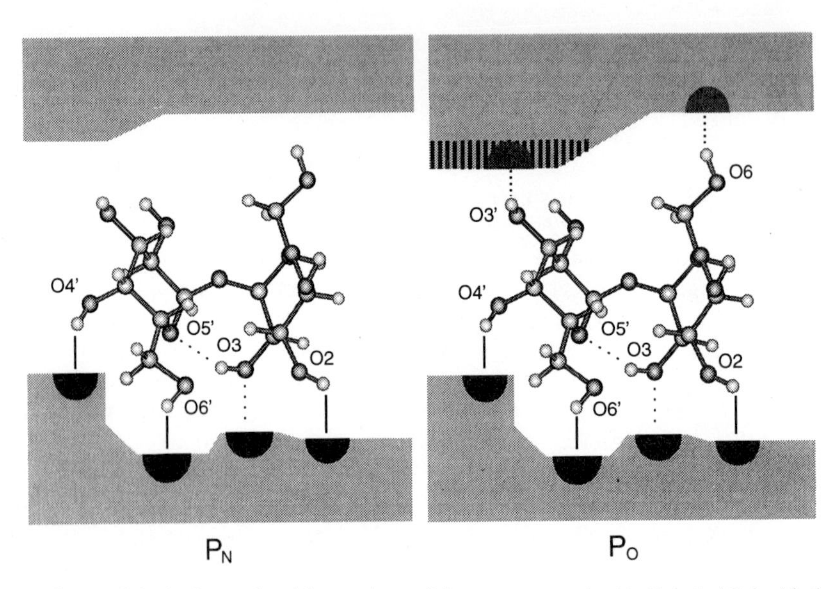

Figure 2. Scheme of the hydrogen bond interactions of the receptor saccharide Galα1–4Gal with the *S. suis* adhesin variants P_N and P_O. The essential hydrogen bonds formed between the receptor disaccharide and the adhesins are indicated by the solid lines and the weak interactions with broken lines. HO-3 may be intramolecularly hydrogen bonded to the ring oxygen O-5′; the inactivity of derivatives lacking HO-3 could result from the breakage of the intra-molecular hydrogen bond or breakage of a hydrogen bond with the adhesin. The hatched area in P_O indicates the crowded region of close contact.

the terminal α-galactose. Since 3-deoxy-fluoro-D-galactose and 3′-deoxy-Galα1–4Galβ-O-Me were as active as D-galactose, HO-3′ is not hydrogen bonded to the adhesin.

HO-3′/type P_O: the 3-deoxy-3-fluoro-D-galactose was inactive and in contrast to type P_N, the 3-epimer was not an inhibitor. The 3′-deoxy-Galα1–4Galβ-O-Me derivative was similarly a weaker inhibitor than Galα1–4Galβ-O-Me, which suggests that HO-3′ may be slightly involved in the interaction with the adhesin.

HO-2′/type P_N: the hydroxyl appears not to be involved in binding since 2-deoxy-D-galactose, 2-deoxy-2-fluoro-D-galactose and 2′-deoxy-Galα1–4Galβ-O-Me were inhibitory. The adhesin may have a binding pocket for N-acetylgalactosamine since 2-acetamido-2-deoxy-D-galactose was a fourfold better inhibitor than D-galactose. D-Talose (2-epimer of galactose) was inactive possibly due to disturbance of the non-polar interactions of the β-side.

HO-2′/type P_O: type P_O differs from type P_N in that it was not inhibited by 2-acetamido-2-deoxy-D-galactose, which suggests that P_O has a narrow combining site in this area. This is further supported by the increase in the inhibitory power of 2′-deoxy-Galα1–4Galβ-O-Me, which suggests that this area may form non-polar interactions with the adhesin.

HO-6/type P_N: the 6-deoxy-Galα1–4Galβ-O-Me was as active as Galα1–4Galβ-O-Me, which indicates that HO-6 is not involved in the interaction.

HO-6/type P_O: 6-deoxy-Galα1–4Galβ-O-Me was a 15 % weaker inhibitor than Galα1–4Galβ-O-Me, which suggests that HO-6 may be slightly involved in the interaction.

HO-2 and HO-3: the 2- and 3-deoxy-Galα1–4Galβ-O-Me derivatives were not inhibitors, which suggests that these hydroxyls are essential for the binding and are hydrogen bonded to the adhesin.

HO-1: the adhesin appears to prefer an aglycon with a β configuration since Galα1–4Gal is less efficient than Galα1–4Galβ-OMe.

IDENTIFICATION AND ISOLATION OF THE *S. SUIS* ADHESIN

Pigeon ovomucoid contains blood group P_1 (Galα1–4Galβ1–4GlcNAcβ1–) terminals in its glycans and is a highly active inhibitor of the hemagglutination *S. suis* (Haataja et al., 1993). Radiolabelled pigeon ovomucoid was used as a probe to identify the adhesin from sonicates of *S. suis* on blots after non-denaturing gel electrophoresis (Fig. 3)(Tikkanen et al., 1995). One single band was detected in all strains. There was no apparent difference in the band mobilities of P_N and P_O strains.

The intensity of the band correlated with the agglutination titers of the strains. As in many other bacteria, hemagglutination of *S. suis* undergoes spontaneous phase variation. In bacteria extracted in the highly agglutinating or low agglutinating phases the intensity of the adhesin bands correlated with the titer (Fig. 3, strain 628). Furthermore, the band intensity correlated with the hemagglutination activity of the strains. Even strains classified as nonhemagglutinating revealed the presence of the adhesin in blot analysis. In fact, all 23 *S. suis* strains so far studied, including several different serotypes, have been found to contain the adhesin (K. Tikkanen, S. Haataja and J. Finne, unpublished results).

The adhesin was purified to homogeneity from sonicated extracts of *S. suis* using fractional ammonium sulfate precipitation and preparative gel electrophoresis (Tikkanen et al., 1995). The adhesin appeared as a single homogenous band of an apparent size of 18 kDa (Fig. 4). The mobility was not changed by the use of a reducing agent, and amino acid analysis revealed the absence of cysteine residues. The N-terminal amino acid sequence had no

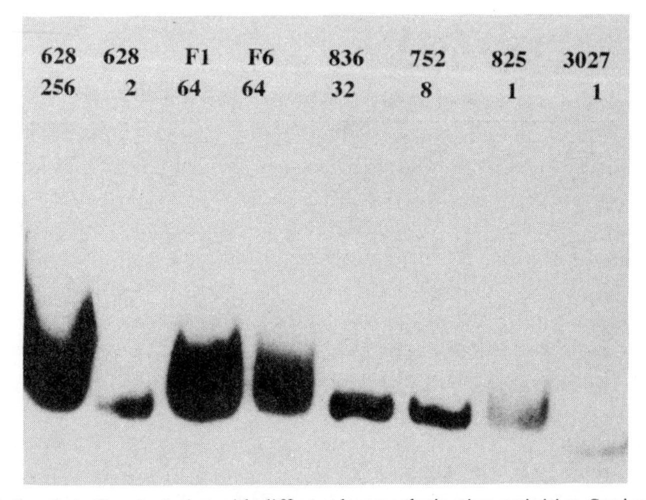

Figure 3. Adhesin levels in *S. suis* strains with different hemagglutinating activities. Sonicated extracts of the *S. suis* strains indicated (top line) with differing hemagglutinating activities (second line; reciprocal of agglutination titer) were separated by electrophoresis in 6 % polyacrylamide gels in the absence of SDS and transferred to PVDF-P membrane which was probed for adhesin activity with ^{125}I-labelled pigeon ovomucoid. Bacteria of strain 628 extracted in high or low hemagglutination titer phases were included in the analysis.

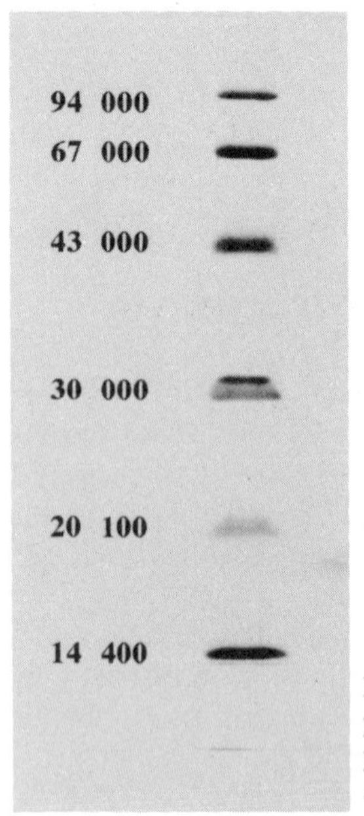

Figure 4. Polyacrylamide gel electrophoresis of purified *S. suis* adhesin. The purified adhesin was run in 15% polyacrylamide gel electrophoresis in the presence of SDS and stained with Serva Blue. The molecular weights of the standard proteins are indicated.

apparent sequence similarity with other bacterial adhesins or surface proteins of gram-positive bacteria, and were identical in the type P_N and P_O strains studied.

The purified adhesin expressed only weak hemagglutination activity, due to its monovalent nature. In contrast the adhesin adsorbed onto latex particles induced a strong hemagglutination which was specifically inhibited with the same inhibitors as hemagglutination induced by whole bacteria (Tikkanen et al., 1995).

CONCLUDING REMARKS

The presence of a galactose-binding adhesin in all *S suis* strains examined suggests that it may be and essential factor in the pathogenesis of *S. suis* infections. *S. suis* binds to frozen sections of pig tonsillar tissues in Galα1–4Gal inhibitable manner (Haataja et al., 1993), but there are at present no further clues to the biological role of the adhesion activity, nor to the presence of the P_N and P_O variant activities. One possibility is that the latter, as well as the sialic-acid binding activity, might be related to different tissue tropisms of the strains.

The adhesin could be extracted from the *S. suis* cells by ultrasonic treatment, and appeared as a homogenous 18 kDa band. However, it is not at present known in which form the adhesin exists on the bacterial surface and whether it is attached to a specific structure such as the fimbriae of gram-negative bacteria.

It is of interest to compare the hydrogen bonding patterns of *S. suis* adhesin and that of the P adhesin of *E. coli*, since this represents the first characterization of the binding mechanisms of two different bacterial species, a gram-negative and a gram-positive, that bind to the same receptor saccharide. The essential hydroxyls required for the binding to the Galα1–4Gal-disaccharide of the *E. coli* $PapG_{396}$ adhesin includes the hydroxyls HO-6′, HO-4′, HO-3′, HO-2′ and HO-6 (Kihlberg et al., 1989). Thus *S. suis* binds a different side of the oligosaccharide (Fig. 5). An additional difference is that HO-6′ is hydrogen bond donor

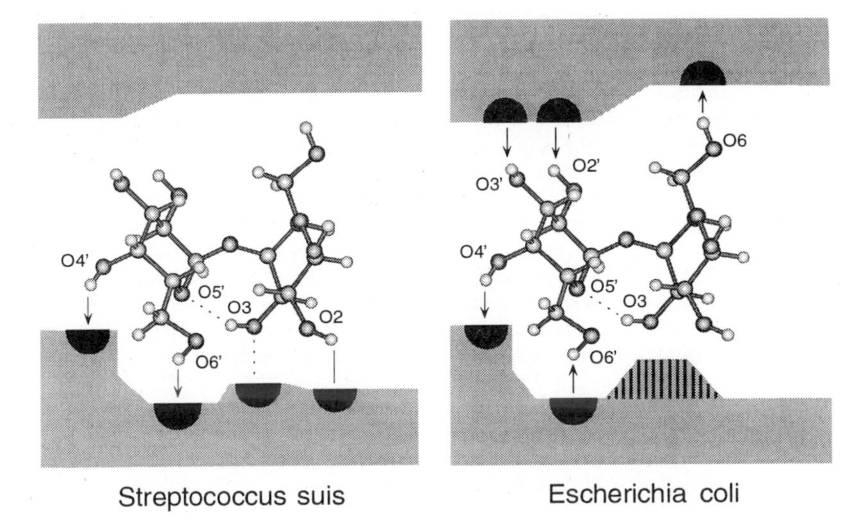

Figure 5. Comparison of the interaction of the receptor saccharide Galα1–4Gal with the *S. suis* type P_N and *E. coli* $PapG_{396}$ adhesins. The essential hydrogen bonds formed between the receptor disaccharide and the adhesins are indicated by the solid lines; the arrows indicate whether the hydroxyl is a hydrogen bond donor or acceptor. The scheme of *E. coli* interaction is based on data published by Kihlberg et al. (1989).

in *S. suis* and acceptor in *E. coli*. The *S. suis* adhesins preferred terminal Galα1–4Gal structures both in oligosaccharide inhibition assays, in the binding to receptors in the solid phase and on natural membranes. The P adhesins of *E. coli* mostly prefer the extended globo-series glycolipids globoside and Forssman glycolipid.

Determination of the binding mechanism of the *S. suis* adhesin with its receptor saccharide awaits the availability of the crystal structure of the adhesin. When it becomes available, the design of inhibitors of the adhesin that could be used as a novel type of antibacterial drug would be facilitated. On the other hand, the identification of the adhesin protein as a probable virulence factor and its presence in all strains examined independent of serotype may open a possibility for the development of a vaccine against *S. suis* infections.

ACKNOWLEDGMENTS

Our work has been supported by grants from the Sigrid Jusélius Foundation and the Academy of Finland.

REFERENCES

J.P. Arends and H.C. Zanen, Meningitis caused by *Streptococcus suis* in humans, *Rev. Infect. Dis.* 10:131–137 (1988).

E.B. Beachey, Bacterial adherence: Adhesin-receptor interactions mediating the attachment of bacteria to mucosal surfaces, *J. Infect. Dis.* 143:325–345 (1981).

M. Cantin, J. Harel, R. Higgins, and M. Gottschalk, Antimicrobial resistance patterns and plasmid profiles of *Streptococcus suis* isolates, *J. Vet. Diagn. Invest.* 4:170–174 (1992).

A.P. Corfield, H. Higa, J.C. Paulson, and R. Schauer, The specificity of viral and bacterial sialidases for alpha(2–3)- and alpha(2–6)-linked sialic acids in glycoproteins, *Biochim. Biophys. Acta* 744:121–126 (1983).

J.P. Duguid, S. Clegg, and M.I. Wilson, The fimbrial and non-fimbrial haemagglutinins of *Escherichia coli*, *J. Med Microbiol.* 12:213–227 (1979).

J. Finne, Polysialic acid – a glycoprotein carbohydrate involved in neural adhesion and bacterial meningitis, *Trends Biochem. Sci.* 10:129–132 (1985).

M.N. Fukuda, M. Fukuda, and S. Hakomori, Cell surface modification by endo-beta-galactosidase. Change of blood group activities and release of oligosaccharides from glycoproteins and glycosphingolipids of human erythrocytes, *J. Biol. Chem.* 254:5458–5465 (1979).

S. Haataja, K. Tikkanen, J. Liukkonen, C. François-Gerard, and J. Finne, Characterization of a novel bacterial adhesion specificity of *Streptococcus suis* recognizing blood group P receptor oligosaccharides, *J. Biol. Chem.* 268:4311–4317 (1993).

S. Haataja, K. Tikkanen, U. Nilsson, G. Magnusson, K.-A. Karlsson, and J. Finne, Oligosaccharide-receptor interaction of the Galα1–4Gal binding adhesin of *Streptococcus suis*. Combining site architecture and characterization of two variant adhesin specificities, *J. Biol. Chem.* 269:27466–27472 (1994).

R. Higgins, M. Gottschalk, M. Boudreau, A. Lebrun, and J. Henrichsen, Description of six new capsular types (29–34) of *Streptococcus suis*, *J. Vet. Diagn. Invest.* 7:405–406 (1995).

K.-A. Karlsson, Animal glycosphingolipids as membrane attachment sites for bacteria, *Annu. Rev. Biochem.* 58:309–350 (1989).

J. Kihlberg, S.J. Hultgren, S. Normark, and G. Magnusson, Probing of the combining site of the PapG adhesin of uropathogenic *Escherichia coli* bacteria by synthetic analogues of galabiose, *J. Am. Chem. Soc.* 111:6364–6368 (1989).

T.K. Korhonen and J. Finne, Agglutination assays for detecting bacterial binding specificities, in: "Enterobacterial surface antigens: methods for molecular characterization", T.K. Korhonen et al., eds., Elsevier Science Publishers, Amsterdam, pp. 301–313 (1985).

J. Liukkonen, S. Haataja, K. Tikkanen, S. Kelm, and J. Finne, Identification of *N*-acetylneuraminyl α2—>3 poly-*N*-acetyllactosamine glycans as the receptors of sialic acid-binding *Streptococcus suis* strains, *J. Biol. Chem.* 267:21105–21111 (1992).

R.Y. Reams, L.T. Glickman, D.D. Harrington, T.L. Bowersock, and H.L. Thacker, *Streptococcus suis* infection in swine: a retrospective study of 256 cases. Part I. Epidemiologic factors and antibiotic susceptibility patterns, *J. Vet. Diagn. Invest.* 5:363–367 (1993).

N. Sharon, Bacterial lectins, cell-cell recognition and infectious disease, *FEBS Lett.* 217:145–157 (1987).

K. Tikkanen, S. Haataja, C. François-Gerard, and J. Finne, Purification of a galactosyl-α1–4-galactose-binding adhesin from the gram-positive meningitis-associated bacterium *Streptococcus suis*, *J. Biol. Chem.* 270:28874–28878 (1995).

A.R. Tunkel and W.M. Scheld, Pathogenesis and pathophysiology of bacterial meningitis, *Annu. Rev. Med.* 44:103–120 (1993).

C.H. Twort, Group R streptococcal meningitis (*Streptococcus suis* type II): a new industrial disease? *Br. Med J. Clin. Res. Ed.* 282:523–524 (1981).

V. Väisänen-Rhen, T.K. Korhonen, and J. Finne, Novel cell-binding activity specific for N-acetyl-D-glucosamine in an *Escherichia coli* strain, *FEBS Lett.* 159:233–236 (1983).

R.S. Windsor, Meningitis in pigs caused by *Streptococcus suis* type II, *Vet. Rec.* 101:378–379 (1977).

4

DEVELOPMENT OF ANTI-ADHESION CARBOHYDRATE DRUGS FOR CLINICAL USE

D. Zopf,[1] P. Simon,[1] R. Barthelson,[1] Diana Cundell,[1]
I. Idanpaan-Heikkila,[2] and E. Tuomanen[2]

[1] Neose Technologies, Inc.
Horsham, Pennsylvania
[2] Rockefeller University
New York, New York

INTRODUCTION

Specific adherence of many human pathogens is mediated by microbial adhesin proteins that recognize sugar chains attached to host cell surface glycoconjugates. More than twenty years ago a strategy was proposed for preventing or interrupting the progression of bacterial and viral infection by using soluble monosaccharides or oligosaccharides as competitive inhibitors of microbial adherence. Since then, investigations from many laboratories into the detailed structures of both microbial adhesin molecules and the carbohydrate epitopes they recognize on mammalian cells have at least partially elucidated the molecular basis for carbohydrate mediated adherence for dozens of microbial species. As new enzyme-based technologies for manufacturing oligosaccharides have emerged during the past few years, commercial scale production of oligosaccharides has become economically feasible. Proceeding from this scientific and technological platform, we are developing a class of drugs known as *carbohydrate adhesion ligand homologs* - inhibitors of microbial adherence comprised of free oligosaccharides chemically identical to sugar chain adhesin ligands carried on surface glycoconjugates of host mucosal cells.

Drug selection begins by testing appropriate oligosaccharide candidates for their abilities to prevent adherence of live fluorescent or radiolabeled microorganisms onto primary cell explants or cell lines derived from appropriate human target tissues. Oligosaccharides that successfully prevent and/or disrupt adherence of the majority of clinical isolates of a pathogen to appropriate target cells undergo further testing in animal models of infection.

Toward Anti-Adhesion Therapy for Microbial Diseases, edited by Kahane and Ofek
Plenum Press, New York, 1996

HELICOBACTER PYLORI

Peptic ulcer disease has been causally linked to chronic infectious gastritis produced by *H. pylori* (Feldman, M., 1994). Several investigations into the chemical entities that mediate adherence of *H. pylori* to gastric cells have been carried out using killed or chemically surface-labeled bacteria binding to fixed human cells or tissues or semipurified cell extracts coated on silica or plastic surfaces (e.g., Saitoh, et al, 1991; Falk, et al, 1993; Slomiany, et al, 1989; Lingwood, et al, 1992, Trust, et al, 1991). In order to define which, if any, of the lectin-like activities previously assigned to *H. pylori* might play a dominant role in binding between live, unmodified *H. pylori* and human gastrointestinal cells, we constructed a test system for comparing oligosaccharides as inhibitors of microbial adhesion to cultured cell lines.

Drug screening was performed using multiple strains of *H. pylori*, including several recent passage clinical isolates. Each strain was added separately to microtiter wells containing monolayers of the human duodenal carcinoma cell line, HuTu-80, in the presence or absence of purified oligosaccharides as competitive inhibitors of bacterial adhesion. After suitable incubation, wells were washed with buffered saline, and a solution containing urea and phenol red was added to each well. Thirty minutes later the number of adherent bacteria was estimated spectrophotometrically according to the degree of color change of the pH indicator dye in response to alkalinization of the medium by microbial urease.

Comparison of inhibitory activities of an oligosaccharides containing a wide variety of structural epitopes yielded a lead compound, termed NE-0080, which contains the active non-reducing epitope, NeuAcα2-3Galβ1.... For most *H. pylori* strains tested, the compound gave an IC50 in the low millimolar range.

We speculated that effect of an antiadhesive agent on *H. pylori* in the stomach will be influenced by several well known physiological processes in the stomach. First, *H. pylori* inhabits a dynamic environment in which the mucosal surface to which it must remain attached for persistent colonization turns over with a half-time in the range of 18 - 24 hours. It is, therefore, necessary for organisms to reattach to new generations of mucosal epithelial cells at frequent time intervals. Second, gastric contents are almost constantly emptied and replenished by ingestion and/or secretion, leading to clearance of unattached microorganisms by peristaltic flow. Given these and other factors, we reasoned that an antiadhesive agent taken orally in a regimen such that it would remain at millimolar concentrations in the stomach for several hours per day should interfere with reattachment of *H. pylori* to the gastric lining and result in progressive diminution of the bacterial population, rendering the infection more susceptible to host natural immune mechanisms, classical antibiotics, or both.

Toxicokinetic investigations of NE-0080 revealed that the compound is at least 90% stable in gastric juice for two hours and is barely absorbed into the circulation, with bioavailability less than 1% in humans. The drug is well tolerated in rats at oral doses more than three hundred times its estimated clinical dose and is completely degraded in the intestine. It is expected to be metabolized according to well-established pathways for simple sugars. Animal trials in gnotobiotic piglet and primate models provided a firm basis for continued human trials. The compound is currently in late Phase 1 human clinical trials.

RESPIRATORY TRACT PATHOGENS

Several carbohydrate targets have been proposed for respiratory pathogens, including one, ...GalNAcβ1-4Gal..., that seems to be a consensus ligand, recognized in common by multiple species of pneumonia-causing bacteria. We screened oligosaccharides to determine whether a single compound might effectively prevent adhesion of three major respiratory pathogens: *Streptococcus pneumoniae, Hemophilus influenzae*, and *Moraxella catarrhalis*. This triad of organisms is responsible for the majority of cases of acute otitis media as well as many cases of community-acquired pneumonia and acute exacerbations of chronic bronchitis.

We evaluated adherence of the above three pathogens in the presence or absence of oligosaccharide inhibitors at various concentrations using either direct microscopic counting techniques or scintillation counting of metabolically radiolabeled organisms bound to target cell monolayers of respiratory origin. The lead compound selected, NE 1530, which contains the active non-reducing epitope NeuAcα2-6Gal..., inhibits binding of 10 *M. catarrhalis* clinical isolates to the Hep2 and A549 cell lines. The same compound inhibits binding of *H. influenzae* to A549 and Chang conjunctival cell lines. Studies carried out at hourly intervals after seeding the organisms into liquid culture, demonstrated varying sensitivity to inhibition of adherence depending upon growth phase. The experiments suggest that expression of an adhesin for NeuAcα2-6Gal... becomes maximal in late log phase, possibly related to nutritional restrictions as the growth medium becomes exhausted, or the release of substances that serve as autoregulatory signals. Growth phase dependent sensitivity to inhibition of adherence by NE 1530 also was noted for *S. pneumoniae in vitro*.

IN VIVO STUDIES: THE RABBIT PNEUMONIA MODEL

When virulent strains of *S. pneumoniae* are instilled intratracheally into the right mainstem bronchus of a rabbit, the animal contracts acute pneumococcal pneumonia and becomes septicemic. Administration of NE 1530 (10 nmol) 24 hours prior to instillation of bacteria reduces by a factor of 100 the mean number of organisms that can be recovered from the right lung by saline lavage, and prevents, or greatly reduces, occurrence of septicemia. The protective effect of the drug also could be observed in animals lavaged at 48 hours. Similar decreases in bacterial counts in lavage fluids were observed when NE 1530 was admixed with the suspension *of S. pneumoniae* prior to instillation into the lung. Greater effects were apparent in animals who received 100 nmol of drug coadmininistered with bacteria. In rabbits that were first infected with *S. pneumoniae*, then treated after 24 hours, a 10- to 100-fold reduction in colony forming units was observed in lavage fluids taken 24 hours post treatment. Several animals that received NE 1530 in all of the above experiments had sterile lung lavages, blood cultures, or both. Histopathological examination and physical tissue imaging techniques confirmed that the lungs of these animals lacked bacterial and inflammatory exudates and were anatomically within normal limits.

CONCLUDING REMARKS

In vitro screening methods for evaluation of microbial antiadherence agents are usually best carried out using live organisms and live host cells. If either the bacteria or epithelial cell is chemically modified , such as by surface labeling of microbes with chemically active fluorescent dyes (e.g., FITC), careful controls should be performed to document that labeling reactions with surface proteins do not alter native properties of adherence. In the course of our pilot experiments we noted that effects such modifications upon binding specificity, especially for *H. pylori*, can be profound.

It is prudent to investigate multiple microbial isolates as well as several potential cell targets to confirm that the inhibitory activity of a lead compound is of general importance. We briefly describe here two oligosaccharide antiadhesive drugs discovered by comparing activities of more than twenty carbohydrate compounds as inhibitors of binding. They were evaluated using at least 10 recent clinical isolates of each bacterial species and not less than three cell lines derived from relevant host tissues. NE-0080, an inhibitor of an *H. pylori* binding to gastrointestinal epithelial cell lines, contains the non-reducing disaccharide epitope, NeuAcα2-3Galβ1.... previously proposed as the target for the *H. pylori* hemagglutinin (Evans, et al, 1988) and as a glycolipid binding receptor in gastric mucosa (Slomiany, et al, 1989; Saitoh, et al, 1991). The oligosaccharide is stable to gastric acid (< 5% hydrolysis per hour), is barely absorbed (bioavailability < 1%), and has an excellent safety profile in preclinical toxicology studies. It therefore appears well suited as a locally acting antiadhesive agent for treatment of *H. pylori* gastritis.

Evaluation of NE 1530 revealed a previously undescribed common adhesion ligand, NeuAcα2-6Gal..., specifically recognized *by S. pneumoniae, H. influenzae*, and *M. catarrhalis*. Evaluations of effectiveness of the compound against other pathogens is in progress. Ongoing experiments suggest that adhesion specificities of *H. influenzae* and *S. pneumoniae* may be differentially expressed in log phase, versus lag phase growth. When administered into the airway, NE-1530 can exert both prophylactic and therapeutic effects, promoting clearance of *S. pneumoniae* from the lung and reducing the incidence of septicemia as a complication of pneumonia. This compound will undergo continuing evaluation as an agent for prevention and/or treatment of upper and lower respiratory tract infections.

REFERENCES

Evans, D.G., Evans, D.J., Moulds, J.J. and Graham, D.Y. 1988. N-Acetylneuramyllactose-binding fibrillar hemgglutinin of *Campylobacter pylori:* a putative colonization factor antigen, Infect. Immun. 56: 2896-2906.

Falk, P., Roth, H..A., Boren. T., Westblom, T.U., Gordon, J.I., and Normark, S. 1993. An in vitro adherence assay reveals that *Helicobacter pylori* exhibits cell lineage-specific tropism in the human gastric epithelium. Proc. Natl. Acad. Sci. 90: 2035-2039.

Feldman, M. 1994. The acid test. Making sense of the consensus conference on *Helicobacter pylori*. J. Am. Med. Assoc. 272: 70-71.

Lingwood, C..A., Huesca, M. and Kuskis, A. 1992. The glycerolipid receptor for *Helicobacter pylori* (and exoenzyme S) is phosphatidylethanolamine. Infect. Immun. 60: 2460-2474.

Saitoh, T., Natomi, H., Zhao, W., Okuzumi, K., Sugano, K., Iwamori, M. and Nagai, Y. 1991. Identification of glycolipid receptors for *Helicobacter pylori* in TLC-immunostaining. FEBS. Lett. 282: 385-387.

Slomiany, B.L., Piotrowski, J., Samanta, A., VanHorn, K., Murty, V.L.N. and Slomiany, A. 1989. *Campylobacter pylori* colonization factor shows specificity for lactosylceramide sulfate and GM3 ganglioside. Biochem. Interntl. 19: 929-936.

Trust, T.J., Doig, P., Emody, L., Dienle, Z., Wadstrom, T. and O'Toole, P. 1991. High-affinity binding of the basement membrane proteins collagen type IV and laminin to the gastric pathogen *Helicobacter pylori*. Infect. Immun. 59: 4398-4404.

5

TOWARDS ANTI-*PSEUDOMONAS AERUGINOSA* ADHESION THERAPY

Nechama Gilboa-Garber

Department of Life Sciences
Bar-Ilan University
Ramat-Gan, Israel

Pseudomonas aeruginosa infections may develop in immunocompromised patients in almost every tissue/organ, including: skin, eyes, ears, nasopharynx, bronchi and lungs, heart, kidneys and urinary tract, intestinal tract, liver, spleen, bones, brain and blood (bacteremia). They are the major cause of death in cystic fibrosis patients and endanger the life of patients suffering from extensive burns, chronic diseases and immunosuppressive ailments and treatments (e.g., for cancer repression or tissue transplantations). Their endurance is dependent on the tissue/host background. They are generally secondary to either immunodeficiency and primary infections or to metabolic abnormalities, which alter cell surface composition, exposing galactose- or N-acetylgalactosamine-bearing glycolipids. In primary infections preceding them, such as influenza, there is either removal of sialic acids or glycopeptides by the infecting factor sialidases or proteases. In metabolic abnormalities such as cystic fibrosis, the reason for the reduced glycolipid sialylation may be either sialylation deficiency (Imundo *et al.*, 1995) or injurious sialo-/proteolysis caused by salivary or leukocytic enzymes.

The result of persistent *P. aeruginosa* infections is extensive lytic and hemorrhagic damage in the target tissues/organs that are destroyed by the bacterial virulence factors (Nicas and Iglewski, 1985), which are generally secondary metabolites (VSMs). The most potent virulence factors are exotoxin A (Liu, 1974) and exoenzyme S, which transfer ADP-ribose of NAD^+ to eukaryotic cell proteins. Exotoxin A blocks protein synthesis by acting on elongation factor EF_2 as a substrate and exoenzyme S acts on Ras and several other GTP-binding proteins.

In addition to them, the *P. aeruginosa* destructive arsenal also includes several families of lytic exoenzymes: metal proteases and elastase, hemolytic and nonhemolytic phospholipase C (lecithinase), phosphatases, amidases, chitinase and leukocidine. It also includes host cell-damaging slime polysaccharides and alginate, hemolytic (heat stable) rhamnolipids and the green pigment pyocyanin (5-methyl-1-hydroxyphenazine), which was shown to disturb lymphocyte proliferation (Sorensen *et al.*, 1983; Mühlradt *et al.*, 1986) and some of the host metabolic pathways,including 5-lipoxygenase activity and leukotriene B_4 biosynthesis (Muller and Sorrell, 1991).

Toward Anti-Adhesion Therapy for Microbial Diseases, edited by Kahane and Ofek
Plenum Press, New York, 1996

Concomitant with the above described VSMs, most *P. aeruginosa* strains also produce a battery of lectins and adhesins which enable the bacterium binding to the host cells by means of their interactions with abundant cell surface components. *P. aeruginosa* cells possess several groups of adhesins (including the sugar-specific lectins) which home onto sialic acids, hexoses (and their derivatives), amino acids, etc. The first *P. aeruginosa* lectin described was PA-I lectin (PA-IL). It was also the first galactophilic bacterial lectin found (Gilboa-Garber, 1972). PA-II lectin (PA-IIL) exhibits a mixed specificity: l-fucose>>d-mannose>fructose (Gilboa-Garber, 1982; Garber *et al.*, 1987). In addition to them, *P. aeruginosa* produces several other galactophilic lectins/adhesins (Prince, 1992), exhibiting preferential affinities for other galactosylated lipids and other galactose derivatives (Krivan *et al.*, 1988; Lanne *et al.*, 1994; Ramphal and Pyle, 1983; Hazlett *et al.*, 1986) as well as sialophilic and other lectins/adhesins and also hydrophobic adhesins (Garber *et al.*, 1985). Alginate (Ceri *et al.*, 1986; Baker and Svanborg-Eden, 1989) polysaccharides (Ramphal and Pier, 1985) and LPS of this bacterium have also been reported to display a considerable tissue adhesiveness. This multitude of adhesive components ensures the tight adherence of the bacterium to host cells with almost no limitation of cell types, age, state and extracellular matrix (ECM) composition.

These lectins/adhesins may support *Pseudomonas* infections by enabling the bacterial persistent attachment and adherence to the host cells, resisting sweeping by flooding body fluids or cilial movements. They may contribute to the wide tissue spectrum on one hand, and evasion of the host resistance and immune factors, as well as protection from them - on the other hand.

The selectivity for the host cells depends on which adhesin gains domination due to its highest compatibility to the host cell surface components and to the composition, as well as to conditions of the microenvironment, ECM, which harbors the bacteria. PA-IL, for example, might contribute to the bacterium adherence to a wide spectrum of sialic acid-deficient cells, owing to its preferential affinity to P system antigens in combination with I and also BI (Gilboa-Garber *et al.*, 1994). The absence of these antigens from the surface of human neutrophils is an advantage of such lectin selectivity. The *in vivo* biodistribution of purified PA-I and PA-II lectins (^{125}I-labeled) in mice also indicates a wide tissue binding with preference of lungs (both), spleen (PA-II>PA-I) and kidney (Fig. 1). The binding specificity of PA-I and PA-II lectins and related *P. aeruginosa* adhesins to asialylated cells may be crucial for the "secondary type" of the *Pseudomonas* infections.

There are indications that the infection aggressiveness is also supported by the bacterium lectins and adhesins. The lectin contribution to it may be both direct and indirect. Purified PA-I and PA-II lectins affect cells and tissues. Examples are clustering and weakening of cell membrane structure and osmotic fragility of blood cells (Gilboa-Garber and Blonder, 1979), and capping (Glick *et al.*, 1981) as well as the mitogenic stimulation in lymphocytes (Sharabi and Gilboa-Garber, 1979; Avichezer and Gilboa-Garber, 1987). Additional examples are the hypertrophy and increased polyamine metabolism in enterocytes (Grant *et al.*, 1995), respiratory cilial demolishing and cell dystrophy (Bajolet-Laudinat *et al.*, 1994), as well as apoptosis in tumor cells (Avichezer and Gilboa-Garber, 1991, Avichezer and Gilboa-Garber, 1995).

However, the more advanced service of the lectins and adhesins to the bacterium aggressiveness is probably their participation in homing the bacterium, which bears them, onto the right cells and targeting its potent exotoxins, exoenzymes and the other virulence factors at the surface of these cells in the right position, ensuring their maximal contact with the cell surface and highest activity on its components (Fig. 2).

The solid proofs of the tight linkage between *P. aeruginosa* lectins/adhesins and its virulence justify investing efforts in searching of anti-lectin/adhesion therapy. This important goal may be achieved by either inhibiting lectin/adhesion production, or by prevention of the lectin interaction with it's complementary receptors on the host cells. For anti adhesive therapy, the following routes and means may be applied (Fig. 3).

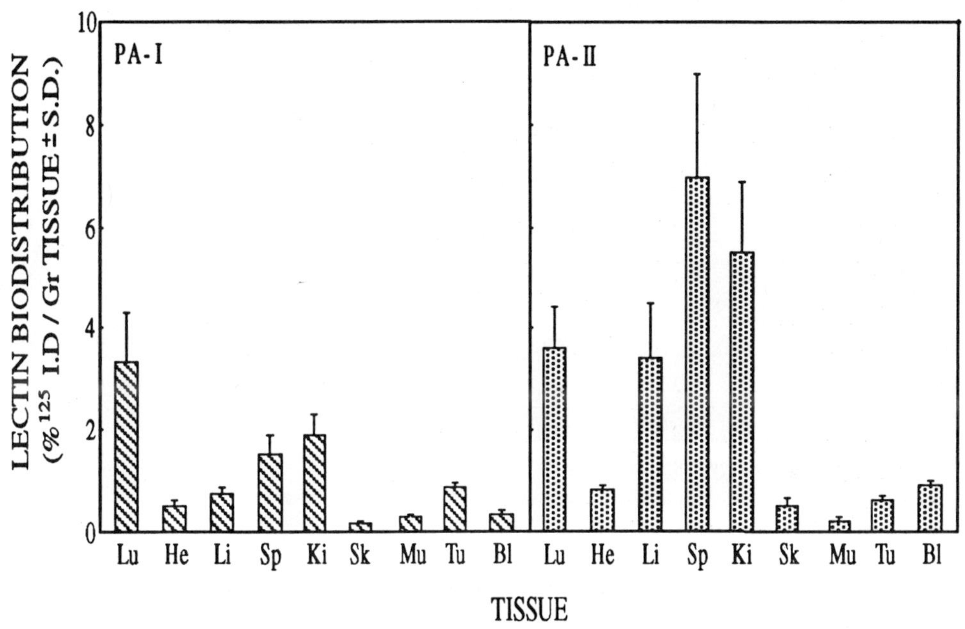

Figure 1. ^{125}I-PA-I and PA-II lectin distribution in ICR mice tissues: Lu = lung; He = heart; Li = liver; Sp = spleen; Ki =kidney; Sk =skin; Mu = muscle; Tu = tumor; Bl = blood (D. Avichezer).

1. PRODUCTION OF ANTIBODIES AND CELLULAR IMMUNITY AGAINST THE PURIFIED LECTINS/ADHESINS

Specific anti-lectin/adhesin antibodies may block lectin binding to the host cell receptors. We have shown that active immunization of mice by purified lectin preparations fully protect them from otherwise lethal infection (Fig. 4, Gilboa-Garber and Sudakevitz, 1982; Sudakevitz and Gilboa-Garber, 1987). As may be seen from this figure, passive immunization, by transfer of antibodies produced in rabbits, was also effective. The rabbit antibodies produced against the purified lectin preparations agglutinate those bacteria which bear the lectins (Fig. 5) and support bacteriolysis by complement (Sofer and Gilboa-Garber, 1995). Such antibodies may also increase phagocytosis of the bacteria, which is possibly reduced by the lectins, due to their low PI=4.9 (Avichezer *et al.*, 1992). Efficient protection of naive mice could also be obtained by immune cell transfer from actively immunized mice (Fig. 6, Avichezer *et al.*, 1989). However, it must be noted that antigenic drift and multiple adhesins produced by *P. aeruginosa* may handicap immunotherapy. At least a mixture of lectins and adhesins must be used as a vaccine.

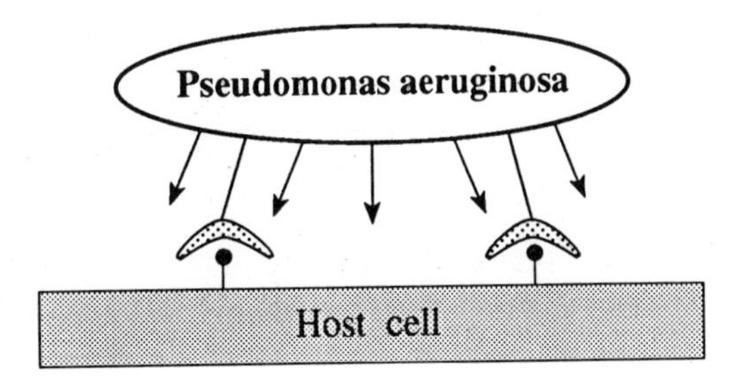

Figure 2. Binding of a *P. aeruginosa* cell to a host cell via specific lectins/adhesins enables focusing of the bacterium toxic factor activities (arrows) on the latter's membrane.

1. ANTI BACTERIAL IMMUNOTHERAPY

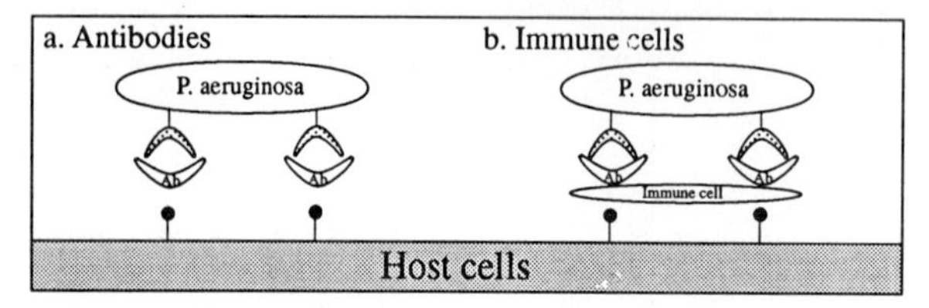

2. SPECIFIC SACCHARIDES (COMPETITION)

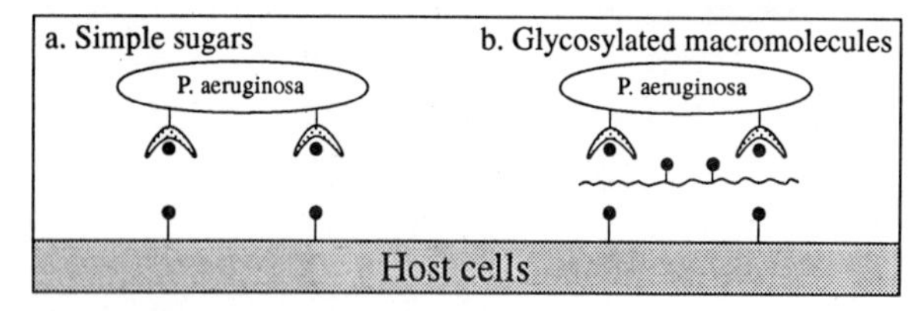

3. RECEPTOR BLOCKING

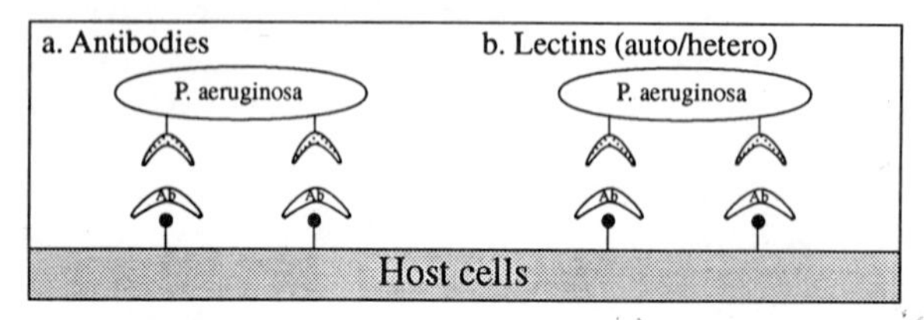

4. INHIBITION OF RECEPTOR/LECTIN SYNTHESIS

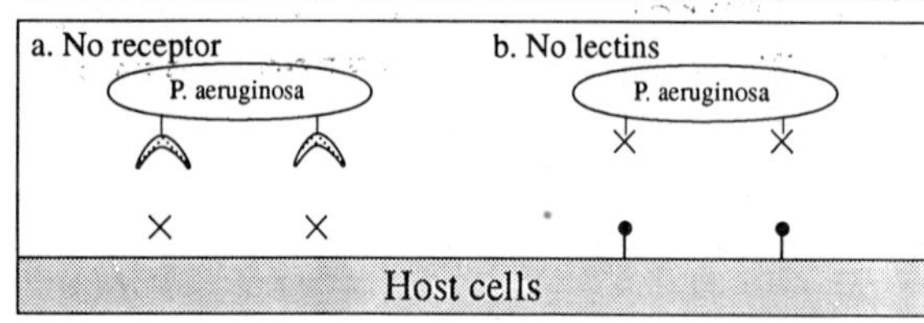

Figure 3. Main means for interference with *P. aeruginosa* lectin/adhesin binding to the respective host cell receptors.

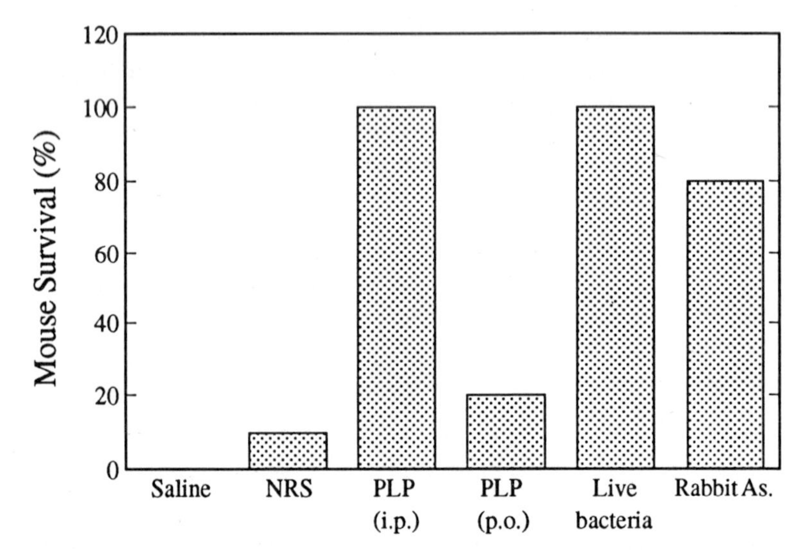

Figure 4. Protection of mice against a lethal dose of *P. aeruginosa* infection by their immunization with purified lectin preparations (PLP) inoculated i.p. or per os (p.o) as compared to live bacteria and passive immunization by rabbit antiserum.

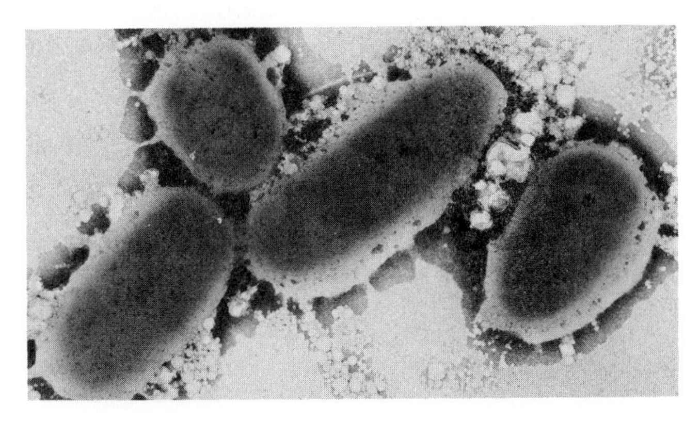

Figure 5. Agglutination of *P. aeruginosa* cells by rabbit antiserum produced against the purified PA-IL preparation.

2. BLOCKING THE LECTIN-BINDING TO THEIR RECEPTORS BY SPECIFIC SACCHARIDES

Simple sugars may be used for local treatment of external *P. aeruginosa* infections where they competitively block the lectin ability to bind to its receptors on the host cells (Steuer *et al.*, 1993). However, in case of internal or dissemenated infections, the sugars are not useful because they are quickly metabolized. There is the possibility of using glycoproteins, as those present in the body fluids, or glycolipids of the P system antigen type (Lanne *et al.*, 1994). However, usage of enzyme-resistant, nontoxic sugar derivatives, especially polymers (Mammen *et al.*, 1995), or analogs, that exhibit increase affinity for the lectin, is preferable. As an example: PA-I lectin of *Pseudomonas aeruginosa* is sensitive to inhibition

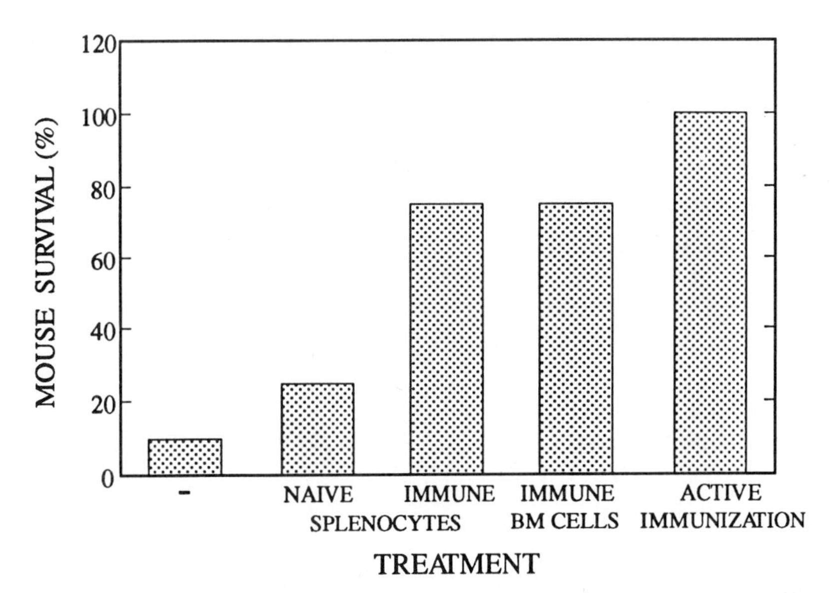

Figure 6. Protection of naive mice against lethal *P. aeruginosa* by immune cells (splenocytes and bone marrow cells from immunized mice) as compared to active immunization by purified PA-I preparations.

by biantennary AB blood group substances present in AB saliva (Gilboa-Garber and Mizrahi, 1985) and by phenyl-β-D-thiogalactoside (Garber *et al.*, 1992). Synthesis of a polymer bearing a series of this compound copies on it may be effective. As an analogous example, recently 2-S-methyl-5-N-thioacetylneuraminic acid was shown to inhibit replication of influenza virus owing to its resistance to sialidase and very high affinity to the virus hemagglutinin (Itoh *et al.*, 1995). Independently, Mammen *et al.* (1995) have shown that synthetic polymers containing multiple copies of sialic acids are much more effective than their monomeric counterparts in inhibition of influenza virus hemagglutinating activity. Since *P. aeruginosa* produces several lectin types concomitantly, mixtures of linked sugars would be needed for inhibition of its adherence. Steuer *et al.* (1993) used a mixture of 5% galactose, 5% mannose and 1% N-acetylneuraminic acid for treatment of *P. aeruginosa*-induced *otitis externa diffusa acuta* and found that this treatment was comparable to standard gentamicin therapy. Of course, such treatments may be applied only externally since the sugars are quickly metabolized. Analogs or stable polymers of these sugars would probably overcome this difficulty, but their intrinsic usage might have side effects on essential host cell-cell interactions.

3. BLOCKING OF LECTIN BINDING TO RECEPTORS BY ANTIBODY FRACTIONS (FAB'S) DIRECTED AGAINST THE RECEPTORS

Antibodies/Fab's directed against the host cell receptors might prevent the bacterial lectin-dependent adhesion to the same receptors. Application of this treatment depends on showing that the Abs or the Fab's do not have adverse effects on the cells or on the host. One of the complications due to such treatments may be stimulation of cytokine secretion and attraction of leukocytes, which may damage the tissues (DiMango *et al.*, 1995). Moreover, there is a need for blocking several receptor types specific for the respective adhesins.

4. REDUCING LECTIN/ADHESIN-DEPENDENT BACTERIAL ADHERENCE BY CELL-FREE ADHESIVE MOLECULES ISOLATED FROM THE BACTERIUM, OR SYNTHETIC PEPTIDES CONTAINING THE APPROPRIATE CARBOHYDRATE-BINDING ACTIVITY

Such preparations of native origin or synthetic peptides may hamper the bacterial-bound lectins from homing the bacteria to the host cell receptors.

5. APPLICATION OF HETEROLOGOUS BACTERIAL ADHESIN MIMICRY FOR COMPETITION WITH THE *P. AERUGINOSA* BOUND ADHESINS

An example for such a treatment is the findings that staphylococcal adhesin which binds to the same host cell receptor may have a competitive potential in inhibition of *P. aeruginosa* adhesion (Imundo *et al.*, 1995).

6. INHIBITION OF SYNTHESIS OF HOST CELL RECEPTORS

Inhibition of, or interference with receptor synthesis is a complicated way which might be performed by enzyme inhibition, gene therapy, or enzyme administration.

7. PREVENTION OF *P. AERUGINOSA* LECTIN/ADHESIN PRODUCTION

Inhibition of *Pseudomonas* lectin/adhesin synthesis is the most reasonable way for achieving the goal of anti-*Pseudomonas aeruginosa* adhesion therapy. The great advantage of this approach is inhibition of most adhesin type synthesis together with some of the important virulence factors. It may be performed by several means:

a. Addition of nonspecific compounds shown experimentally to inhibit lectin production. This is not very complicated since many compounds (including sugars, salts, ions or various organic acids) and temperature greatly affect lectin levels (Sudakevitz *et al.*, 1979; Gilboa-Garber, 1982). The treatment by such non-specific compounds is useful especially for external, local infections.
b. Antibiotic treatment. This treatment, even at subgrowth inhibitory concentration may be very useful for inhibition of lectin production (Gilboa-Garber, 1982; Fig. 7). This treatment inhibits not only the lectin production, but also that of some of the *Pseuomonas* virulence factors, such as protease (Fig. 8) (Shibl, 1980) and other virulence factors and adhesins (Molinari *et al.*, 1993; Piatti, 1994).

We have described as early as 15 years ago the close linkage in *P. aeruginosa* between the lectins PA-I and PA-II, protease, hemolysin and pyocyanin production (Gilboa-Garber *et al.*, 1981, 1982; Gilboa-Garber, 1986; Gilboa-Garber and Garber, 1989) based on examination of lectin-deficient mutants and strains. Since 1981 we have reported that lectin-deficient mutants and strains also exhibit protease and pyocyanin deficiency and lower

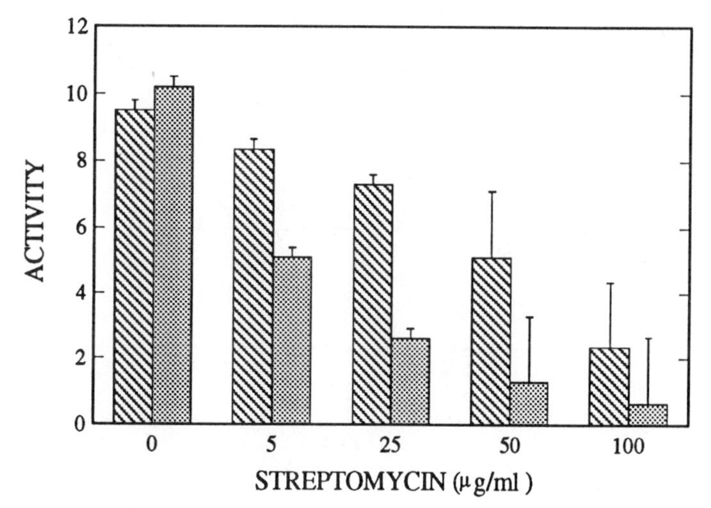

Figure 7. Inhibition of lectin and exoprotease production by streptomycin at different concentrations (Gilboa-Garber, 1982).

hemolysin levels, suggesting that these virulence factors are corregulated (Fig 8). These strains were also less virulent [considerably higher LD_{50} (Fig. 8)].

These findings of lectin-enzyme corregulation are now supported by the discovery of *P. aeruginosa* autoinducers and regulatory mechanisms. These systems include autoinducers analogous to the VAI, which is involved in the bioluminescence of *Photobacterium* (*Vibrio*) *fischeri* (Eberhard *et al.*, 1981; Bainton *et al.*, 1992; Passador *et al.*, 1993) and regulatory proteins (Meighen, 1991; Kaplan and Greenberg, 1987; Gambello and Iglewski, 1991; Gambello *et al.*, 1993; Pearson *et al.*, 1994, Brint and Ohman, 1995). The autoinducers, which are N-acyl derivatives of L-homoserine lactone, bind to the regulatory proteins endowing them with the ability to bind to promoters of structural genes coding secondary metabolism enzymes. The secondary metabolism enzymes are directly, or indirectly, involved in virulence - by synthesis of toxic factors (metabolites). The binding of the autoinducer - regulatory protein complex to the promoter activates the translation of the adjacent genes and lead to expression of the virulence factors (Gambello *et al.*, 1993), including elastase and alkaline protease. Latifi *et al.* (1995) showed that PAO1 mutants, which are elastase- and chitinase-less (PAN067) gain the ability to produce both enzymes as well as cyanide and pyocyanin when cultured in the presence of added autoinducers. Garber's group (Garber *et al.*, 1995) has similarly shown that our old mutants of *P. aeruginosa* ATCC 33347 strain, lacking PA-I and PA-II lectins together with protease and pyocyanin, are also defective in autoinducer production (Garber *et al.*, 1995). Autoinducer addition was shown to increase both lectin production (Hammer-Münz *et al.*, 1996) and the *in vivo* virulence of these mutants for mice. A cooperative research of both groups showed that addition of the autoinducer to the PAN067 mutants induced in them production of both lectins and virulence secondary metabolites: elastase, pyocyanin, chitinase and HCN (Camara *et al.*, 1995).

The antibiotic positive effect may also result from inhibition of the regulatory system. Its advantage, of course, is the elimination of both the lectins and the virulence factor production. Supporting this aspect are results indicating positive effects of long-term

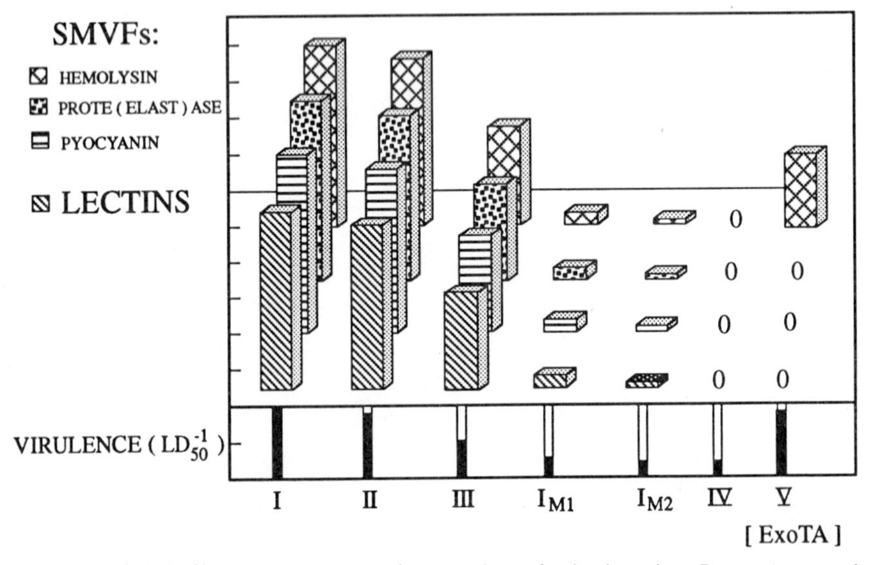

Figure 8. Corregulation of lectin, exoenzymes and pyocyanin production in various *P. aeruginosa* strains and mutants and their LD_{50} for mice.

erythromycin therapy in patients with chronic *P. aeruginosa* infection. Fujii *et al.* (1995) have recently reported that such treatment considerably improved respiratory functions irrespective of the presence or absence of *P. aeruginosa* in the sputum. The treatment also resulted in a reduction in the neutrophils in the respiratory fluid without differences between patients in whom the bacteria cleared and those with presistent bacteria, with regard to the degree of improvement of respiratory function, arterial blood gas tensions and the surrounding fluid composition. Fujii *et al.* (1995) suggested that the efficacy of the erythromycin treatment could be due to an anti-inflammatory effect, independent of the bacterial infection or clearence. However, our suggestion is, that the antibiotic effect could result from its inhibition of the VSM and lectin/adhesin production (Gilboa-Garber, 1982; Kita *et al.*, 1991; Sofer *et al.* 1996).The inhibition is probably including preceding autoinducer synthesis [also shown to induce IL-8 production and neutrophil invasion in respiratory epithelial cells (DiMango *et al.*, 1995)].

c. Specific knocking out the specific pathways providing the autoinducer or the regulatory protein leading (together) to positive regulation of the lectin and VSM production - by negative feedback regulation, or by competitive inhibition, of the autoinducer binding to that protein, using N-acyl-homoserine analogs.

CONCLUDING REMARKS

Several ways lead towards anti-bacterial adhesion therapy. Most of them prevent lectin/adhesin binding to the host cells, by blocking either the adhesive molecules or their host cell receptors, using antibodies or competitive inhibitors (glycoconjugates, lectins, etc.). These manipulations may be successful in treatment of external infections and of infections caused by organisms possessing uniform (one type) adhesins. However, most *P. aeruginosa* strains bear several types of lectins/adhesins, easily overcoming conventional blocking. Therefore, the most attractive and promising treatment is repression of the synthesis of autoinducers and positive regulators of lectin/adhesin production. This approach combines two huge advantages, not shared by the lectin - blocking treatments:

a. repression of the adhesive potential of the bacterium together with at least some of its virulence factors.
b. selective activity toward the bacterial target virulence - regulating system without harming the host.

ACKNOWLEDGMENTS

This project is a result of a multiannual cooperative work with the following colleagues and research assistants: Prof. N.C. Garber, Dr. D. Sudakevitz, Dr. D. Avichezer, Dr. C. Levene, Ms. A. Belz, Ms. H. Mymon, Ms. D. Sofer and Ms. B. Lerrer. The author thanks Ms. Ella Gindi and Ms. Sharon Victor for great help in the graphic presentations and preparation of the manuscript. This research is supported in part by Bar-Ilan University Research Foundation.

REFERENCES

Avichezer, D., and Gilboa-Gilboa, N., 1987, PA-II, the l-fucose and d-mannose binding lectin of *Pseudomonas aeruginosa* stimulates peripheral lymphocytes and murine splenocytes, *FEBS Lett.* 216:62-66.

Avichezer, D., Gilboa-Garber, N., Mumcuoglu, M., and Slavin, S., 1989, Adoptive transfer of resistance to *Pseudomonas aeruginosa* infection by splenocytes and bone marrow cells from BALB/c mice immunized by *Pseudomonas aeruginosa* lectin preparations, *Infection* 17:407-410.

Avichezer, D., and Gilboa-Garber, N., 1991, Anti-tumoral effects of *Pseudomonas aeruginosa* lectins on Lewis lung carcinoma cells cultured *in vitro* without and with murine splenocytes, *Toxicon* 29:1305-1313.

Avichezer, D., Katcoff, D. J., Garber, N. C., and Gilboa-Garber, N., 1992, Analysis of the amino acid sequence of the *Pseudomonas aeruginosa* galactophilic PA-I lectin, *J. Biol. Chem.* 267:23023-23027.

Avichezer, D., and Gilboa-Garber, N., 1995, Effects of *Pseudomonas aeruginosa* lectins on human ovarian, breast and oral epidermoid carcinoma cells, *Proc. 1st FISEB Meeting, Eilat (Israel)*, p. 192.

Bainton, N. J., Bycroft, B. W., Chhabra, S. R., Stead, P., Gledhill, L., Hill, P. J., Rees, C. E. D., Winson, M. K., Salmond, G. P. C., Stewart, G. S. A. B., and Williams, P., 1992, A general role for the lux autoinducer in bacterial cell signalling: control of antibiotic synthesis in Erwinia, *Gene* 116:87-91.

Baker, N. R., and Svanborg-Eden, C., 1989, Role of alginate in the adherence of *Pseudomonas aeruginosa*, in: *Pseudomonas aeruginosa Infection* (N. Hoiby, S. S. Pedersen, G. H. Sland *et al.*, eds.), Karger AG, Basel, pp. 72-79.

Bajolet-Laudinat, O., Girod-de Bentzmann, S., and Tournier, J. M., 1994, Cytotoxicity of *Pseudomonas aeruginosa* internal lectin PA-I to respiratory epithelial cells in primary culture, *Infect. Immun.* 62:4481-4487.

Brint, M; and Ohman, D. E., 1995, Synthesis of multiple exoproducts in Pseudomonas aeruginosa in the control of Rh1R-Rh1I, another set of regulators in strain PAO1 with homology to the auto-inducer responsive LuxR-LuxI family. J. Bacteriol. 177: 7155-7163.

Camara, M., Winson, M. K., Latifi, A., Falconer, C., Briggs, G. S., Chhabra, S. R., Foglino, M., Garber, N. C., Gilboa-Garber, N., Belz, A., Bycroft, B. W., Lazdunski, A., Stewart, G. S. A. B., and Williams, P., 1995, Multiple quorum sensing modulons interactively regulate production of virulence determinants and secondary metabolites in *Pseudomonas aeruginosa*, *Proc. "Pseudomonas 1995" 5th Int. Symp. on Pseudomonas Biotechnol. and Mol. Biol., Tsukuba, Japan*, p. 159.

Ceri, H., McArthur, H. A. I., and Whitfield, C., 1986, Association of alginate from *Pseudomonas aeruginosa* with two forms of heparin-binding lectin isolated from rat lung, *Infect. Immun.* 51:1-5.

DiMango, E., Zar, H. J., Bryan, R., and Prince, A., 1995, Diverse *Pseudomonas aeruginosa* gene products stimulate respiratory epithelial cells to produce interleukin-8, *J. Clin. Invest.* 96:2204-2210.

Eberhard, A., Burlingame, A. L., Eberhard, C., Kenyon, G. L., Nealson, K. H., and Oppenheimer, N. J., 1981, Structural identification of autoinducer of *Photobacterium fischeri* luciferase, *Biochemistry* 20:2444-2449.

Fujii, T., Kadota, J., Kawakami, K., Iida, K., Shirai, R., Kaseda, M., Kawamoto, S., and Kohno, S., 1995, Long term effect of erythromycin therapy in patients with chronic *Pseudomonas aeruginosa* infection, *Thorax* 50:1246-1252.

Gambello, M. J., and Iglewski, B. H., 1991, Cloning and characterization of the *Pseudomonas aeruginosa* las R gene, a transcriptional activator of elastase expression, *J. Bacteriol.* 173:3000-3009.

Gambello, M. J., Kaye, S., and Iglewski, B. H., 1993, Las R of *Pseudomonas aeruginosa* is a transcriptional activator of the alkaline protease gene (apr) and an enhancer of exotoxin A expression, *Infect. Immun.* 61:1180-1184.

Garber, N., Sharon, N., Shohet, D., Lam, J. S., and Doyle, R. J., 1985, Contribution of hydrophobicity to the hemagglutination reactions of *Pseudomonas aeruginosa*, *Infect. Immun.* 50:336-337.

Garber, N., Guempel, U., Gilboa-Garber, N., and Doyle, R. J., 1987, Specificity of fucose-binding lectin of *Pseudomonas aeruginosa*, *FEMS Microbiol. Lett.* 48:331-334.

Garber, N., Guempel, U., Belz, A., Gilboa-Garber, N., and Doyle, R. J., 1992, On the specificity of the d-galactose-binding lectin (PA-I) of *Pseudomonas aeruginosa* and its strong binding to hydrophobic derivatives of d-galactose and thiogalactose, *Biochim. Biophys. Acta* 1116:331-333.

Garber, N. C., Hammer-Müntz, O., Belz, A., and Krakower, Y., 1995, The lux autoinducer stimulates the production of the lectins of *Pseudomonas aeruginosa* , *ISM Lett.* 15:164.

Gilboa-Garber, N., 1972, Inhibition of broad spectrum hemagglutinin from *Pseudomonas aeruginosa* by d-galactose and its derivatives, *FEBS Lett.* 20:242-244.

Gilboa-Garber, N., 1982, *Pseudomonas aeruginosa* lectins, in: *Methods in Enzymology*, Vol. 83, "Complex Carbohydrates, Part D", [Colowick and Kaplan (Ginsburg), eds.], Academic Press, Inc., New York, pp. 378-385.

Gilboa-Garber, N., 1986, Lectins of *Pseudomonas aeruginosa*: properties, biological effects and applications, in: *Microbial Lectin and Agglutinins: Properties and Biological Activity* (D. Mirelman, ed.), John Wiley & Sons, New York, pp. 255-269.

Gilboa-Garber, N., and Blonder, E., 1979, Augmented osmotic hemolysis of human erythrocytes exposed to the galactosephilic lectin of *Pseudomonas aeruginosa, Israel J. Med. Sci.* 15:537-539.

Gilboa-Garber, N., Buxenbaum, R., Mizrahi, L., and Avichezer, D., 1981, Correlation between lectins and protease activities in *Pseudomonas aeruginosa, XI World Cong. Pathol., Jerusalem, Israel*, p. 98.

Gilboa-Garber, N., Mizrahi, L., Buxenbaum, R., and Sudakevitz, D., 1982, The *Pseudomonas aeruginosa* lectins are linked to the production of protease and other exocellular enzymes, *Israel J. Med. Sci.* 18:19.

Gilboa-Garber, N., and Sudakevitz, D., 1982, The use of *Pseudomonas aeruginosa* lectin preparations as a vaccine, in: *Advances in Pathology*, Vol. 1 (E. Levy, ed.), Pergamon Press, Ltd., Oxford, pp. 31-33.

Gilboa-Garber, N., and Mizrahi, L., 1985, *Pseudomonas* lectin PA-I detects hybrid product of blood group AB genes in saliva. *Experientia* 41:681-682.

Gilboa-Garber, N., and Garber, N., 1989, Microbial lectin cofunction with lytic activities as a model for a general basic lectin role, *FEMS Microbiol. Rev.* 63:211-222.

Gilboa-Garber, N., Sudakevitz, D., Sheffi, M., Sela, R., and Levene, C., 1994, PA-I and PA-II lectin interactions with the ABO(H) and P blood group glycosphingolipid antigens may contribute to the broad spectrum adherence of *Pseudomonas aeruginosa* to human tissues in secondary infections, *Glycoconjugate J.* 11:414-417.

Glick, J., Malik, Z., and Garber, N., 1981, Lectin-bearing protoplasts of *Pseudomonas aeruginosa* induce capping in human peripheral blood lymphocytes, *Microbios.,* 32:181-188.

Grant, G., Bardocz, S., Ewen, S. W. B., Brown, D. S., Duguid, T. J., Pusztai, A., Avichezer, D., Sudakevitz, D., Belz, A., Garber, N. C., and Gilboa-Garber, N., 1995, Purified *Pseudomonas aeruginosa* PA-I lectin induces gut growth when orally ingested by rats, *FEMS Immun. Med. Microbiol.* 11:191-196.

Hammer-Münz, O., Krakower, Y., and Garber, N.C., 1996, N-(3-oxohexanoyl)-L-homoserine lactone increases both lectin production in *Pseudomonas aeruginosa* and its virulence to mice, *ISM News* 20:21.

Hazlett, L. D., Moon, M., and Berk, R. S., 1986, *In vivo* identification of sialic acid as the ocular receptor of *Pseudomonas aeruginosa, Infect. Immun.* 51:687-689.

Imundo, L., Barasch, J., Prince, A., and Al-Awqati, Q., 1995, Cystic fibrosis epithelial cells have a receptor for pathogenic bacteria on their apical surface, *Proc. Natl. Acad. Sci. USA* 92:3019-3023.

Itoh, M., Hetterich, P., Isecke, R., Brossmer, R., and Klenk, H. -D., 1995, Suppression of influenza virus infection by an N-thioacetylneuraminic acid acrylamide copolymer resistant to neuraminidase, *Virology* 212:340-347.

Lanne, B., Ciopraga, J., Bergstrom, J., Motas, C., and Karlsson, K.-A., 1994, Binding of the galactose-specific *Pseudomonas aeruginosa* lectin, PA-I, to glycosphingolipids and other glycoconjugates, *Glycoconjugate J.* 11:292-298.

Latifi, A., Winson, M. K., Foglino, M., Bycroft, B. W., Stewart, S. A. B., Lazdunski, A., and Williams, P., 1995, Multiple homologues of Lux R and Lux I control expression of virulence determinants and secondary metabolites through quorum sensing in *Pseudomonas aeruginosa* PAO1, *Mol. Microbiol.* 17:333-343.

Liu, P.V., 1974, Extracellular toxins of *Pseudomonas aeruginosa, J. Infect. Dis.* 130(suppl.):94-99.

Kaplan, H. B., and Greenberg, E. P., 1987, Overproduction and purification of the lux R gene product: transcriptional activation of the *Vibrio fischeri* luminescence system. *Proc. Natl. Acad. Sci. USA* 84:6639-6643.

Kita, E., Sawaki, M., Oku, D., Hamuro, A., Mikasa, K., Konishi, M., Emoto, M., Takeuchi, S., Narita, N., and Kashiba, S., 1991, Suppression of virulence factors of *Pseudomonas aeruginosa* by erythromycin, *J. Antimicrob. Chemther.* 27:273-284.

Krivan, H. C., Ginsburg, V., and Roberts, D. D., 1988, *Pseudomonas aeruginosa* and *Pseudomonas cepacia* isolated from cystic fibrosis patients bind specifically to gangliotetraosylceramide (asialo GM_1) and gangliotriaosyl-ceramide (asialo GM_2), *Arch. Biochem. Biophys.* 260:493-496.

Mammen, M., Dahmann, G., and Whitesides, G. M., 1995, Effective inhibitors of hemagglutination by influenza virus synthesized from polymers having active ester groups. Insight into mechanism of inhibition, *J. Med. Chem.* 38:4179-4190.

Meighen, E. A., 1991, Molecular biology of bacterial bioluminescence, *Microbiol. Rev.* 55:123-142.

Molinari, G., Guzman, C. A., Pesce, A., Schito, G. C., 1993, Inhibition of *Pseudomonas aeruginosa* virulence factors by subinhibitory concentrations of azithromycin and other macrolide antibiotics, *J. Antimicrob. Chemother.* 31:681-688.

Mühlradt, P.F., Tsai, H., and Conradt, P., 1986, Effects of pyocyanine, a blue pigment from *Pseudomonas aeruginosa*, on separate steps of T cell activation: interleukin 2 (IL2) production, IL2 receptor formation, proliferation and induction of cytolytic activity, *Eur. J. Immunol.* 16:434-440.

Muller, M., and Sorrell, T. C., 1991, Production of leukotriene B_4 and 5-hydroxyeicosatetraenoic acid by human neutrophils is inhibited by *Pseudomonas aeruginosa* phenazine derivatives, *Infect. Immun.* 59:3316-3318.

Nicas, T. I., and Iglewski, B. H., 1985, The contribution of exoproducts to virulence of *Pseudomonas aeruginosa*, *Can. J. Microbiol.* 31:387-392.

Passador, L., Cook, J. M., Gambello, M. J., Rust, L., and Iglewski, B. H., 1993, Expression of *Pseudomonas aeruginosa* virulence genes required cell-to-cell communication, *Science* 260:1127-1130.

Pearson, J. P., Gray, K. M., Passador, L., Tucker, K. D., Eberhard, J. P., Iglewski, B. H. and Greenberg, E. P., 1994, Structure of autoinducer required for expression of *Pseudomonas aeruginosa* virulence genes, *Proc. Natl. Acad. Sci. USA* 91:197-201.

Piatti, G., 1994, Bacterial adhesion to respiratory mucosa and its modulation by antibiotics at sub-inhibitory concentrations, *Pharmacol. Res.* 30:289-299.

Prince, A., 1992, Adhesins and receptors of *Pseudomonas aeruginosa* associated with infection of the respiratory tract, *Microbiol. Pathog.* 13:251-260.

Ramphal, R., and Pyle, M., 1983, Evidence for mucins and sialic acid as receptors for *Pseudomonas aeruginosa* in the lower respiratory tract, *Infect. Immun.* 41:339-344.

Ramphal, R., and Pier, G. B., 1985, Role of *Pseudomonas aeruginosa* mucoid exopolysaccharide in adherence to tracheal cells, *Infect. Immun.* 47:1-4.

Sharabi, Y., and Gilboa-Garber, N., 1979, Mitogenic stimulation of human lymphocytes by *Pseudomonas aeruginosa* galactosephilic lectin, *FEMS Microbiol. Lett.* 5:273-276.

Sheth, H. B., Lee, K. K., Wong, W. Y., Srivastava, G., Hindsgaul, O., Leung, O., Krepinsky, G., Hodges, R.S., Paranchych, W., and Irvin, R.T., 1994, The pili of *Pseudomonas aeruginosa* strains PAK and PAO bind specifically to the carbohydrate sequence βGalNAc(1-4)βGal found in glycosphingolipids asialo-GM_1 and asialo-GM_2, *Mol. Microbiol.* 11:715-723.

Shibl, A. M., and Al-Sowaygh, I. A., 1980, Antibiotic inhibition of protease production by *Pseudomonas aeruginosa*, *J. Med. Microbiol.* 13:345-349.

Sofer, D., and Gilboa-Garber, N., 1995, Effect of antibodies to *Pseudomonas aeruginosa* lectins in the presence of complement on the bacterium growth *in vitro* and its harm *in vivo*, *Proc. 1st F.I.S.E.B. Meeting, Eilat, Israel*, p. 134.

Sofer, D., Gilboa-Garber, N. and Garber, N.C., 1996, Erythomycin, which does not affect *Pseudomonas aeruginosa* growth, blocks the corregulated production of virulence factors and lectins by inhibiting synthesis of corregulatory autoinducer molecules, *Proc. 8th IUMS Congress of Bacteriol. and Applied Microbiol., Jerusalem, Israel.*

Sorensen, R.U., Klinger, J.D., Cash, H.A., Chase, P.A., and Dearborn, G.D., 1983, In vitro inhibition of lymphocyte proliferation by *Pseudomonas aeruginosa* phenazine pigments, *Infect. Immun.* 41:321-330.

Steuer, M. K., Herbst, H., Beuth, J., Steuer, M., Pulverer, G., Matthias, R., 1993, Inhibition of lectin mediated bacterial adherence by receptor blocking carbohydrates in patients with *Pseudomonas aeruginosa* induced otitis externa: A prospective phase-II study, *Otorhinolaryngol. Nova* 3:19-25.

Sudakevitz, D., Gilboa-Garber, N., and Mizrahi, L., 1979, Regulation of lectin production in *Pseudomonas aeruginosa* by culture medium composition, *Isr. J. Med. Sci.* 15:97.

Sudakevitz, D., and Gilboa-Garber, N., 1987, Immunization of mice against various strains of *Pseudomonas aeruginosa* by using *Pseudomonas* lectin vaccine, *FEMS Microbiol. Lett.* 43:313-315.

6

INHIBITION OF BACTERIAL ADHESION AND INFECTIONS BY LECTIN BLOCKING

J. Beuth, B. Stoffel, G. Pulverer

Institute of Medical Microbiology and Hygiene
University of Köln
Goldenfelsstr. 19-21
50935 Köln, Germany

Bacterial attachment (e.g. to mucosal and solid surfaces) may be considered to be a prerequisite for the colonization and infection of numerous tissues and foreign devices. In general, attachment of microorganisms to surfaces of epithelial cells is the first of a series of events that may include organ colonization and invasion of the host.

Bacterial agglutinins and/or hemagglutinins were first discovered in 1902, one year after discovery of human blood groups by Landsteiner. In the mean time, considerable evidence has been accumulated showing the fundamental role that lectin-carbohydrate interactions play in cell-cell recognition, as well as in microbial pathogenicity. On the one hand, these lectins increase the virulence of bacteria by enhancing microbial adhesion to epithelial cells or solid surfaces; on the other, they decrease virulence by enhancing phagocytosis. It may be supposed that, in the near future, much more will be learned about bacterial lectins, and many others will probably be isolated, purified and characterized. A detailed knowledge of the binding sites of these lectins and their receptors should lead to the design of potent inhibitors of adhesion, which will hopefully be suitable for testing in humans. We shall also gain a more detailed understanding of lectinophago-cytosis/opsono-phagocytosis and its role in natural defense. This outlook, indeed, is promising. By understanding the molecular biology of adhesion and the different forms of nonspecific adherence, we can alter Paul Ehrlich's postulates originally meant for "chemical corpora" into "bacteria non agunt, nisi fixata". Accordingly, it is the task for the future to find out which events, signals, and messengers follow the fixation and what happens after the recognition of the specific and/or nonspecific receptor.

BIOLOGY OF LECTINS

Living cells and tissues generally express surface carbohydrates that participate in intercellular interactions and in reactions of the cells with a variety of molecules. Accordingly, antibodies, agglutinins, toxins, or transmitter substances interact with cell-bound carbohydrates, triggering cell surface alterations, signal transmission or metabolic activities.

Toward Anti-Adhesion Therapy for Microbial Diseases, edited by Kahane and Ofek
Plenum Press, New York, 1996

Because of their relative abundance at the cell surface, carbohydrates are preferentially selected as microbial receptors rather than peptides. This form of adhesion depends upon lectins (adhesins) on the surface of microorganisms and on the tip of filamentous appendages (fimbriae, fibrillae), respectively. It is considered to be highly specific and responsible for the tissue tropism of pathogenic and symbiotic microorganisms [1-3].

Since the turn of the century, it has been known that cell agglutinating proteins (hemagglutinins) are widely distributed in nature. Such proteins were first found in plants and were therefore known as phytagglutinins, phytohemagglutinins or, more recently, lectins. Lectins are ubiquitious proteins/glycoproteins that exhibit a specific and reversible carbohydrate-binding activity [1-5]. They react with glycosylated macromolecules or cells, may coaggregate them or lead to their lysis or other alterations [3, 6]. The term "lectin" (Latin: legere, to select) was based on the observation that some seed extracts could discriminate among human blood groups. The first discription of hemagglutinating activity of a plant protein from Ricinus communis was presented by Stillmark in 1888 [7]. This hemagglutination was shown to be reversible and inhibitable by D-galactose. During the last decade, many lectins were detected in plants and in micro-/macroorganisms [3, 8]. Their specific binding was compared to that of enzymes and antibodies, which were both excluded from lectins by definition [2, 3]. In the course of investigations, lectins have been shown to be useful tools for various scientific approaches, including detection and identification of blood groups and bacteria, mitogenic stimulation of mononuclear immune cells, detection of carbohydrates in solutions, on macromolecules and cells, protein purification and cell fractionation [9-11]. Furthermore, they could be shown to be involved in specific adhesion of symbiotic and pathogenic microorganisms to host tissue, in specific adhesion of tumor cells to organ cells in metastatic spread, in certain interactions with the cellular immune system and in removal of senescent molecules or cells [5, 12, 13].

PLANT LECTINS: CLINICALLY RELEVANT IMMUNOMODULATORS

Recent investigations have shown that defined non-toxic doses of galactoside-specific mistletoe lectin (Viscum album agglutinin-1, VAA-1) have immunomodulatory potency [12]. The well known ability of certain lectins (e.g. VAA-1, Con A) to activate (non)specific defense mechanisms supports the assumption that lectin-carbohydrate interactions may induce clinically beneficial immunomodulatory activities [14]. In vitro investigations with VAA-1 showed enhanced expression of lymphocyte, monocyte and granulocyte activities, evident cytokin release of human mononuclear immune cells and no enhancement of tumor cell proliferation [15, 16]. In vivo, regular subcutaneous administration of the optimal immunomodulating VAA-1 dose (1 ng/kg BW, twice a week) yielded: enhanced thymocyte proliferation, maturation and emigration as well as significant antimetastatic effects in different murine tumor models; statistically increased counts and activity of peripheral blood lymphocytes and natural killer (NK)-cells; greatly increased serum levels of acute phase reactants as indicators of cellular and humoral immunopotentiation in cancer patients; significantly increased serum levels of ß-endorphins correlating to improved quality of life in breast cancer patients [17-19]. Currently, various prospectively randomized clinical multicenter trials are in progress to evaluate the effects of VAA-1 treatment in cancer patients on tumor progress, relapse rate, metastasis, and on survival of patients [20].

ORGAN LECTINS: ADHESION MOLECULES IN METASTASIS

Adhesion and inhibition experiments with parenchymal cells and tumor cells indicated that lectins mediate specific cellular interactions. Both in vitro and in vivo (murine models) the adhesion of tumor cells to parenchymal cells (e.g. of the liver and lung) could be inhibited by lectin-blocking glycoconjugates while non-specific glycoconjugates did not interfere with the adhesion process [5]. Accordingly, when organ-characteristic lectins were blocked with competitive receptor analogues, tumor colonization to murine liver and lung could be significantly reduced [21, 22]. These experimental approaches encouraged clinical trials to prevent metastatic liver colonization by hepatic lectin blocking with receptor analogue carbohydrate (D-galactose) in colorectal cancer patients. Since initial prospectively randomized clinical studies were very promising (with benefical effects for patients), multicenter studies are currently considered to substantiate these data.

BACTERIAL LECTINS IN INFECTIOUS DISEASES

Consideration of the host-parasite interaction encompasses a wide range of phenomena, from adhesion to mucosal and/or solid surfaces to interactions with cells of the immune system. Bacterial attachment is a means of colonizing the appropriate ecological niche to avoid being swept away by mucosal secretions. It may be considered to be the first step in infectious diseases [23-25].

Mucosal surfaces are characterized by an extensive carbohydrate coat contributed by glycoproteins and glycolipids, as well as by more loosely associated mucin glycoproteins. Accordingly, it is not surprising that most bacterial adhesins have evolved to act as lectins, using carbohydrates as receptor site [26]. Certain pathogens display a predilection for specific tissues, which was previously shown in cases of microbial colonization, e.g. of the gastrointestinal tract by *Vibrio cholerae*, of lungs and meninges by *Streptococcus pneumoniae*, of urinary tract by *Staphylococcus saprophyticus* and others [23, 27]. However, it was a long time until the concept of receptor site became recognized in microbiology. Recent work has demonstrated that carbohydrate specific bacterial adhesins (lectins, hemagglutinins) may recognize the carbohydrate receptor both in an internal and terminal position [3, 28]. However, chemical groups distal as well as proximal to the oligosaccharide receptor site in the glycoprotein/glycolipid may in some cases enhance binding and in others sterically interfere with receptor binding [3, 29]. Accordingly, two lectins recognizing different epitopes on the same oligosaccharide may have a considerably different tissue-binding distribution (organotropy) due to different response of neighbouring groups. Previously, considerable evidence was accumulated showing that bacteria interact with components of the intercellular matrix, as well as with cell membrane receptors [30, 31]. Apart from lectin-mediated specific adhesion, non-specific adherence mechanisms (e.g. electrostatic forces, lipophilic/hydrophobic interactions, binding to fibronectin, vitronectin, laminin, and collagen) are common properties of bacteria [3, 13].

To demonstrate the lectin-carbohydrate mechanism of bacterial adhesion, to prove that anti-adhesion therapeutics may be introduced for clinical administration and to check the hypothesis that the blood group status may determine the susceptibility to bacterial infections, laboratory and clinical studies were performed with patients suffering from *S. saprophyticus*-induced urinary tract infections (UTIs) and from *P. aeruginosa*-induced otitis externa diffusa acuta.

S. saprophyticus is a coagulase-negative, novobiocin-resistant microorganism that frequantly causes UTIs, especially in young women [32]. Hemagglutination of sheep red blood cells and adherence to various cell types are associated with this microorganism [33]. Several possible virulence factors have been suggested for *S. saprophyticus*, including surface lectins, urease, lipoteichoic acid and it was speculated that staphylococcal lectins apparently mediate the specific adhesion to host tissue [34, 35]. The surface lectin specificity of *S. saprophyticus* isolates from UTIs and their possible interaction with blood group characteristic antigens (carbohydrates) were checked and an attempt was made to correlate *S. saprophyticus*-induced UTIs with patients blood groups. The medical records of 55 individuals with moderate or heavy growth ($\geq 10^4$ colony forming unit per ml, pure culture) of *S. saprophyticus* in midstream urine samples were studied. The findings can be summarized as follows: *S. saprophyticus*-induced UTIs seem to be sex and age dependent since they were found almost exclusively among female patients (94.5%) not older than 40 years (76.4%); acute cystourethritis was the predominant clinical correlate of UTIs associated with *S. saprophyticus*; preexisting disease was not a condition for *S. saprophyticus*-induced UTIs. Furthermore, this study proved that UTIs caused by *S. saprophyticus* can be positively correlated with the patients' blood group. Thus, the A/AB phenotypes of the ABO blood group system could be determined in 94.5% of patients. This percentage was significantly different from the regular distribution of the A/AB phenotypes among the population of Western Europe (48.2%). This obviously correlates with the N-acetylgalactosamin (GalNAc) lectin-specificity of *S. saprophyticus* isolates from UTIs [36] as GalNAc represents the terminal carbohydrate moiety and the most important antigenic determinant of blood group A. Apparently, the blood group antigens (terminal carbohydrates) represent receptors recognized by *S. saprophyticus* surface lectins [37].

Pseudomonas aeruginosa, a gram-negative rod-shaped microorganism has emerged as one of the predominant pathogens in the antibiotic era, in which antibiotic pressure obviously is responsible for the change in pathogens. As an opportunistic human pathogen, it is currently the major cause of morbidity and mortality in seriously ill hospitalized patients suffering from immunosuppression, cancer, cystic fibrosis or otitis externa [5, 38].

Adherence and organotropism of *P. aeruginosa* to host tissue is strain dependent. It is also influenced by numerous factors like cell surface carbohydrates and/or microbial lectins as well as by alterations of host tissues [39, 40]. Thus, adherence of *P. aeruginosa* could be shown to be directly related to decreased amounts of cell surface fibronectin and increased levels of salivary proteases. Strong adherence to injured tissue could be demonstrated by scanning electron microscopy with uninfected and virus-infected murine tracheas [41, 42]. Apparently, specific receptors and adhesion molecules (lectins) are necessary to influence this characteristic behaviour which could be shown experimentally by blocking the adherence of *P. aeruginosa* to host cells or tissues [5, 39, 40]. Promising preclinical experimental studies encouraged to perform a clinical trial in humans, proving the efficacy of alternative drugs to antibiotics to minimize side effects of local application of antibiotics (e.g. development of resistance, induction of allergic reactions). Basing on the fact that surface lectins of *P. aeruginosa* could be shown to be specific for D-galactose (Gal), D-mannose (Man) and N-acetyneuraminic acid (NeuAc) and since this bacterium is found in more than 50% of patients suffering from acute otitis externa, a double blind trial was started to compare the efficacy and tolerability of *P. aeruginosa* lectin specific carbohydrate solution with the *P. aeruginosa* sensitive aminoglycosid gentamicin. The carbohydrate solution for local adminstration contained 5% Gal, 5% Man, and 1% NeuAc. A standard gentamicin preparation was locally administered to a control group of patients with *P. aeruginosa* induced otitis externa. Of 58 patients originally enrolled into this study, 36 (62%) were culturally positive for *P. aeruginosa*. Summing up the study an average of 7.42 days was required for patients locally treated with the specific carbohydrate preparation (5.92 days for the gentamicin-treated group) to return to normal appearance of the external auditory canal. The subjective degree of pain was the decisive parameter for the number of therapeutic days, whereas swelling and secretion disappeared faster in both groups. Since both therapeutical approaches presented with comparable results, local administration of microbial lectin-specific carbohydrate solution should be considered to be a therapeutical alternative without side effects, disadvantages and at lower costs [43]. In another prospective clinical study with 133 patients enrolled, the correlation between *P. aeruginosa*-induced otitis externa acuta and ABO blood group distribution was investigated [44]. Indications of such an association had already been observed in a clinical trial demonstrating that the adhesion of *P. aeruginosa* to outer ear canal epithelium cells could be blocked by administration of microbial lectin-specific carbohydrate solution [45, 46]. In this study with 133 patients, *P. aeruginosa* was the most frequently isolated pathogen (25% in mixed infections; 32% in monoinfections) leading to otitis externa acuta. The blood group distribution of the 43 patients suffering from *P.aeruginosa* monoinfections was as followes: 36 patients (83.7%) presented with blood groups A/AB, 7 patients (16.3%) presented with blood group O, no patient presented with blood group B. Statistical evaluation proved a highly significant correlation between *P. aeruginosa*-induced otitis externa acuta and the blood group A/AB phenotypes. These results support the hypothesis, that *P. aeruginosa* adhers to distinct receptors of the outer ear canal epithelium.

Apparently, patients with blood group A/AB have a genetic disposition to *P. aeruginosa*-associated otitis externa acuta and *S. saprophyticus*-induced UTIs. Experimental studies are currently in progress to further substantiate the role of ABO blood groups in infectious diseases and the therapeutical possibilities to combat bacterial adhesion (leading to organ colonization and infectious diseases) by blocking of microbial surface lectins. Initial clinical experience is very promising.

REFERENCES

1. Beachey, E.H., Bacterial adherence Chapman and Hall, London, 1980.
2. Liener, E., Sharon, N., Goldstein, I.J., The lectins, Academic Press, London, 1986.
3. Mirelman, D.,Microbial lectins and agglutinins, John Wiley & Sons, New York, 1986.
4. Boedecker, E.C., Attachment of organisms to the gut mucosa, CRC Press, Boca Raton, 1984.
5. Beuth, J., Pulverer, G., Lectin blocking: New strategies for the prevention and therapy of tumor metastatis and infectious diseases, G. Fischer Verlag, Stuttgart, Jena, New York, 1994.
6. Nowell, P.C., Phytohemagglutinin: an initiator of mitosis in cultures of normal human leukocytes. Cancer Res. 20, 462, 1960.
7. Stillmark, H., Ueber Ricin, ein giftiges Ferment aus dem Samen von Ricinus comm. und einigen anderen Euphorbiaceen, Inaug. Diss., Dorpat, 1888.
8. Gold, E.R., Balding, P., Receptor specific proteins. Plant and animal lectins, Excerpta Medica, Amsterdam, 1975.
9. Barondes, S.H., Lectins: their multiple endogenous cellular functions, Ann. Rev. Biochem. 50, 207, 1981.
10. Uhlenbruck, G., Die Biologie der Lektine: Eine biologische Lektion, Funkt. Biol. Med. 2, 40, 1983.
11. Prokop, O., Uhlenbruck, G., A new source of antibody-like substances having anti-blood group specificity, Vox Sang. 14, 321, 1968.
12. Gabius, H.-J., Gabius, S., Lectins and glycobiology, Springer Verlag, Berlin, 1993.
13. Richardson, P.D., Steiner, M., Priciples of cell adhesion, CRC Press, Boca Raton, 1995.
14. Beuth, J., Stoffel, B., Ko, H.L., Jeljaszewicz, J., Pulverer, G., Mistellektin-1: neue therapeutische Perspektiven in der Onkologie, Onkologie 18, 36, 1995.
15. Hajto, T., Hostanska, K., Gabius, H.-J., Modulatory potency of the β-galactoside-specific lectin from mistletoe extracts (Iscador) on the defence system in rabbits and patients, Cancer Res. 49, 4803, 1989.
16. Braun, J.M., Gemmell, C.G., Beuth, J., Ko, H.L., Pulverer, G., Respiratory burst of human polymophonuclear leukocytes in response to the galactoside-specific mistletoe lectin, Zbl. Bakt. 283, 90, 1995.
17. Beuth, J., Ko, H.L., Tunggal, L., Steuer, M.K., Geisel, J., Jeljaszewicz, J., Pulverer, G., Thymocyte proliferation and maturation in response to galactoside-specific mistletoe lectin-1, *In vivo* 7, 407, 1993.
18. Heiny, B.M., Beuth, J., Mistletoe extract standardized for the galactoside-specific lectin (ML-1) induces β-endorphin release and immunopotentiation in breast cancer patients, Anticancer Res. 14, 1339, 1994.
19. Beuth, J., Ko, H.L., Gabius, H.-J., Pulverer, G., Influence of treatment with the immunomodulatory effective dose of the β-galactoside-specific lectin from mistletoe on tumor colonization in BALB/c-mice for two experimental model systems, *In Vivo* 5, 29, 1991.

20. Steuer, M.K., Bonkowsky, V., Strutz, J., Beuth, J., Kau, R.J., Ambrosch, P., Neiß, A., Kühn, A., Arnold, W., Pulverer, G., Ko, H.L., Immundefizienz bei Patienten mit Kopf-Hals-Plattenepithelkarzinomen: Eine rationale Grundlage für den adjuvanten Einsatz des Immunmodulators ML-1, Otorhinolaryngol. Nova 4, 152, 1994.
21. Beuth, J., Ko, H.L., Schirrmacher, V., Uhlenbruck, G., Pulverer, G., Inhibition of liver tumor cell colonization in two animal models by lectin-blocking with D-galactose and arabinogalactan. Clin. Exp. Metastasis 6, 115, 1988.
22. Beuth, J., Ko, H.L., Pulverer, G., Uhlenbruck, G., Pichlmaier, H., Importance of lectins for the prevention of bacterial infections and cancer metastases, Glycoconjugate J., 12, 1, 1995.
23. Ofek, I., Beachey, E.H., Sharon, N., Surface sugars of animal cells as determinants of recognition in bacterial adherence, Trends Biochem. Sci. 3, 159, 1978.
24. Beachey, E.H., Bacterial adhesion: adhesin-receptor interactions mediating the attachment of bacteria to mucosal sursfaces, J. Infect. Dis. 143, 325, 1981.
25. Wadström, T., Trust, T.J., Microbial surface lectins, Med. Microbiol. 4, 287, 1984.
26. Sharon, N., Bacterial lectins, cell-cell recognition and infectious disease, FEBS Lett. 217, 145, 1987.
27. Beuth, J., Ko, H.L., Pulverer, G., Lectin-mediated bacterial adhesion to human tissue, Europ. J. Clin. Microbiol. 6, 591, 1987.
28. Weir, D.M., Carbohydrates as recognition molecules in infection and immunity, FEMS Microbiol. Immunol. 47, 331, 1989.
29. Jackson, G., Schlumberger, H.D., Zeiler, H.J., Current topics in infectious diseases and clinical microbiology, Vieweg Verlag, Wiesbaden, 1989.
30. Ryden, C., Rubin, K., Speziale, P., Höök, M., Lindberg, M., Wadström, T., Fibronectin receptors from *Staphylococcus aureus*, J. Biol. Chem. 258, 3396, 1983.
31. Hasky, D.L., Simpson, W.P., Effects of fibronectin and other salivary macromolecules on the adherence of *Escherichia coli* to buccal epithelial cells, Infect. Immun. 55, 2103, 1987.
32. Jordan, P.A., Iravani, A., Richard, G.A., Baer, H., Urinary tract infection caused by *Staphylococcus saprophyticus*, J. Infect. Dis. 142, 510, 1980.
33. Beuth, J., Ko, H.L., Schumacher-Perdreau, F., Peters, G., Heczko, P., Pulverer, G., Hemagglutination by *Staphylococcus saprophyticus* and other coagulase-negative staphylococci, Microbial Pathogenesis 4, 379, 1988.
34. Gatermann, S., John, S., Marre, R., *Staphylococcus saprophyticus* urease: characterization and contribution to uropathogenicity in unobstructed urinary tract infections in rats, Infect. Immun. 57, 110, 1989.
35. Beuth, J., Ko, H.L., Ohshima, Y., Yassin, A., Uhlenbruck, G., Pulverer, G., The role of lectins and lipoteichoic acid in adherence of *Staphylococcus saprophyticus*, Zbl. Bakt. 268, 357, 1988.
36. Beuth, J., Ko, H.L., Tunggal, L., Pulverer, G., Harnwegsinfektionen durch *Staphylococcus saprophyticus*: Gehäuftes Auftreten in Abhängigkeit von der Blutgruppe. Dtsch. med. Wschrft. 117, 687, 1992.
37. Geisel, J., Steuer, M.K., Ko, H.L., Beuth, J., The role of ABO blood groups in infections induced by *Staphylococcus saprophyticus* and *Pseudomonas aeruginosa*, Zbl. Bakt. 282, 427, 1995.
38. Dibb, W.L., Microbial etiology of otitis externa, J. Infect. Dis. 22, 233, 1991.
39. Gilboa-Garber, N., Garber, N.C., Microbial lectin cofunction with lytic activities as a model for a general basic lectin role, FEMS Microbiol Rev. 63, 211, 1989.
40. Gilboa-Garber, N., *Pseudomonas aeruginosa* lectins, Meth. Enzymol. 83, 278, 1982.
41. Ramphal, R.I., Sadoff, J.C., Pyle, M., Silipigni, J.D., Role of pili in adherence of *Pseudomonas aeruginosa* to injured tracheal epithelium, Infect. Immun. 44, 38, 1984.
42. Woods, D.E., Straus, D.C., Johanson, W.G., Bass, J. A., Factors influencing the adherence of *Pseudomonas aeruginosa* to mammalian buccal epithelial cells, Rev. Infect. Dis. 5, 23023, 1983.
43. Steuer, M.K., Herbst, H., Beuth, J., Steuer, M., Pulverer, G., Matthias, R., Inhibition of lectin mediated bacterial adhesion by receptor blocking carbohydrates in patients with *Pseudomonas aeruginosa* induced otitis externa: A prospective phase II study, Otorhinolaryngol. Nova 3, 19, 1993.
44. Steuer, M.K., Beuth, J., Kwok, P., Strutz, J., Pulverer, G., Lessmeister, R., Ist die durch *Pseudomonas aeruginosa* induzierte Otitis externa diffusa acuta mit dem Blutgruppen-A-Antigen assoziiert? Otorhinolaryngol. Nova 4, 76, 1994.
45. Steuer, M.K., Hofstädter, F., Pröbster, L., Beuth, J., Strutz, J., Are ABH antigenic determinants on human outer ear canal epithelium responsible for *Pseudomonas aeruginosa* infections? Otorhinolaryngol. Nova 57, 148, 1995.
46. Steuer, M.K., Beuth, J., Hofstädter, F., Pröbster, L., Ko, H.L., Pulverer, G., Strutz, J., Blood group phenotype determines lectin-mediated adhesion of *Pseudomonas aeruginosa* to human outer ear canal epithelium, Zbl. Bakt. 282, 287, 1995.

7

ADHESION AND INVASION OF *ESCHERICHIA COLI*

Studies on Function and Regulation

T. A. Oelschlaeger, J. Morschhäuser, C. Meier, C. Schipper, and J. Hacker

Institut für Molekulare Infektionsbiologie
Universität Würzburg
Würzburg, Germany

INTRODUCTION

Escherichia coli are well known as part of the intestinal flora of healthy individuals. However, there are several groups of *E. coli* strains which are important pathogens. These *E. coli* strains cause intestinal diseases as enteritis, diarrhea, or dysentery. Furthermore, certain *E. coli* strains are responsible for extraintestinal diseases like urinary tract infections (UTI), sepsis, or meningitis. *E. coli* are among the most frequent causes for intestinal diseases and urinary tract infections. In order to prevent infection by these microorganisms, it is necessary to understand on a molecular level the pathophysiology of diseases caused by *E. coli*. It seems reasonable to try to block already the very first steps of infection in order to prevent developement of disease. These very first steps are adhesion and/or invasion to/into host cells. This article will focus on S-fimbriae of UTI and meningitis causing *E. coli* (MENEC) as an example of an adhesin as a virulence factor as well as on the ability of MENEC and enterohemorrhagic *E. coli* (EHEC) to invade nonprofessional phagocytes.

S-FIMBRIAE

One of the very first steps in establishing infection is adherence to host surfaces. Most pathogenic *E. coli* strains produce fimbrial and/or afimbrial adhesins responsible for the ability to adhere to host tissue or mucoid slime. Moreover, adherence is essential for a variety of bacterial effects on host cells as action of LT and ST enterotoxins, destruction of microvilli, reorganisation of the host cell cytoskeleton etc. Intestinal *E. coli* strains often carry transferable plasmids with gene clusters encoding fimbrial adhesins like K88 and K99 or CFAI or II. In contrast, the fimbrial gene clusters of extraintestinal *E. coli* strains, such as P, S, or F1C fimbriae, are located on the chromosome. Some fimbrial determinants of extraintestinal *E. coli* are part of so-called pathogenicity islands (pai's), DNA regions of up to 200kb that

Toward Anti-Adhesion Therapy for Microbial Diseases, edited by Kahane and Ofek
Plenum Press, New York, 1996

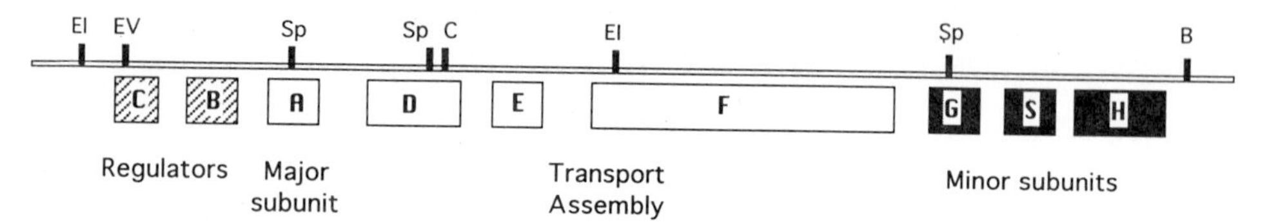

Figure 1. Physical map of the *sfa* locus consisting of 9 genes. The gene products of *sfaB* and *sfaC* function as regulators of *sfa* expression and those of *sfaD*, *sfaE*, and *sfaF* are responsible for transport and assembly of the fimbrial subunits. The major fimbrial subunit is encoded by *sfaA* and the minor subunits by *sfaG*, *sfaS*, and *sfaH*. The indicated restricitons sites are *Bam*HI (B), *Cla*I (C), *Eco*RI (EI), *Eco*RV (EV), and *Sph*I (Sp).

are spontaneously lost. Such deletion mutants are no longer virulent (1). The genetic determinant of S-fimbriae (*sfa*) is composed of 9 genes, who's arrangement and function is very similar to that of other fimbrial genes (Figure 1)(2; 3). Two regulatory genes (*sfaB*, *sfaC*) are followed by *sfaA*, encoding the major subunit of S-fimbriae. The products of the next three genes (*sfaD*, *E*, *F*) are responsible for transport and assembly of the subunits into fimbriae. The last three genes (*sfaG*, *S*, *H*) encode minor subunits. SfaS is the adhesin proper, responsible for binding of the S-fimbriae to sialyl-(2-3)-galactose containing receptors and anchored by SfaG and SfaH to the tip of the fimbriae (4; 5)(Figure 1).

Expression of S-fimbriae is not constitutive but regulated by a variety of factors. It is not only enhanced by SfaB and SfaC, but also by the products of the genes *prfb* and *prfI*. These genes are located on paiII and belong to the gene cluster of the P-related fimbriae (prf). Expression of *sfa* genes is therefore dramatically reduced after deletion of paiII, even so no *sfa* genes are located on paiII. Deletion of *prfI* but not of *prfB* also strongly reduces S-fimbriae expression, however, not as dramatic as deletion of the whole paiII, indicating the involvement of additional regulatory pai encoded factors (Figure 2) (6). However, there are further cellular components influencing S fimbriae expression, as RNaseE processing *sfa* mRNA and H-NS (=OsmZ) acting as a repressor (7).

Environmental factors are also influencing S fimbriae expression: 1) Aerobic conditions repress *sfa* expression in the logarithmic growth phase via Fnr; 2) Solid surface contact induces expression of *sfa;* 3*)* Temperature optimum is at 37°C for S fimbriae expression; 4) *sfa* expression is also regulated by catabolic repression (8) (Table 1). Most interesting are the effects of sialyl-galactose derivates and subinhibitory concentrations of antibiotics on expression of S fimbriae. After growth of S fimbriated *E. coli* in medium containing receptor analogs like sialyl-lactose, expression of *sfa* genes is twice as high compared to the control. In contrast, subinhibitory concentrations (i.e. 25% of the minimal inhibitory concentration) of antibiotics active against the test strain like Aztreonam, Gentamicin, Chloramphenicol, Trimethoprim, and Norfloxacin reduce *sfa* expression by up to 90%. Even antibiotics not active against the test strain as Clindamycin and Novobiocin reduce *sfa* expression (9) (Table 1).

INVASION/INTERNALIZATION

In general, intestinal as well as extraintestinal *E. coli* are thought, until recently, to be extracellular pathogens. However, studies in several laboratories revealed, that intestinal *E. coli* strains like enterotoxic (ETEC) (10), enteropathogenic (EPEC)(11; 12) and enterohemorrhagic (EHEC)(13) *E. coli*, besides the well known facultative intracellular enteroinvasive *E. coli* (EIEC), are able to invade human nonprofessional phagocytes. Similarly, invasion ability was demonstrated for extraintestinal *E. coli* like uropathogenic and menin-

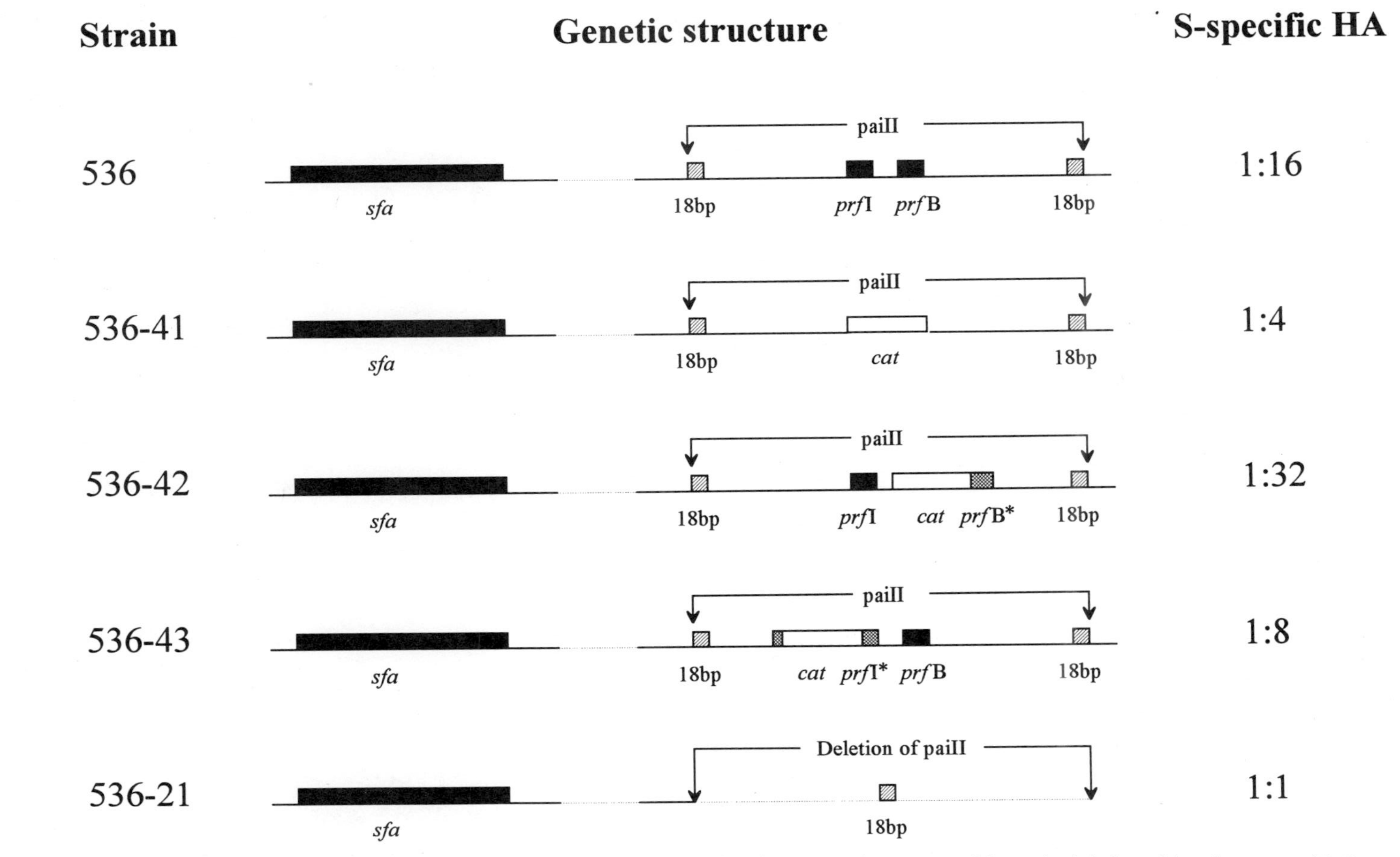

Figure 2. Expression of S fimbriae in mutant derivatives of the wild-type strain 536. The relevant genetic structure of the strains is indicated (not drawn to scale). Intact *sfa* and *prf* genes (black bars), the fragments containing a *cat* gene (open bars), and genes inactivated by insertions (*prfI**) or partial deletions (*prfB**)(cross-hatched bars) are noted. The 18bp direct repeats at the left and right ends of paiII are represented by hatched boxes. Gaps and deletions are indicated by stippled lines. The expression of S fimbriae was quantified in an S-specific haemagglutination assay. (This Figure is taken from ref. 6).

Table 1. Environmental factors influencing S fimbrae expression

Environmental Factor	Effect on *sfa* Expression
Aerobic conditions	Repression [a]
Solid surface	Induction
Temperature	Optimum at 37°C
Glucose	Catabolic repression
Receptor analogues (e. g. sialyl lactose)	Induction
Certain antibiotics	Repression

[a]Repression during the logarithmic growth phase under aerobic conditions

gitis causing *E. coli* (14;15;16). Internalization of these bacteria might be the second step after adherence or might even occure without prior adherence. In contrast to the well established genetic basis of adherence, only limited information is available about the genetic basis of the invasion ability. The only exception are enteroinvasive *E. coli*. Their invasion system is identical to the invasion system of *Shigella* spp. and contains more than 30 genes of a complex regulon (17). A simple system consisting of only one gene was cloned from enterotoxigenic *E. coli*. However, these *E. coli* strains encode at least two of such invasion systems (10). The complexity of the invasion system of enteropathogenic *E. coli* is between those of the ETEC and EIEC. Invasion related genes/proteins of extraintestinal *E. coli* were only identified for MENEC. The outer membrane protein OmpA as well as the product of the *ibe10* (invasion of brain endothelial cells) gene are essential for efficient invasion in vitro and in vivo in the newborn rat meningitis model (16; 18). Mutants in *ompA* or *ibe10* were still as adherent as the isogenic wild type strain although they were no longer invasive (18). These findings clearly demonstate that adherence and invasion are independent processes. EHEC internalization seems also not to depend on adherence, even so there are several adherence factors known for EHEC, as a fimbrial antigen encoded on the 60MDa plasmid of EHEC and Intimin, the product of the chromosomal *eaeA* gene (19; 20). Isogenic EHEC strains whith or without the 60MDa plasmid as well as wild type EHEC and deletion mutants of *eaeA* showed almost identical invasion efficiencies (Ölschläger and Schipper, unpublished results). Furthermore, the plasmidless invasive EHEC strain was negative in adhesion tests

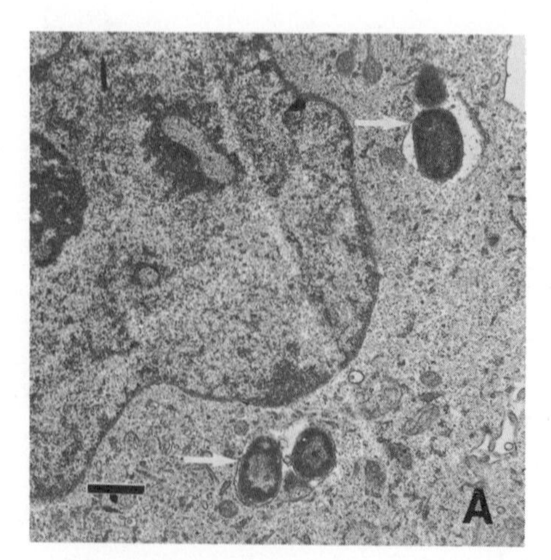

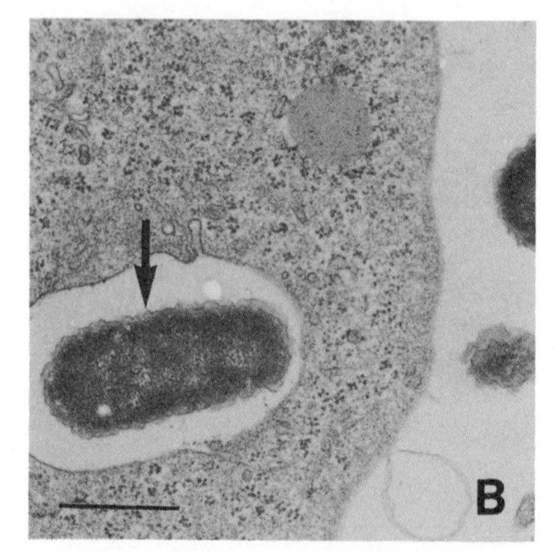

Figure 3. Transmission electron micrographs of T24 cells infected with the enterohemorrhagic *E. coli* strain B6914 (A) and the *E. coli* meningitis isolate IHE3034 (B). Bacteria (arrows) are located intracellularly in endosomes. Bar represents 1μm.

for type I mediated adhesion and did not show mannose resistant hemagglutination (Ölschläger and Schipper, unpublished results). However, it can not be ruled out, that there are other, yet unidentified adhesins essential for EHEC invasion.

HOST CELL STRUCTURES ESSENTIAL FOR ADHERENCE OR INVASION

Besides bacterial requirements for adherence and invasion, host cell structures are investigated, which might be involved in these processes. The essential structures for fimbriae mediated adherence are mainly different carbohydrate epitopes of glycolipids and glycoproteins on the host cell surface. For S fimbriae these are α-sialyl-(2-3)-β-galactose residues of glycoproteins. These residues are present on almost all cell types with the exception of phagocytes. They are also present on the Tamm-Horsefall glycoprotein in urine and on factors in human milk. In both cases these factors might be designed to prevent binding of S fimbriated *E. coli* to host tissue. In fact, sialyl-galactose residues in solution, if allowed to bind to S fimbriae, block binding of S fimbriated *E. coli* in vitro to host cells (e.g. erythrocytes).

Eukaryotic cell structures and processes involved in bacterial internalization are kinases, microfilaments and microtubules (21). Microfilaments and microtubules are also essential for efficient internalization of MENEC and EHEC (13; 15). As for adherence there are also certain eukaryotic cell surface structures used as specific receptors for bacterial invasion. These receptors are recognized by bacterial ligands termed invasion proteins, to trigger internalization of bacteria. Recently, the epitopes of such receptors for MENEC and EHEC internalization have been identified. Both *E. coli* groups need to interact with GlcNAc-β-(1-4)-GlcNAc for efficient invasion (22; Ölschläger and Schipper, unpublished results). The biological significance of the in vitro data for MENEC was demonstrated in the newborn rat meningitis model by the inhibition of the entry of these *E. coli* into the central nervous system by soluble GlcNAc-β-(1-4)-GlcNAc oligomers (22).

CONCLUSION

The very first steps in establishing infection are adherence and invasion. Already the current, still limited knowledge of these processes can form the basis for new preventive and therapeutic measures. Such new steps in prevention and/or treatment of infections caused by bacteria could be blocking of eukaryotic cell surface receptors, essential for adherence/invasion of bacteria, by 1) isolated ligands, 2) antibodies, 3) treatment with appropriate glycosidases. Furthermore, the bacterial ligands for adherence (i.e. adhesins) or invasion (i.e. invasion proteins) could be blocked by receptor analogues or antibodies. However, a more detailed understanding of adherence and invasion on the molecular level is necessary in order to achive clinically significant anti-adhesion and anti-invasion strategies.

REFERENCES

1. Ritter, A., Blum, G., Emödy, L., Kerenyi, M., Böck, A., Neuhierl, B., Rabsch, W., Scheutz, F., and Hacker, J. 1995. tRNA genes and pathogenicity islands: influence on virulence and metabolic properties of uropathogenic *Escherichia coli*. Mol. Microbiol. 17:109-121.
2. Hacker, J., Schmidt, G., Hughes, C., Knapp, S., Marget, M., and Goebel., W. 1985. Cloning and characterization of genes involved in production of mannose-resistant, neuraminidase-suseptible (X) fimbriae from a uropathogenic O6:K15:H31 *Escherichia coli* strain. Infect. Immun. 47:434-440.

3. Schmoll, T., Morschhäuser, J., Ott, M., Ludwig, B., van Die, I., and Hacker, J. 1990. Complete genetic organization and functional aspects of *Escherichia coli* S fimbrial adhesin determinant: nucleotide sequence of the genes *sfa B, C, D, E, F*. Microb. Pathog. 9:331-343.
4. Moch, T., Hoschützky, H., Jann, K., and Hacker, J. 1987. Isolation and characterization of the alpha-sialyl-beta-2,3-galactosyl-specific adhesin from fimbriated *Escherichia coli*. Proc. Natl. Acad. Sci. USA 84:3462-3466.
5. Schmoll, T., Hoschützky, H., Morschhäuser, J., Lottspeich, F., Jann, K., and Hacker, J. 1989. Analysis of genes coding for the sialic acid-binding adhesin and two other minor fimbrial subunits of the S fimbrial adhesin determinant of *Escherichia coli*. Mol. Microbiol. 3:1735-1744.
6. Morschhäuser, J., Vetter, V., Emödy, L., and Hacker, J. 1994. Adhesin regulatory genes within large, unstable DNA regions of pathogenic *Escherichia coli*: cross-talk between different adhesin gene clusters. Mol. Microbiol. 11:555-566.
7. Morschhäuser, J., Vetter, V., Korhonen, T., Uhlin, B.E., and Hacker, J. 1993. Regulation and binding properties of S fimbriae cloned from *E. coli* strains causing urinary tract infection and meningitis. Zbl. Bakt. 278:165-176.
8. Schmoll, T., Ott, M., Oudega, B., and Hacker, J. 1990. Use of a wild-type gene fusion to determine the influence of enviromental conditions on expression of the S fimbrial adhesin in an *Escherichia coli* pathogen. J. Bacteriol. 172:5103-5111.
9. Hacker, J., Ott, M., and Hof, H. 1993. Effects of low, subinhibitory concentrations of antibiotics on expression of a virulence gene cluster of pathogenic *Escherichia coli* by using a wild-type gene fusion. Int. J. Antimicrob. Agent 2:263-270.
10. Elsinghorst, E.A. and Kopecko, D.J. 1992. Molecular cloning of epithelial cell invasion determinants from enterotoxigenic *Escherichia coli*. Infect. Immun. 60:2409-2417.
11. Miliotis, M., Koornhof, H.J., and Phillips, J.I. 1989. Invasive potential of noncytotoxic enteropathogenic *Escherichia coli* in an in vitro Henle 407 cell model. Infect. Immun. 57:1928-1935.
12. Donnenberg, M.S., Donohue-Rolfe, A., and Keusch, G.T. 1989. Epithelial cell invasion: an overlooked propertiy of enteropathogenic *Escherichia coli* associated with the EPEC adherence factor. J. Infect. Dis. 160:452-459.
13. Oelschlaeger, T.A., Berrett, T.J., and Kopecko, D.J. 1994. Some structures and processes of human epithelial cells involved in uptake of enterohemorrhagic *Escherichia coli* O157:H7 strains. Infect. Immun. 62:5142-5150.
14. Straube, E., Schmidt, G., Marre, R., and Hacker, J. 1993. Adhesion and internalization of *E. coli* strains expressing various pathogenic determinants. Zbl. Bakt. 278:218-228.
15. Meier, C., Oelschlaeger, T.A., Merkert, H., Korhonen, T.K., and Hacker, J. 1996. Ability of the new born meningitis isolate *Escherichia coli* IHE 3034 (O18:K1:H7) to invade epithelial and endothelial cells. Infect. Immun., in press.
16. Huang, S.-H., Wass, C., Fu, Q., Prasadarao, N.V., Stins, M., and Kim, K.S. 1995. *Escherichia coli* invasion of brain microvascular endothelial cells in vitro and in vivo: molecular cloning and characterization of invasion gene *ibe10*. Infect. Immun. 63:4470-4475.
17. Hsia, R.C., Small, P.L.C., and Bavoil, P.M. 1993. Characterization of virulence genes of enteroinvasive *Escherichia coli* by Tn*phoA* mutagenesis: identification of *invX*, a gene required for entry into HEp-2 cells. J. Bacteriol. 175:4817-1823.
18. Prasadarao, N.V., Wass, C.A., Weiser, J.N. Stins, M.F. Huang, S.-H., and Kim. K.S. 1996. Outer membrane protein A of *Escherichia coli* contributes to invasion of brain microvascular endothelial cells. Infect. Immun. 64:146-153.
19. Karch, H., Heesemann, J., Laufs, R., O'Brian, A.D., Tacket, C.O., and Levine, M.M. 1987. A plasmid of enterohemorrhagic *Escherichia coli* O157:H7 is required for expression of a new fimbrial antigen and for adhesion to epithelial cells. Infect. Immun. 55:455-461.
20. Yu, J., and Kaper, J.B. 1992. Cloning and characterization of the *eae* gene of enterohemorrhagic *Escherichia coli* O157:H7. Mol. Microbiol. 6:411-417.21.
21. Oelschlaeger, T.A., Guerry, P., and Kopecko, D.J. 1993. Unusual microtubule-dependent endocytosis mechanisms triggered by *Campylobacter jejuni* and *Citrobacter freundii*. Proc. Natl. Acad. Sci. USA 90:6884-6888.
22. Prasadarao, N.V., Wass, C.A., and Kim, K.S. 1996. Endothelial cell GlcNAcβ1-4GlcNAc epitopes for outer membrane protein A enhance traversal of *Escherichia coli* across the blood-brain barrier. Infect. Immun. 64:154-160.

8

NONFIMBRIAL ADHESINS OF *ESCHERICHIA COLI*

Janina Goldhar

Department of Human Microbiology
Sackler Faculty of Medicine
Tel-Aviv University
Tel-Aviv, Israel

INTRODUCTION

Pathogenic *Escherichia coli* exhibit a variety of adhesins classified according to their morphology, antigenic structure or receptor specificity.

Morphologically, *E. coli* adhesins are fimbrial, fibrillar and nonfimbrial. This classification is based on varying appearances in the electron microscope (E.M). Fimbrial adhesins are relatively thick and rigid (5-7nm diameter), while the fibrillar types are much thinner, more flexible and curly (2-3 nm diameter). Both can be demonstrated directly by negative staining procedure. A third group of extracellular adhesive structures can be visualized only after stabilization with specific antibodies and then they may possess a capsule-like appearance. (Jann & Hochutzky, 1990, Schmidt, 1994).The latter, generally described as nonfimbrial (NFAs) or A-fimbrial (AFAs), will be described in the following.

STRUCTURE AND GENETICS

Table 1 lists NFAs of *E. coli* strains related to human infections. Included are also outer membrane proteins that function as adhesins: MIAT (Ofek & Doyle, 1994) and intimine (Law, 1994).

Although NFAs represent a heterogeneous group of structures, most of them share a few common characteristics:

a) Expression is regulated by growth conditions and, like fimbrial mannose-resistant (MR) adhesins, NFAs are synthesized only above temperature of 20°C; b) they are readily released from a bacterial surface; c) they consist of noncovalently linked subunits with a molecular weight ranging between 15-30kDa; d) the nonfimbrial adhesins isolated and purified in various laboratories by different techniques, tend to form aggregates of apparent molecular weight in excess of 10^6 Da. These purified adhesive polymers in soluble form cause MR hemagglutination (HA) of human erythrocytes only. Because the adhesins consist

Toward Anti-Adhesion Therapy for Microbial Diseases, edited by Kahane and Ofek
Plenum Press, New York, 1996

Table 1. Nonfimbrial Adhesins of *Escherichia coli*[1]

Adhesin	Size (KDa)	Receptor	Source of isolation
GV-12	13.4	MR[2]	clinical
AFA-I	16	Dr[3]	UTI[4]
AFA-III	15	Dr[3]	UTI, diarrhea
M	21	M	UTI
NFA-1	21	MR	UTI, sepsis
NFA-2	19	MR	UTI
NFA-3	17.5	N[3]	sepsis
NFA-4	28	M[3]	UTI
NFA-5	23	M[3]	UTI
NFA-6(Z1)	14.4	MR	diarrhea
2230	16	None	diarrhea
8786	16.3	MR	diarrhea
AIDA-1	104.8	MR	diarrhea
INTIMIN	94.0	None	diarrhea
MIAT	n.d.	MS[2]	UTI

[1]Adapted from Schmidt, 1994.
[2]MR -mannose-resistant; MS- mannose-sensitive
[3]Blood group specificity
[4]UTI-Urinary Tract Infection

of identical subunits, each expressing a binding epitope, they retain multivalency after extraction and purification unlike purified fimbriae (type 1, P, S, or CFA/I). In the latter the adhesive protein is located on the tip of the fimbrial structure. This type of fimbriae retained adhesive capacity, but HA capacity was observed only after aggregation of fimbriae at low pH, in the presence of divalent ions, or with specific antibodies (Jann & Hoschutzky, 1991).

Structural studies revealed that NFA-1 exhibits a polymeric structure, which disintegrate in elevated temperatures into a 19 kDa monomers with relatively stable dimers. Measurements by light scattering suggest that NFA-1 also has a fibrillar structure. The morphological appearance in the E.M. depends on the linear mass densities of adhesin polymers, which are as follows: P-fimbriae- 60,000 D/nm, K88- 20,000 D/nm, NFA-1- 8,000 D/nm (Ahrens et al., 1993). Thus the distinction between fimbrial and nonfimbrial morphology may be superficial and in fact, be due to the limitation of E. M. resolution. For example, the adhesins 444-3 and 469-3 of enteropathogenic *E. coli*, originally described as nonfimbrial, were later found to form very fine fibrillae on bacterial surface (Ofek & Doyle, 1994).

Treatment of bacteria expressing NFAs with anti-adhesin antibodies cause stabilization of the adhesive protein, followed by what appears to be capsule-like structure. Such capsules were demonstrated in bacteria expressing NFA-1, NFA-4 and NFA-6 (Jann & Hoschutzky, 1990) In *E. coli* strains expressing NFA-2 and NFA-3 some extracellular material could be detected with adhesin-specific antibodies. However, the capsule was not so well developed as in NFA-1, 4 and 6 (Goldhar et al., 1987; Grunberg et al., 1988).

Since strains of *E. coli* expressing NFAs also possess polysaccharide capsules (K antigens) it was important to clarify how the two types of capsules are coexpressed. In the case of *E. coli* expressing NFA-4 and NFA-6, it was demonstrated that both the polysaccharide capsule and the nonfimbrial adhesive capsule were expressed on the same bacterial cell (Kroncke et al.,1990). The adhesins were stabilized with specific monoclonal antibodies and the polysaccharide capsule was visualized with antibodies labeled with gold. The adhesin appears as the outermost layer surrounding the polysaccharide capsule, connected to the cell surface by thread-like structures. In the tested bacterial population most bacterial cells expressed the capsular polysaccharide, while only approximately 20% expressed the NFA-4 adhesin, as well. In some bacteria the K antigen was visible - extending beyond the adhesin.

It is still not clear, how the phenomenon of the coexpression of the capsule and NFA influences phagocytosis.

1. Antigenic Structure

Monoclonal (MAbs) antibodies against NFAs were prepared. Mabs possessing anti-adhesive activity were used in ELISA studies showing that the NFAs are distinct antigens, although some serological cross-reactivity can be observed (Table 2). The adhesins did not cross-react with any of fimbrial adhesins F1-F14 (Jann & Hoschutzky, 1991; Kahana et al., 1994).

2. Genetic Organization

The genetic organization of nonfimbrial adhesins was best documented for NFA-1, M agglutinin and the AFA family. NFA-1 was cloned (Hales et al., 1988) and later the gene cluster coding for NFA-1 was further characterized (Ahrens et al., 1993). Despite the fact that the gene coding for the structural protein of NFA-1 does not exhibit any homology to loci responsible for subunits of the other adhesins, the gene structures of the corresponding determinants share common features. In all cases five genes are responsible for full expression of adhesin phenotype. These genes are located on a stretch of DNA of approximately 6 to 8 kb. The genes encoding the corresponding adhesin structural proteins are located at the distal (3′) end of the gene clusters. Interestingly, large proteins such as the nfaC gene products are found in all nonfimbrial adhesins, as well as in fimbrial adhesin complexes. The proteins are involved in the biogenesis of fimbriae and fibrillae and possess a similar function in biogenesis of NFAs.

The operon afa encodes for five polypeptides (afaA to afaE). AFA_I active adhesin protein is encoded by afaE. Hybridization experiments revealed the presence of the related gene clusters in various *E. coli* isolates. The gene clusters contain a conserved DNA fragment, harboring afaB, afaC and afaD genes, while afaE was found variable. From these studies four distinct adhesins of the AFA family were identified and designated AFA-I, AFA-II, AFA-III, and AFA-IV (Labigne-Roussel & Falkow, 1988; LeBougenec et al.,1993). Members of the family showed serological cross-reactions, were mostly expressed by *E. coli* belonging to serogroup O75 and recognized receptors on Dr blood group antigens (see below). Because of the receptor specificity, the AFA family was later named Dr family of adhesins. Of four Dr adhesins, AFA-I and AFA-III are nonfimbrial, while AFA-II and AFA-IV correspond to Dr hemagglutinin (fibrillar) and F1845 (fimbrial) adhesins, respectively. AFA-III is encoded by genes located on plasmid (Nowicki et al., 1990).

Table 2. Reactions of NFAs with monoclonal antibodies (ELISA)[1]

	M Abs			
NFA	NFA-1	NFA-2	NFA-3	NFA-4
NFA-1	+++	-	-	-
NFA-2	+	++	-	-
NFA-3	+	-	+++	-
NFA-4	+	-	-	++

[1]Antibodies inhibited HA caused by homologous NFA (Jann & Hoschutzky, 1991; Kahana et al., 1994)

The bma genes associated with M-agglutinin are also contained in DNA chromosomal fragments of uropathogenic *E. coli* (Rhen et al., 1986; Schmidt, 1994). As for AFA-I five gene are essential for adhesin expression on the surface of a recombinant strain. Gene bmaE encodes for a major 21 kDa protein identified as M-agglutinin. The organization of genes coding for expression AFA-1 (afa), NFA-1 (nfa) and M agglutinin (bma) is similar to that of fimbrial F1845 (daa) and fibrillar Dr adhesin (dra) operons (Schmidt, 1994).

Despite similarity in gene organization, comparison of N-terminal amino acid sequences shows a diversity of nonfimbrial adhesins and lack of homology with any fimbrial or fibrillar adhesins. One single exception is the K88, a fibrillar adhesin of *E. coli* enteropathogenic for calves.

Different in gene organization is the large adhesin AIDA-I, isolated and cloned from diarrheogenic *E. coli* of EPEC serotype. Only two genes located on plasmid are necessary for expression of the large precursor leading to the mature outer membrane protein functioning as adhesin. The C-terminal part of the precursor protein shows a 36% similarity to the VirG protein, which is involved in the intra- and inter-cellular spread of *Shigella flexneri*. The protein is serologically related to adhesive proteins of EPEC strains which cause diffuse adherence (DA) (Bentz et al., 1992; Schmidt, 1994).

Two additional plasmid encoded nonfimbrial adhesins were isolated from diarrheogenic *E. coli*: The 2230 adhesin that does not possess HA activity, but was found to mediate the binding of the bacteria to brush borders of enterocytes in the experimental model (Forestier et al., 1987); and another NFA isolated from an EPEC strain 8786 (Aubel et al., 1991).

RECEPTOR SPECIFICITY AND BIOLOGICAL SIGNIFICANCE

All adhesins described above were shown to mediate binding of the bacteria to human epithelial cells, concomitant to the capacity to cause hemagglutination of human erythrocytes.

NFA-1 and NFA-2 carrying bacteria and isolated adhesins were shown to bind to Human Kidney (HK) cell line monolayer in a linear dose response. The binding of the bacteria could be inhibited by isolated adhesins that seem to share a common receptor (non-identified) (Goldhar et al., 1987).

The receptor specificities of only a limited number of NFAs were determined. Four nonfimbrial adhesins recognize receptors of the M/N human blood group antigen system. Both determinants are expressed on glycophorin A, a major sialo-glycoprotein of human erythrocytes. Specificities M and N differ by the sequence of five amino acids on the N-terminal of glycophorin A. The two blood group antigenic determinants are cross-reacting and may be differentiated by monoclonal antibodies. A plant lectin (*Vicia graminea*) was known as the reagent specific for blood group N antigen. Human erythrocytes carry either both specificities (heterozygotic, MN) or homozygotic MM or NN. Individuals lacking both determinants are extremely rare (Anstee, 1981).

Nonfimbrial M-specific hemagglutinin was first characterized (Rhen, et al., 1986) in a uropathogenic *E. coli* isolate belonging to serogroup O2. The same strain also expressed N-acetyl-D-glucosamine-specific G- fimbriae. The M-specific adhesin agglutinated erythrocytes MM and MN, but not NN. The hemagglutination was inhibited by glycophorin A^M and by serine - a terminal amino-acid of the glycoprotein chain. Bacteria expressing M-hemagglutinin were able to bind glycophorin A^M.

Two other M-specific NFAs of two uropathogenic *E. coli* isolates, belonging to two different O7 serotypes were characterized by Hoschutzky et al. (1989) and Nimmich (1990). The adhesins, differing in their amino acid N-terminal sequence, were named NFA-4 and

NFA-5. A certain degree of homology was found between M-agglutinin and NFA-5, while no relation has been found between NFA-4 and M-adhesin, nor to other NFAs. Nevertheless, certain homologies were discovered between NFA-4 and K88 fibrillar adhesin.

An *E. coli* strain carrying N-specific adhesin was isolated from a systemic infection case and named NFA-3 (Grunberg et al., 1988). Its N-specificity, among others, was demonstrated with the help of human erythrocytes of "Antigenic constitution Matrix Panel". It was found that erythrocytes lacking N-antigen could not be agglutinated by NFA-3. Glycophorin A^{NN} (but not A^{MM}) inhibited the HA of RBC-NN and the purified adhesin bound to glycophorin A^{NN} in a dose-dependent manner. The binding of NFA-3 to glycophorin A^{MM} was significantly lower than that to glycophorin A^{NN}.

Bacteria expressing NFA-3 bound to epithelial cell line monolayers, but the rate of binding was lower than that of NFA-1. In contrast, *E. coli* carrying NFA-3 bound well to erythroleukemic cell line K562, expressing glycophorin A (Kahana et al., 1994).

Further studies indicated that NFA-3 also mediates binding of the bacteria to human granulocytes (PMNs). The interaction was examined between NFA-3 carrying bacteria, its isolated adhesin and with PMNs obtained from donors of different blood groups (Grunberg et al., 1994). As shown in figure 1, PMNs from donors of blood group NN (PMN-NN) bound significantly more NFA-3 carrying bacteria than PMNs from donors of blood group MM

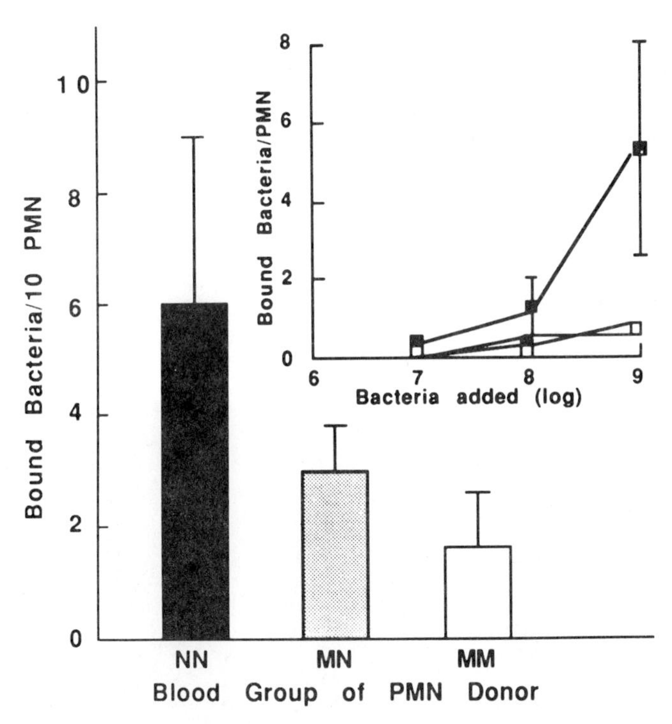

Figure 1. Binding of E. coli 9 expressing NFA-3 to PMNs of donors with blood groups MM, MN and NN. Bacteria (1x10^9 /ml) were incubated with fresh PMNs (1x10^6) at 4°C for 30 min, or bacteria in various concentrations were incubated with monolayers of dried PMNs at 37°C for 30 min (insert). The number of bacteria bound to PMNs was estimated using the antiserum against *E. coli* 9 in ELISA. Closed bars - PMN-NN, open bars - PMN-MM, shadowed bars - PMN-MN. Vertical lines indicate standard deviation (SD). Values for PMN-MN and PMN-MM were significantly lower than those for PMN-NN ($p<0.01$) according to Students t-test, calculated for experiments with fresh PMNs and according to Analysis of variance with repeated measures for experiments with dried PMNs. Data from Grunberg et al., 1994.

Table 3. Effect of pretreatment with enzymes and addition of inhibitors on binding of bacteria carrying NFA-3 to PMN, or on hemagglutination caused by NFA-3

Enzyme pretreatment	Inhibitor added	Binding of *E. coli* 9 to PMN-NN[1] % of control	Hemagglutination caused by NFA-3 % of control[2]
None	None	100	100
Ficine 200 mg/ml	None	0	0
Sialidase			
0.5u	None	70	12.5
1.0u	None	45	n.d
5.0u	None	n.d	0
None	Glycophorin A^{NN}	25	<25
	Glycophorin A^{MM}	68	100
	V.graminea[3]	8	+
	Concanavalin A[3]	100	-

[1]Binding of the bacteria to PMN-NN determined by ELISA.[2]Haemagglutination titer was determined using either erythrocytes pretreated with indicated enzymes or untreated erythrocytes (RBC-NN) in the presence of indicated inhibitors. 100% Hemagglutination = titer 1:32.[3]Binding to PMNs was estimated by microscopic examination.+ indicates inhibition- indicates lack of inhibition Data from Grunberg et al., 1994.

(PMN-MM). Table 3 shows results of a number of experiments suggesting that PMNs express the N-like determinant that serves as attachment site for *E. coli* via its NFA-3 adhesin. The NFA-3 mediated attachment of *E. coli* to PMNs was followed by ingestion and killing of the organisms in the serum free system (Figure 2) (Grunberg et al., 1994).

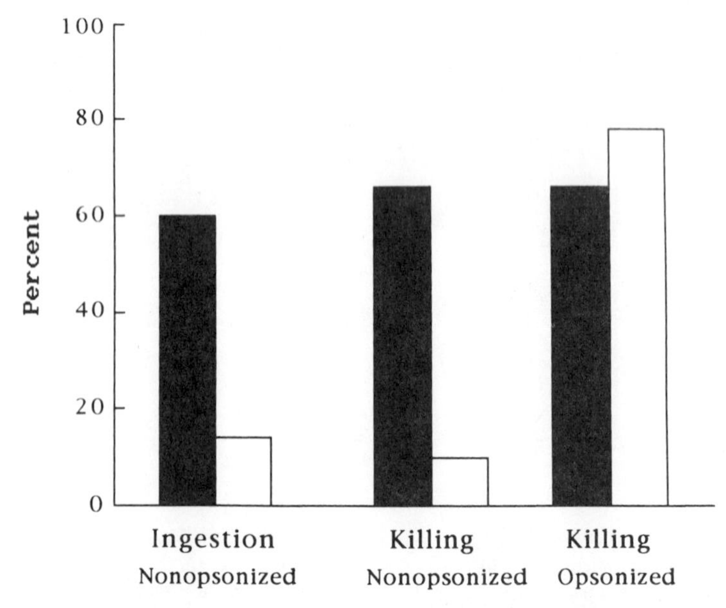

Figure 2. Ingestion and killing of *E. coli* 9 expressing NFA-3 by PMNs of donors with blood groups MM and NN. Nonopsonized or opsonized bacteria were incubated with PMNs (closed bars) or PMN-MM (open bars) ($5x10^8$ bacteria/ $5x10^5$ PMN) at 4°C for 30 min and after washings the mixture was shifted to 37°C for further incubation. Percent ingestion was calculated by estimation of decrease in number of bacteria remaining bound to PMNs in samples taken after 60 min of incubation at 37°C, using ELISA. Parallel samples were used to determine percent killing by estimating the decrease in the number of colony forming units after 60 min of incubation at 37°C. Data from Grunberg et al., 1994.

In another study it was demonstrated that NFA-1 also mediated binding of bacteria to human PMNs - a conclusion based on data showing that soluble NFA-1 inhibited both the attachment of NFA-1 carrying *E. coli* to PMNs and the stimulation of PMNs induced by bacteria. Binding to PMNs was followed by ingestion and killing of the organisms. Similar results were obtained with both the wild strain and recombinant expressing NFA-1 (Goldhar et al., 1991). It was found that association of NFA-1 carrying bacteria with PMNs triggered an oxidative burst in the phagocytic cells. The rate of ingestion and killing of *E. coli* bound via NFA-1 was found to be considerably slower than that of *E. coli* bound via type 1 (MS) fimbrial adhesin (Figure 3). For example, during 20 min of incubation at 37°C only 15% of NFA-1 expressing bacteria were ingested, as compared with 60% of the same strain bacteria expressing type 1 fimbriae. Furthermore, after 60 min of incubation 80% of the attached MS *E. coli* were killed, as compared with only 50% of the attached NFA-1 carrying bacteria. These distinctions may reflect differences in the nature and activity of the receptor molecules involved in the binding of bacteria to PMNs.

As mentioned above, the nonfimbrial AFA-I and AFA-III and Dr fibrillar hemagglutinin together with the fimbrial F1845 adhesin of diarrheogenic *E. coli*, all possess a common genetic basis and recognize the Dr blood group antigen located on decay accelerating factor (DAF) present on human erythrocytes. DAF is a phosphatidyl-inositol linked RBC membrane protein involved in the regulation of complement cascade. It is present on the surface of granulocytes and various other types of human cells, including urinary tract and intestinal epithelium. *E. coli* strains expressing Dr-specific adhesins agglutinate Dr-positive RBC, but not Dr-negative ones (extremely rare). The distinct receptor epitopes of the Dr adhesins family were characterized (Nowicki et al., 1990). It was found that binding of Dr hemagglutinin could be inhibited by chloramphenicol, as compared with other adhesins of the Dr family. Treatment of erythrocytes by various proteases showed distinct epitopes for each of AFA-I, AFA-III, Dr hemagglutinin and F1845. *E. coli* strains carrying Dr hemagglutinin have been reported to bind to uroepithelial cells and to Hep-2, HeLa and Caco-2 human epithelial cell lines, as well as to basement membranes of the kidney glomerules and tubules. It was suggested that besides DAF, also type IV collagen may serve as receptor to Dr-specific adhesins.

Recently Johnson et al (1995) reported, that the Dr family of adhesins mediated adherence of *E. coli* to PMNs. The study showed parallel inhibitions of interaction between Dr - adhesins with erythrocytes and PMNs, suggesting the similarity of Dr receptors on

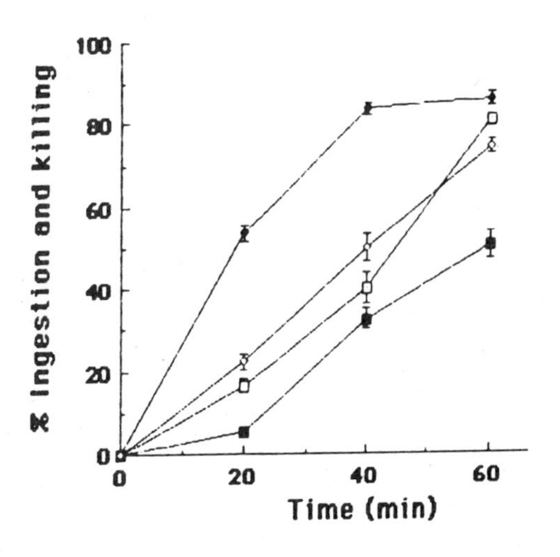

Figure 3. Ingestion and killing of *E.coli* 827 expressing MS type 1 fimbriae (MS) or NFA-1. Bacteria grown in broth to express type 1 fimbriae or on agar to express NFA-1 were incubated ($5x10^6$/ml) with PMNs suspensions ($5x10^6$/ml) at 4°C for 30 min. After washing the mixtures were shifted to 37°C and samples were withdrawn at various time intervals to determine the percent of ingested bacteria (by ELISA) and percent of killed bacteria (by CFU counting). The number of bacteria attached, or CFU per PMNs at time zero was considered as 0% ingestion or killing, respectively (Goldhar et al., 1991). Ingestion (□) and killing (■) of *E.coli* 827 carrying NFA-1; Ingestion (●) and killing (O) of E.coli 827 carrying type 1 fimbriae. Data from Goldhar et al.,1991.

various types of human cells. Unlike NFA-1 and NFA-3, however, the binding mediated by Dr-specific adhesins was not followed by ingestion and killing of the organisms.

The above reported data indicate that a variety of factors are involved in interactions of *E. coli* bacteria with PMNs. The rate of binding depends on quantitative differences in expression of adhesins and on the type of receptors and their quantitative accessibility on the surface of PMNs. Subsequent events, such as rate of ingestion, extra- or intra-cellular killing, vary according to adhesin. It is difficult to conclude whether binding to PMNs followed by phagocytosis is related to nonfimbrial morphology of MR adhesins. It may rather be speculated, that expressing of nonfimbrial adhesins reflects evolutionary adaptation of bacterial pathogens to host tissue receptors and defense mechanisms - resulting in the diversity of receptor specificity, antigenic structure and morphology of adhesins.

Despite a number of the known genetical and functional similarities of the fimbrial and nonfimbrial adhesins, the facts that they are readily released from the bacterial surface, that they form a capsule-like surface, and that they mediate "intimate" binding, may cause different biological behavior.

Some suggestion as to the biological significance of NFAs may be based on studies concerning the frequency of isolation of strains carrying the nonfimbrial adhesins.

The first study to determine relative frequency of *E. coli* clinical isolates containing afa operon was published by Labigne-Roussel and Falkow (1988). The authors screened 138 strains by colony hybridization and found 13% of UTI isolates containing DNA sequences related to afa operon, as compared to only 5% of fecal isolates. According to a recent publication (Donnenberg & Welch, 1996) the pooled results from a number of molecular epidemiological studies, indicate that positive signals were found in 16% of *E. coli* strains isolated from pyelonephritis (generally in association with P or S-fimbriae) and that the highest prevalence of strains reacting with probes of members of the AFA family was found in the cystitis strains (19%). It was suggested that the presence of the family of Dr receptors identical in human intestinal and urinary tract cells enables the bacteria harboring adhesins of the Dr family to colonize the intestinal tract - the prerequisite step to ascending urinary tract infection. AFA adhesin complex alone, however, seems to be insufficient to cause infection in a normal host (Nowicki et al., 1989). As mentioned, AFA were generally detected in strains expressing P or/and S fimbrial adhesins.

Maslow et al. (1993) screened *E. coli* strains isolated from bacteremia cases. They used appropriate probes for the detection of strains carrying P and S fimbriae, AFA adhesin and hemolysin. In a more recent study, in which clonal relationships among blood isolates of patients from Boston and Nairobi were determined, the bma operon was also included. Relatively small numbers of isolates were found afa and bma positive (lower percentage than in pyelonephritis) and in all cases together with other virulence factors (Maslow et al., 1995).

A more comprehensive study on the relative frequency of *E. coli* strains carrying nonfimbrial adhesins require the use of probes of all NFAs since phenotypic detection is possible only in strains not expressing any fimbrial MR adhesin.

CONCLUDING REMARKS

- The highly heterogeneous nonfimbrial adhesins reflect an evolutionary process whereby human bacterial pathogens can adapt, during infection, to a variety of environmental conditions.
- Virulent *E. coli* strains are able to express various adhesins in different compartments of the host.
- The expression of each of these adhesins is regulated by environmental signals.

- In certain situations, nonfimbrial adhesins contribute to the pathogenic process. It is possible that adhesion mediated by NFAs is more effective for signal transduction than that mediated by fimbrial adhesins. We have yet to clarify the conditions which are advantageous for bacteria expressing nonfimbrial adhesins.

REFERENCES

Ahrens, R., M. Ott, M. Ritter, H. Hoschutzky et al. Genetic analysis of the gene cluster encoding nonfimbrial adhesin I from *Escherichia coli* uropathogen. Infect. Immun. 1993, 61: 2505-12.

Anstee, D. J. The blood group MNSs active sialoglycoproteins. Semin. Hematol. 1981, 18: 13-27.

Aubel, D., A. Darfeuille-Michaud and B. Joly. New adhesive factor (antigen 8786) on a human enteropathogenic *Escherichia coli* O117:H4 strain isolated in Africa. Infect. Immun. 1991, 59: 1290-99.

Bentz, I. and A. Schmidt. Isolation and serologic characterization of AIDA-I, the adhesin mediating the Diffuse Adherence phenotype of the diarrhea-associated *Escherichia coli* strain 2787 (O126: H27). Infect. Immun. 1992, 60: 13-18.

Donnenberg, M. S. and R.A. Welch. Virulence determinants of uropathogenic *Escherichia coli*. 1996. In: Urinary Tract Infections. Molecular Pathogenesis and Clinical Management. H. L. T. Mobley and J. W. Warren, eds. ASM Press, Washington. 135-174.

Forestier, C., K. W. Welinder, A. Darfeuille-Michaud, and P. Klemm. Afimbrial adhesin from *Escherichia coli* strain 2230: purification, characterisation and partial covalent structure. FEMS Microbiol. Letters. 1987, 40: 47-50.

Goldhar, J., M. Yavzori, Y. Keisari and I. Ofek. Phagocytosis of *Escherichia coli* mediated by mannose resistant non_fimbrial hemagglutinin (NFA-1). Microbial Pathogenesis. 1991, 11: 171-78.

Goldhar, J., R. Perry, J. R. Golecki, H. Hoschutzky, B. Jann and K. Jann. Nonfimbrial, mannose-rasistant adhesins from uropathogenic *Escherichia coli* O83:K1:H4 and O14:K?:H11. Infect. Immun. 1987, 55: 1837-42.

Grunberg, J., I. Ofek, R. Perry, M. Wiselka, G. Boulnois and J. Goldhar. Blood group NN-dependent phagocytosis mediated by NFA-3 haemagglutinin of *Escherichia coli*. Immunol. & Infect. Dis. 1994, 4: 28-32.

Grunberg, J., R. Perry, H. Hoschutzky, B. Jann, K. Jann and J. Goldhar. Nonfimbrial blood group N-specific adhesin (NFA-3) from *Escherichia coli* O20: KX104:H-, causing systemic infection. FEMS Microbiol. Letters. 1988, 56: 241-46.

Hales, B. A., H. Beverly-Clarke, N. J. High, K. Jann et al. Molecular cloning and characterization of the genes for a non-fimbrial adhesin from *Escherichia coli*. Microbial Pathogenesis. 1988, 5: 9-17.

Hoschutzky, H., W. Nimmich, F. Lottspeich, and K. Jann. Isolation and characterisation of the non-fimbrial adhesin NFA-4 from uropathogenic *Escherichia coli* O7: K98: H6. Microbial Pathogenesis. 1989, 6: 351-59.

Jann K. and H. Hoschutzky. 1990. Nature and Organization of Adhesins. In: Current Topics in Microbiology and Immunology. Vol. 151. K. Jann and B. Jann, eds. Springer Verlag, Berlin. 55-70.

Jann, K. and H. Hoschutzky. 1991. Characterization and surface organization of *E. coli* adhesins. In: Microbial surface components and toxins in relation to pathogenesis. E.Z. Ron and S. Rotem, eds. Plenum Press, New York. 3-9.

Johnson, J. R., K. M. Skubitz, B. J. Nowicki, K.Jacques-Palaz and R. M. Rakita. Nonlethal adherence to human neutrophils mediated by Dr antigen-specific adhesins of *Escherichia coli*. Infect. Immun. 1995, 63: 309-16.

Kahana, H., J. Grunberg, Y. Bartov, R. Perry et al. Binding of *Escherichia coli* recognising N-blood group antigen to the erythroleukemic cell line K562 expressing glycophorin A. Immunol & Infect. Dis. 1994, 4: 161-165.

Kroncke, K-D., F. Orskov, I. Orskov, B.Jann, K. Jann. Electron microscpe study of coexpression of adhesive protein capsules and polysaccharide capsules in *Escherichia coli*. Infect. Immun. 1990, 58: 270-275.

Labigne-Roussel,A. and S. Falkow. Distribution and degree of heterogeneity of the Afimbrial-adhesin-encoding operon (afa) among uropathogenic *Escherichia coli* isolates. Infect. Immun. 1988, 56: 640-48.

Law, D. Adhesion and its role in the virulence of enteropathogenic *Escherichia coli*. Clin. Microbiol. Rev. 1994, 7: 152-73.

LeBouguenec, C., M. I. Garcia, V. Ouin, J.-M. Dessperrier, P. Gounon, and A. Labigne. Characterization of plasmid-borne afa-3 gene clusters encoding afimbrial adhesins expressed by *Escherichia coli* strains associated with intestinal or urinary tract infections. Infect. Immun. 1993, 61: 5106-14.

Maslow, J. N., M. E. Mulligan, K. S. Adams, J. C. Justis and R. D. Arbeit. Bacterial adhesins and host factors: Role in the development and autcome of *Escherichia coli* bacteremia. Clin. Infect. Dis. 1993, 17: 89-97.

Maslow, J. N.,T. S. Whittam, C. F. Gliks et al. Clonal relationships among bloodstream isolates of *Escherichia coli*. Infect. Immun. 1995, 63: 2409-17.

Nimmich, W. Adhesins of *Escherichia coli*. Acta Biotechnol. 1990, 2: 151-61.

Nowicki, B., A. Labigne, S. Mosley, R. Hull, S. Hull, and J.Moulds. The Dr hemagglutinin, afimbrial adhesins AFA-I and AFA_III, and F1845 fimbriae of uropathogenic and diarrhea-associated *Escherichia coli* belong to a family of hemagglutinins with Dr receptor recognition. Infect. Immun. 1990, 56: 279-81.

Nowicki, B., C. Svanborg-Eden, R. Hull, and S. Hull. Molecular analysis and epidemiology of the Dr hemagglutinin of uropathogenic *Escherichia coli*. Infect. Immun. 1989, 57: 446-451.

Ofek, I. and R. Doyle. 1994. Bacterial Adhesion to Cells and Tissues. Chapman and Hall, N.Y. 321-377.

Ofek, I., J. Goldhar, Y. Kesari and N. Sharon. Nonopsonic phagocytosis of Microorganisms. Ann. Rev. Microbiol. 1995, 49: 239-76.

Rhen, M., P. Klemm and T. Korhonen. Identification of two new heamagglutinins of *Escherichia coli* N-acetyl-D- Glucosamine specific fimbriae and a blood group M-specific agglutinin, by cloning the corresponding genes in *Escherichia coli* K-12. J. Bacteriol. 1986, 16: 1234-42.

Schmidt, A. M. 1994. Nonfimbrial adhesins of *Escherichia coli*. In: Fimbriae, Adhesion, Genetics, Biogenesis and Vaccines. P. Klemm, ed. C.R.C. Press Inc. 85-96.

9

FIMBRIAE–MEDIATED ADHERENCE INDUCES MUCOSAL INFLAMMATION AND BACTERIAL CLEARANCE

Consequences for Anti-Adhesion Therapy

Hugh Connell,[1] William Agace,[1] Maria Hedlund,[1] Per Klemm,[2] Mark Shembri,[2] and Catharina Svanborg[1]

[1] Department of Medical Microbiology
Section of Clinical Immunology
Lund University
Lund, Sweden
[2] Department of Microbiology
The Technical University of Denmark
Lyngby, Denmark

INTRODUCTION

Escherichia coli strains express a variety of fimbrial and non-fimbrial adherence factors which bind via lectin-receptor interactions to host cell glycoconjugate receptors. The attached state provides several advantages; it allows bacteria to resist elimination by the flow of secretions, it enhances their ability to trap nutrients, to multiply and to colonize the mucosa. Attachment may indeed be the endpoint for microbes that form part of the indigenous flora at different mucosal sites. For the pathogens however, adherence is only the first step in a complex series of events that lead to disease. The pathogens may activate the mucosal cells to which they bind, may invade into and through those cells, and may disrupt the integrity of the mucosal cell layer to reach underlying tissue compartments. Common to these events is the induction of an inflammatory response in the mucosa.

We have studied the relationship between bacterial adherence, epithelial cell activation and inflammation, using urinary tract infection (UTI) as a model. *E. coli* predominate as a cause of UTI. The bacteria ascend into the urinary tract, and establish a population of $>10^5$ cfu/ml. Acute pyelonephritis involves the kidney, while acute cystitis is limited to the lower urinary tract. Both disease states are characterized by inflammatory changes in the host. Acute pyelonethritis is diagnosed by fever, elevated acute-phase reactants, and recruitment of inflammatory cells to the urinary tract. Neutrophil recruitment occurs also in acute cystitis, together with other signs of local inflammation in the urinary bladder.

Toward Anti-Adhesion Therapy for Microbial Diseases, edited by Kahane and Ofek
Plenum Press, New York, 1996

Fimbriae-mediated adherence is important for the virulence of *E. coli* in the urinary tract (Svanborg et.al., 1994). The mechanisms of adherence have been defined. Uropathogenic *E.coli* express P, S, Dr and type 1 fimbriae that bind glycoconjugates with receptor epitopes specified by sialic acid, Galα1-4Galβ and α-mannose (Johnson, 1991).

P fimbriae are expressed by most *E.coli* strains causing acute pyelonephritis (~90%) but by few strains causing asymptomatic bacteriuria (~20%). P fimbriae mediate attachment to urinary tract epithelial cells and enhance cytokine responses *in vitro* and *in vivo* (Hedges et.al., 1995). Mutations in the *pap* gene cluster encoding P fimbriae, reduced bacterial persistence in the mouse urinary tract (Hagberg et.al., 1983). Mutational inactivation of the *pap*G adhesin in a urinary tract pathogen was recently shown to dramatically decrease colonisation and inflammation in the kidneys of monkeys (Roberts et.al., 1995). Studies in animal models have suggested that type 1 fimbriation increases the survival of *E.coli* in the urinary tract (Hagberg et.al., 1983; Iwahi et.al., 1983; Abraham et.al., 1985; Hultgren et.al., 1985; Keith et.al., 1986), however, epidemiological studies have failed to reveal a correlation between type 1 fimbriation and virulence (Hagberg et.al., 1981). While P fimbriae enhance the virulence of uropathogenic strains through specific adherence and increased induction of mucosal inflammation (Svanborg et.al., 1994), the role of type 1 fimbriae in virulence remains undefined (Klemm and Krogfeldt, 1994).

MUCOSAL CYTOKINE RESPONSES DURING UTI

Infections of the urinary tract elicit a cytokine response. This was first observed in mice, that secreted Interleukin-6 (IL-6) into the urine shortly after experimental infection. The IL-6 response was impaired in LPS non responder mice. Deliberate colonization of the human urinary tract with *Escherichia coli* bacteria was also found to elicit a cytokine response. IL-6 and IL-8 levels in urine increased from undetectable levels prior to colonization, to significant levels during the first few hours. There was no concomitant increase in circulating cytokine levels, suggesting that these molecules were produced locally in the urinary tract. Subsequently, patients with UTI were shown to mount a cytokine response in urine, and in the case of acute pyelonephritis also in serum.

The cellular source of the cytokines was examined, *in vitro*. Epithelial cells were selected for study, since they dominate the mucosal surface, and are likely to be the first cells to encounter bacteria during infection. Epithelial cells exfoliated from the urinary tract, and epithelial cell lines were exposed to uropathogenic *Escherichia coli* and the cytokine response was measured. There was an increase in cytokine specific mRNA levels, in the intracellular cytokine content and in cytokine secretion by those cells. Stimulation of IL-1α, IL-1β, IL-6 and IL-8 was detected.

ROLE OF BACTERIAL ADHERENCE FOR UROEPITHELIAL CELL CYTOKINE RESPONSES

The role of bacterial adherence for the epithelial cell cytokine response was examined using several approaches.

1. Isogenic *Escherichia coli* strains differing in P or type 1 fimbrial expression were used to stimulate epithelial cell lines, and the cytokine responses were compared. Both types of fimbriae were shown to enhance the cytokine response.
2. Purified P fimbriae were shown to induce low but significant cytokine responses. Adhesion negative P fimbriae were inactive.

3. Receptor glycolipid expression was inhibited, using PDMP; a structural analogue of ceramide that inhibits its glycosylation PDMP treated cells had reduced receptor expression, reduced adherence and reduced cytokine responses to P fimbriated but not to type 1 fimbriated *Escherichia coli*.
4. Receptor analogues were used to inhibit adherence of P or type 1 fimbriated *Escherichia coli* to uroepithelial cell lines. Adhesion inhibition was accompanied by a reduction in the cytokine response by those cells.

These results suggested that fimbriae-receptor interactions triggered transmembrane signalling events that result in cytokine activation, either directly through the receptor, or through associated mechanisms.

P FIMBRIAE INDUCED CYTOKINE RESPONSES THROUGH THE CERAMIDE SIGNALLING PATHWAY

P-fimbriae bind Galα1-4Galβ-containing receptor epitopes on the globoseries of glycolipids. The globoseries of glycolipids consist of an oligosaccharide chain bound to ceramide that is localized in the outer leaflet of the lipid bilayer of the epithelial cell membrane (Leffler and Svanborg-Edén, 1980; Bock et.al., 1985). Ceramide has recently been identified as a second messenger in cell signalling. Exogenous ligands like TNF, FAS, IL-1 bind to their respective receptors and activate endogenous sphingomyelinases, that release ceramide from sphingomyelin (Schütze et.al, 1992; Mathias et.al., 1993). Ceramide can in turn, activate protein kinases belonging to the Ser/Thr family of protein kinases and phosphatases with Ser/Thr specificty (Mathias et.al., 1991; Dobrowsky and Hannun, 1993). Down-stream signalling results in cell activation and cytokine production, or alternatively in apoptotic cell death.

We tested the hypothesis that P-fimbriated *E.coli* might release ceramide, and that the ceramide signalling pathway may be involved in epithelial cell cytokine responses (Hedlund et.al., 1996). The human kidney epithelial cell line, A498, that expresses the globoseries of glycolipids, binds P-fimbriated *E.coli* and responds with cytokine production to stimulation with P-fimbriated *E.coli,* was used.

Released ceramide was detected in lipid extracts of A498 cells by *in vitro* phosphorylation using the diacylglycerolkinase assay. The labelled product was run on thin layer chromatogram plates, and quantitated by autoradiography. The P-fimbriated strains caused ceramide release that peaked after approx. 20 minutes of stimulation. The isogenic non-fimbriated control strain did not cause detectable ceramide release in the A498 cells.

The phosphorylation of ceramide to ceramide-1-phosphate was quantitated in cells that had been prelabelled with $^{32}P_i$ for 72 hours. P-fimbriated bacteria caused an increase in ceramide-1-phosphate that peaked around 20 minutes after stimulation. Increases in ceramide-1-phosphate-levels were not detected in cells exposed to non-fimbriated strains.

Ceramide-activated protein kinases are Ser/Thr kinases with specificity for proline (Mathias et.al., 1991). We tested the effects of protein kinase inhibitors on the IL-6 response of A498 cells to P-fimbriated *E.coli.* Staurosporin and K252-a (inhibitors of Ser/Thr kinases) inhibited the IL-6 response to P-fimbriated *E.coli* and PMA (positive control), but had no effect on the IL-6 response to the non-fimbriated *E.coli* control. Genestein and tyrphostin (inhibitors of tyrosine kinase), had no effect on the PMA or P-fimbriae induced response.

These results demonstrated that P-fimbriated *E.coli* fimbriae caused the release of ceramide, and suggested that the ceramide signalling pathway participates in the cytokine responses in epithelial cells. The mechanism(s) of the P-fimbriae induced ceramide release need to be defined and the origin of the ceramide involved in the induction of the IL-6 response needs to be established. Several hypotheses may be discussed and tested.

1. Binding of P-fimbriae to the oligosaccharide portion of globotetraosyl-ceramide or other Galα1-4Galβ-containing glycolipids may cause the release of ceramide from those molecules. Studies to examine this hypothesis are ongoing.
2. Sphingomyelinase was shown to activate IL-6 production in the A498 cells. There was a dose-dependant increase in IL-6 secretion with kinetics similar to those observed with bacterial stimulation.
3. Binding of the P-fimbriae to the globoseries of glycolipids may activate endogenous sphingomyelinases, that cleave ceramide from sphingomyelin.
4. Binding of P-fimbriae to the glycolipid receptors may increase the concentration of other bacterial components at the cell surface. Activation of cytokine responses may be through these components rather than, or in addition to the adhesin-receptor interaction.

(a). The bacteria may secrete sphingomyelinases that activate the release of ceramide from sphingomyelin.

(b). LPS was recently shown to be a structural analogue of ceramide (Joseph et.al., 1994). Thus LPS might, bypass ceramide and directly activate CAPK involved in the down-stream activation of the cytokine response. This pathway was shown to exist in CD-14 positive cells and to require Lipopolysaccharide binding protein. We have shown that purified LPS is a poor activator of epithelial cell cytokine responses (Linder, 1988; de Man, 1989; Hedges, 1992).

P-fimbriae have been proposed to contain LPS as an integral part of the pap_{IA2} adhesin, adjacent to the receptor-binding domain (Linder, 1991). If so, the cells may see the bacterial lectin and the LPS at the same time, and a dual signal may be delivered. The fimbriae, may compensate for the lack of an LPS receptor on epithelial cells, by delivering LPS to the epithelial cell surface in a molecular context that leads to cell activation.

TYPE 1- AND P-FIMBRIATED *E.COLI* ACTIVATE EPITHELIAL CELL CYTOKINE RESPONSES VIA DIFFERENT TRANSMEMBRANE SIGNALLING PATHWAYS

The A498 cells express receptors for *E.coli* type 1-fimbriae. The type 1-fimbriated strains attach to these cells, and elicit higher cytokine responses than non-fimbriated controls. The receptor has not been identified, but the attachment is inhibited by α-Methyl-D-mannoside, suggesting that the fimbriae recognize a mannosylated glycoprotein(s). Mannose residues occur on glycoproteins rather than glycolipids. Consequently, type 1-fimbriated bacteria did not bind to glycolipid extracts from the A498 cells. There was no evidence of ceramide release or phosphorylation of ceramide in cells exposed to type 1-fimbriated bacteria. Furthermore, the type 1-fimbriated *E.coli* induced cytokine response was insensitive to treatment with inhibitors of Ser/Thr kinases. These results strongly support the notion that the fimbrial receptor specificity directs the transmembrane signalling pathways involved in cell activation.

TYPE 1 FIMBRIAL ADHESION ENHANCES *ESCHERICHIA COLI* VIRULENCE FOR THE URINARY TRACT

The presence of *fim* DNA sequences is common among *E.coli* strains. Greater than 90% of clinical isolates, both virulent and avirulent from the gastrointestinal and urinary tract carry the *fim* sequences and can be induced to express type 1 fimbriae. There has been no evidence from epidemiological studies of an association between type 1 fimbriae and the severity of infection (Hagberg et.al., 1981). The fact that a high percentage of *E.coli* strains carry the *fim* DNA sequences makes it difficult to correlate fimbrial expression to disease.

Type 1 fimbriae are encoded by the *fim* cluster which contains 9 genes. *fim*A encodes the major structural subunit and *fim*H encodes the adhesin. The FimH adhesin is located at the fimbrial tip and interspersed along the shaft of the fimbriae and recognizes terminally located D-mannose moieties on cell-bound and secreted glycoproteins (Krogfeldt et.al., 1990; Giampapa et.al., 1988; Wold et.al., 1990). It was also recently shown to mediate mannose sensitive binding to non-glycosylated peptide epitopes (Sokurenko et.al., 1994; Sokurenko et.al., 1995). Interactions with such receptors enables type 1 fimbriated bacteria to bind to a range of cells, including erythrocytes, epithelial cells, granulocytes, macrophages and mast cells (Duguid et.al., 1955; Bar-Shavit et.al., 1977; Ofek et.al., 1977; Malaviya et.al, 1994). This interaction can result in the activation of the respiratory burst in granulocytes, degranulation of mast cells and cytokine release from epithelial cells *in vitro* (Malaviya et.al, 1994; Steadman et.al., 1988; Agace et.al., 1993).

Type 1 mediated adherence has been proposed to play a role in the induction of inflammation. Early *in vitro* studies showed that type 1 fimbriated bacteria bind to phagocytic cells and induce a respiratory burst (Sharon and Ofek, 1986). Further studies have shown that type 1 fimbriae and FimH induce many of the inflammatory effects associated with type 1 fimbriated *E.coli* including an oxidative burst in neutrophils (Steadman et.al., 1988; Tewari et.al., 1993) and proliferation and differentiation of human B cells to immunoglobulin secretion (Ponniah et.al., 1989; Ponniah et.al., 1992). More recently, type 1 fimbriated *E.coli* and FimH coated latex beads were shown to induce mast cell degranulation and histamine release in mice (Malaviya et.al., 1994). Despite these observations the role of type 1 fimbriae as inducers of inflammation in UTI remains undefined.

We have shown that type 1 fimbriae increase the virulence of *E.coli* for the urinary tract by promoting bacterial persistence and by enhancing the inflammatory response to infection. In a clinical study, it was observed that disease severity was greater in children infected with *E.coli* O1:K1:H7 isolates expressing type 1 fimbriae than in those infected with type 1 negative isolates of the same serotype (Mårild et.al., 1989). The children infected with *E.coli* O1:K1:H7 type 1 positive isolates showed a more rapid onset of infection, higher fever, longer fever duration, and higher blood leucocyte counts.

The *E.coli* O1:K1:H7 isolates had the same electrophoretic type, were hemolysin negative, expressed functional P fimbriae, and carried the *fim* DNA sequences. When tested in a mouse UTI model, the type 1 positive *E.coli* O1:K1:H7 isolate (*E.coli* 1177) survived in higher numbers, induced a higher urinary interleukin-6 response and a greater neutrophil influx into the urine, than O1:K1:H7 type 1 negative isolate (*E.coli* 845).

To confirm a role for type 1 fimbriae in urinary tract infection a *fim*H null mutant (*E.coli* CN1016) was constructed from an O1:K1:H7 type 1 positive parent (*E.coli* 1177). Site direct mutagenesis was carried out using a suicide plasmid (pGP704) carrying *fim*D to *fim*H with a neomycin phophatase gene (kanamycin resistance) inserted in the *fim*H gene. Selection was made for double crossover recombinants expressing kanamycin resistance and which had lost the ability to agglutinate guinea pig erythrocytes in the absence of mannose

(*E.coli* CN1016). The construction was subsequently confirmed by Southern blot analysis using the *fim* gene cluster, *fim*H and the *npt* genes as probes.

E.coli CN1016 had reduced survival and inflammatogenicity in the mouse UTI model. Bacterial numbers in the kidneys and bladders of mice infected with *E.coli* CN1016 were significantly lower compared to mice infected with *E.coli* 1177. The numbers of neutrophils in the urine of mice infected with *E.coli* CN1016 were also significantly less than in mice infected with *E.coli* 1177.

To fulfill the molecular Koch's postulates we tranformed *Ecoli* CN1016 with the plasmid pPKL04 (Klemm et.al., 1985) containing the entire *fim* gene cluster (*E.coli* CN1018). Mannose sensitive agglutination of guinea pig erythrocytes was restored in *E.coli* CN1018. Mice infected with *E.coli* CN1018 had similar numbers of bacteria in the kidneys and bladders as mice infected with *E.coli* 1177 and differed significantly from mice infected with *E.coli* CN1016. The numbers of neutrophils in the urine of mice infected with *E.coli* CN1018 were the same as *E.coli* 1177 and differed significantly from *E.coli* CN1016.

Several studies in experimental UTI models have indicated that type 1 fimbriae can aid in the persistence of *E.coli* in the urinary tract. Aronson et. al. showed that the addition of α methyl-D-mannoside to an inoculum of type 1 fimbriated *E.coli* significantly reduced bacteriuria in mice (Aronson et.al., 1979). Similarly, antibodies against type 1 fimbriae prevented ascending colonization of the kidneys in a rat model of *E.coli* induced pyelonephritis (Silverblatt and Cohen, 1979). Hagberg et. al. showed that a type 1 negative mutant of a wild type uropathogenic strain, produced by chemical mutagenesis with nitrosoguanidine, survived in significantly lower numbers in the mouse bladder than the type 1 positive parent (Hagberg et.al., 1983). In contrast, survival in the human urinary tract was reduced after transformation of a wild type *E.coli* strain with a plasmid encoding the *fim* sequences (Anderson et.al., 1991).

This study differed from previous studies since it was based on a type 1 related difference in disease severity in patients with UTI, and it examined type 1 fimbriae in the genetic background of a fully virulent strain isolated from these patients. Furthermore, the adhesive capacity of type 1 fimbriae was specifically inactivated by insertion of a kanamycin resistance gene in the *fim*H gene without further disruption of the wildtype genome. Our results demonstrate that type 1 fimbriae contribute to bacterial persistence in the urinary tract when expressed in the background of a fully virulent uropathogen. The results presented here show that the virulence of a P fimbriated uropathogenic *E.coli* strain can be reduced by inactivation of a second fimbrial type. This illustrates how the different fimbriae, and probably other virulence factors act in concert to achieve the virulent phenotype.

MUCOSAL INFLAMMATION AND RESISTANCE TO UTI

The influx of inflammatory cells into the urine has long been used as a diagnostic tool in patients with UTI. Several observations suggest that the inflammatory response, and specifically the neutrophils are important for the clearance of infection. First, the onset of the inflammatory response coincides with the clearance of infection. Second, defects in the inflammatory response are associated with an impaired resistance to infection. LPS non responder mice have low cytokine responses to infection and a poor neutrophil recruitment. Infections persist in those mice for several weeks. In contrast, LPS responder mice quickly activate inflammation, and clear the infection within 48-72 hours. Third, treatment of normal mice with anti-neutrophil serum or with anti-inflammatory agents, impairs their resistance to infection.

The role of inflammation for resistence to infection needs to be considered in discussions on anti-adhesive therapy. Anti-adherence may turn out to be a two-edged sword.

Blocking of attachment is likely to reduce bacterial persistence but also the inflammatory response that may be essential to eliminate remaining bacteria. The balance between adhesion, antiadhesion and inflammation needs further study.

ACKNOWLEDGMENTS

These studies were supported by: The Swedish Medical Association; the Medical Faculty, University of Lund; the Swedish Medical Research Council (grant 7934); the Royal Physiographical Society of Lund; and, the Österlund, NORFA and Crafoord foundations.

REFERENCES

Abraham, S., Babu, J, Giampapa, C., Hasty, D., Simpson, W., and Beachey, E. (1985) *Infect. Immun.* 48, 625-628.

Agace, W., Hedges, S., Ceska, M., and Svanborg, C. (1993) *J. Clin Invest.* 92, 780-785.

Anderson, P., Engberg, I., Lidin-Janson, G., Lincoln, K., Hull, R.,Hull, S. and Svanborg-Edén, C. (1991) *Infect. Immun.* 59, 2915-2921

Aronson, M., Medalia, O., Schori, L., Mirelman, D., Sharon, N., and Ofek, I. (1979) *J. Infect. Diseas.* 139, 329-332.

Bar-Shavit, Z., Ofek, I., Goldman, R., Mirelman, D., and Sharon, N. (1977) *Biochem. Biophys. Res. Commun.* 78, 455-460.

Bock, K., Breimer, M., Bringole, A., Hansson, G., Karlsson, K-A., Larsson, G., Leffler, H., Samuelsson, B., Strömberg, N., Svanborg-Edén, C., and Thurin, J. (1985) *J. Biol. Chem.* 260:8545-8551.

de Man, P., van Kooten, C., Aarden, L., Engberg, I., and Svanborg-Edén, C. (1989) *Infect. Immun.* 57:3383-3388.

Dobrowsky, R., and Hannun, Y. (1993) *Adv. Lipid Res.* 25:91-104.

Duguid, J., Smith, I., Dempster, G., and Edmunds, P. (1955) *J. Path. Bact.* 70, 335-348.

Giampapa, C., Abraham, S., Chiang, T., and Beachey, E. (1988) *J. Biol. Chem.* 263, 5362-5367.

Hagberg, L., Jodal, U., Korhonen, T., Lidin-Janson, G., Lindberg, U., and Svanborg-Edén, C. (1981) *Infect. Immun.* 31, 564-570.

Hagberg, L., Hull, R., Hull, S., Falkow, S., Freter, R., and Svanborg-Edén, C. (1983) *Infect. Immun.* 40, 265-272.

Hedges, S., Svensson, M., and Svanborg, C. (1992) *Infect. Immun.* 60:1295-1301.

Hedges, S, Agace, W., and Svanborg, C. (1995). *Trends in Microbiology.* 3. 266-270.

Hedlund, M., Svensson, M., Nilsson, Å., Duan, R-D., and Svanborg, C. (1996) *J. Exp. Med.* 183:in press.

Hultgren, S., Porter, T., Schaeffer, A., and Duncan, J. (1985) *Infect. Immun.* 50, 370-377.

Iwahi, T., Abe, Y., Imada, A., and K. Tsuchiya. (1983) *Infect. Immun.* 39, 1307-1315.

Johnson, J. (1991). *Clin. Microbial. Rev.* 4, 80-128.

Joseph, C., Wright, S., Bornmann, W., Randolph, J., Kumar, E., Bittman, R., Liu, J., and Kolesnick, R. (1994) *J. Biol. Chem.* 269:17606-17610.

Keith, B., Maurer, L., Spears, P., and Orndorff, P. (1986) *Infect. Immun.* 53, 693-696.

Klemm, P., Jørgensen, B., van Die, I., de Ree, H., and Bergmans, H. (1985) *Mol. Gen. Genet.* 199, 410-414.

Klemm, P., and Krogfelt, K. (1994) in *Fimbriae; Adhesion, Genetics, Biogenesis, and Vaccines.* ed. Klemm, P. (CRC Press, Boca Raton), pp. 9-26.

Krogfelt, K., Bergmans, H., and Klemm, P. (1990) *Infect. Immun.* 58, 1995-1998.

Leffler, H., and Svanborg-Edén, C. (1980) *FEMS Lett.* 8:127-134.

Linder, H., Engberg, I., Mattsby-Baltzer, I., Jann, K., and Svanborg-Edén, C. (1988) *Infect. Immun.* 56:1039-1313.

Linder, H., Engberg, I., Hoschüttzky, H., Mattsby-Baltzer, I., and Svanborg, C. (1991) *infect. Immun.* 59:4357-4362.

Malaviya, R., Ross, E., Jakschik, B., and Abraham, S. (1994) *J. Clin Invest.* 93, 1645-1653.

Mårild, S., Jodal, U., Ørskov, I., Ørskov, F., Svanborg Edén, C. (1989) *J. Pediatr.* 115, 40-45.

Mathias, S., Dressler, K., and Kolesnick, R. (1991) *Proc. Natl. Acad. Sci. USA.* 88:10009-10013.

Mathias, S., Younes, A., Kan, C., Orlow, I., Joseph, C., and Kolesnick, R. (1993) *Science (Wash DC)* 259:519-522.

Ofek, I., Mirelman, D., and Sharon, N. (1977) *Nature (London)*. 265, 623-625.

Ponniah, S., Abraham, S., Docktor, M., Wall, C., and Endres, R. (1989) *J. Immunol.* 142, 992-998.

Ponniah, S., Abraham, S., and Endres, R. (1992) *Infect. Immun.* 60, 5197-5203.

Roberts, J. Marklund, B., Ilver, D., Haslam, D., Kaak, M., Baskin, B., Louis, M., Möllby, R., Winberg, K., and Normark, S. (1995). *Proc. Natl. Acad. Sci.* 91, 11889-11893.

Schütze, S., Potthoff, K., Machleidt, T., Berkovic, C., Wiegman, K., and Krönke, M. (1992) *Cell* 71:765-776.

Sharon, N., and Ofek, I. (1986) In *Microbial lectins and agglutinins: properties and biological activity*, ed. Mirelman, D. (Wiley and sons, New York), pp. 55-82.

Silverblatt, F., and Cohen, L. (1979) *J. Clin. Invest.* 64, 333-336.

Sokurenko, E., Courtney, H., Ohman, D., Klemm, P., and Hasty, D. (1994) *J. Bacteriol.* 176, 748-755.

Sokurenko, E., Courtney, H., Maslow, J., Siitonen, A., and Hasty, D. (1995) *J. Bacteriol.* (In press)

Steadman, R., Topley, N., Jenner, D. E., Davies, M., and Williams, J. (1988) *Infect Immun.* 56, 815-822.

Svanborg, C., Ørskov, F., and Ørskov, I. (1994) in *Fimbriae; Adhesion, Genetics, Biogenesis, and Vaccines*. ed. Klemm, P. (CRC Press, Boca Raton), pp. 239-254.

Tewari, R., MacGregor, J., Ikeda, T., Little, R., Hultgren, S., and Abraham, S. (1993) *J. Biol. Chem.* 268, 3009-3015.

Wold, A., Mestecky, J., Tomana, M., Kobata, A., Ohbayashi, H., Endo, T., and Svanborg-Edén, C. (1990) *Infect. Immun.* 58, 3073-3077.

10

GROUP A STREPTOCOCCAL ADHESION

All of the Theories Are Correct

David L. Hasty[1,2] and Harry S. Courtney[1,3]

[1] Research Service
VA Medical Center
Memphis, Tennessee
[2] Departments of Anatomy and Neurobiology and of [3]Medicine
University of Tennessee
Memphis, Tennessee

1. INTRODUCTION

1.1. General

Streptococcus pyogenes, the group A streptococcus (GAS), is a common human pathogen that causes a wide array of infections, ranging from uncomplicated pharyngitis and pyoderma to invasive infections including sepsis, pyomyositis and fasciitis, pneumonia and toxic shock syndrome. Two nonsuppurative sequelae, acute rheumatic fever (ARF) and acute glomerulonephritis (AGN), are the most significant health problems stemming from GAS infection worldwide and account in large part for the intense clinical interest in this pathogen over the years. In the United States and other developed countries, where the incidence and severity of these diseases had declined over much of the 20th century,[15,19] there has been a remarkable resurgence in the number of ARF cases over the last 10 years.[60,69,70] In addition, a new clinical entity, streptococcal toxic shock syndrome has been described.[58]

Although the emergence of antibiotic-resistant GAS strains is not yet a critical problem and GAS remain exquisitely sensitive to penicillin, it is clear that a vaccine that would prevent these infections is urgently needed. While penicillin therapy at the appropriate time during pharyngitis can effectively prevent the development of ARF,[59] some individuals develop ARF without having realized they needed medical care. In the recent outbreak of ARF in the state of Utah, only a small fraction of the patients sought health care prior to the onset of symptoms and a large number of patients did not even recall having had a sore throat.[69] In other instances, especially in developing countries, medical care is simply not available.[31] Another reason a vaccine is needed is that certain serious GAS infections often progress so rapidly that antibiotics may have little or no impact on the outcome.[58]

There is little question that the prime candidate for a protective GAS vaccine is currently M protein. It is the major virulence factor and protective antigen of group A

Toward Anti-Adhesion Therapy for Microbial Diseases, edited by Kahane and Ofek
Plenum Press, New York, 1996

streptococci and it elicits bactericidal antibodies.[36,38] Although a number of other candidate immunogens have been proposed for a vaccine that may protect against infection, an overwhelming amount of evidence supports the focus on M protein. M protein antibodies provide long-lasting, type-specific immunity and bactericidal antibodies have been detected as long as 30 years following documented infections.[37] Recurrent infections with the same serotype are rare.

While anti-M protein antibodies are opsonic and protective, there have been a number of obstacles to M protein vaccine development. First is the large number of M serotypes (currently over 90). Even if one focuses only on the primary "rheumatogenic" serotypes, there are 12-15 types that would have to be included in a vaccine.[6,7,59] A second, perhaps more serious problem is the fact that certain regions of M proteins elicit antibodies that crossreact with human tissues, primarily the myocardium.[19,34]

We have hoped to contribute toward more effective treatments for GAS infections by identifying an adhesin that could be included in a GAS vaccine. This could be important in at least two ways. First, evoking anti-adhesive antibodies could prevent GAS from ever gaining a foothold in the oropharyngeal cavity. Second, it is possible that an adhesin candidate could be found that would be more broadly represented among serotypes, reducing the number of immunogenic sequences that would be required.

Understanding GAS adhesion sufficiently well to contribute to a vaccine or other therapeutic procedure obviously requires a detailed knowledge of the molecular architecture of GAS surfaces. Defining the most relevant adhesins of the serotypes most commonly involved in ARF and AGN is essential. A thorough knowledge of the host niche that becomes infected and the identification of relevant receptors is also important in the process of trying to determine which are the relevant GAS adhesins.

Significant progress has been made on the issue of GAS adhesins and receptors. Since the first paper on GAS adhesion almost 25 years ago,[22] at least 11 adhesin candidates have been described: M protein,[14-16,20,22,23,47,74] lipoteichoic acid (LTA),[43,45,52,54,73] protein F/Sfb,[28,62] a 28 kDa fibronectin (Fn)-binding protein (FBP),[13] glyceraldehyde-3-phosphate-dehydrogenase (G3PDH),[49] a 70 kDa galactose-binding protein,[24,72] C-carbohydrate,[8] a vitronectin (Vn)-binding protein,[66] a 57 kDa collagen-binding protein,[71] serum opacity factor (SOF)[33] and a 54 kDa Fn-binding protein, FBP54[15,19] (Table I). The data supporting a role for most of the adhesin candidates other than LTA, M protein and protein F/Sfb are scant, at least to date. For G3PDH and the 28 kDa Fn-binding protein, the only supporting evidence is their ability to bind Fn or, in the case of G3PDH, other host proteins as well.[8,49] A Vn-binding substance on GAS has not been identified but is assumed to be present because Vn binds to GAS and blocks adhesion.[66] The 57 kDa collagen-binding protein isolated from one strain of GAS blocks adhesion of GAS to collagen.[71] Sequencing the gene for SOF has revealed a Fn-binding motif.[33] The evidence in favor of LTA, M protein and two of the Fn-binding proteins, protein F/Sfb and FBP54, as adhesins is more substantial and can serve to illustrate some of the issues involved (see below).

We are currently of the opinion that many of the adhesins so far described, and/or others yet to be described, will probably be found to be involved in GAS adhesion in important ways. Thus, we expect that the various theories regarding adhesins and receptors put forth by each of the laboratories examining this topic are probably also correct, at least in part. It is certainly true that GAS are capable of utilizing more than one adhesin. Whether a strain is able to attach to a mucosal surface is dependent upon both the nature of the adhesins expressed by a particular GAS strain and on the nature of the receptors expressed in a particular niche of the host. Other layers of complexity must also be taken into account, such as genotypic variation and environmental effects on regulation of expression of adhesins. Since strains do not always have genes for each of the adhesins,[41] GAS strains cannot express identical arrays of adhesins. Since environmental factors affect adhesin gene expression,[68]

Table I. GAS Adhesins and Receptors

Adhesin/receptor	Target substratum	References
LTA/Fn or ?	Human buccal cells	3,14,42
	Human pharyngeal cells	8, 14
	Mouse pharyngeal cells	20
	HEp-2 cells	14,27
	Human/mouse fibroblasts	39a
	Human leukocytes	9a
FBP*/fibronectin or ?	Human fibronectin	57, 21,62
	Human buccal cells	18
	HEp-2 cells	62,63,67
	Epidermal Langerhans' cells	47
M protein/galactose fucose, fibrinogen, or ?	Human pharyngeal cells	22,23, 65,72
	Human keratinocytes	47,48
	HEp-2 cells	14-18,74
	virus infected MDCK cells	51
VBP/vitronectin	Human pharyngeal cells	8
G3PDH	-	49
Serum Opacity Factor	-	33

*FBP= fibronectin binding proteins; VBP= vitronectin binding protein

it is probably equally unlikely that any single strain will express the same adhesins in every host niche. The modulation of mucosal surfaces by variations in the composition of mucosal secretions,[12] the presence of dietary components[25] and the effects of other microbes[32] must also be kept in mind. Thus, we expect that any effective anti-adhesion vaccine may require inclusion of several adhesins or adhesin fragments. Work remains to be done to fully explain basic mechanisms of GAS adhesion, but more focus should now be given to evaluating the potential of the described adhesin candidates for inclusion in vaccine constructs designed to prevent GAS infections.

1.2. The Initial Conflict

In 1972 and 1974, Richard Ellen and Ron Gibbons published papers in which they proposed that the GAS adhesin was associated with the M protein-containing "fuzz" at the cell surface.[22,23] M-positive strains adhered and also persisted in the oropharyngeal cavity of mice, while M-negative strains did not. In 1975, Itzhak Ofek, Ed Beachey and their colleagues published the first in a long series of papers from which they concluded that the GAS adhesin was not M protein, but lipoteichoic acid (LTA).[43] LTA is an amphipathic molecule which has mono-, di- and tri-acylated forms associated with the cell membrane and released through the surface "fuzz" and into the medium. LTA was the only GAS surface component that inhibited adhesion. Among other molecules tested, a pepsin-extracted amino terminal fragment of M protein roughly half the size of the intact M molecule had no inhibitory effect. The initial studies by Ofek and Beachey were limited to one host cell type, buccal epithelial cells.

1.3. LTA as Adhesin; Fibronectin as Receptor

Studies of LTA numerically overwhelmed studies of any other GAS adhesin candidate for a number of years. In 1982, Ofek, Andy Simpson and Beachey published a model for how LTA might become oriented at the surface of the bacteria by way of the interaction of its negatively charged PGP backbone with positively charged residues of surface proteins, its lipid moiety projecting outward toward the periphery of the organism where it can interact with fatty acid binding sites of host cell surface proteins.[45] The evidence supporting the role of LTA as an adhesin is substantial (Table II). However, several laboratories have described studies in which LTA has not been an effective inhibitor of adhesion.[9,62] It is not possible to directly compare the studies in which LTA inhibited adhesion to those in which it didn't, because of significant differences in the characteristics of LTAs used (*e.g.* impure or diacylated *vs* monoacylated) and technical differences in the ways experiments were conducted (*e.g.* high heat, presence of detergents and LTA-binding proteins). Whether or not the model proposed by Ofek, Simpson and Beachey is precisely correct, it has certainly generated relatively enthusiastic groups of supporters and detractors and, thereby, has had a notable and long-lasting impact on the field.

As the "tree model" for anchorage of LTA on the GAS surface was being developed, Beachey, Ofek and Simpson began a series of studies searching for an LTA receptor molecule.[52-54] The well-characterized fatty acid binding sites of bovine serum albumin proved to be a good LTA receptor model.[53] Although these investigators did not believe albumin was the receptor for GAS adhesion, it did provide a useful model to determine the appropriate biochemical characteristics for a receptor. These studies eventually led to a series of publications firmly establishing Fn as one important LTA-reactive GAS receptor on buccal epithelial cells.[54] Although the role of LTA continues to be questioned, Fn has apparently become rather firmly established as a prominent GAS receptor.[1,15,21,33,35,39,42,54,55,57,62,63,67] However, there is still probably more data in support of LTA as an important adhesin than there is that Fn is the primary receptor.

The Beachey, Ofek and Simpson research group that we joined in the late 1970s was a strong and effective proponent of LTA as an important GAS adhesin. However, even then we realized that LTA would not be a good vaccine candidate because of its cytotoxicity[9a] and because of its virtual ubiquity among gram positive organisms, many of which are commensal strains that are extremely important in maintaining the balance of oral flora. For these

Table II. LTA as an adhesin/ Fn as a receptor

Evidence that LTA is an adhesin	Evidence that Fn is a Receptor
1. LTA binds to specific receptors on host cells	1. Fibronectin is present on host cell surfaces
2. LTA inhibits streptococcal adhesion	2. Fibronectin inhibits adhesion
3. The degree of streptococcal adhesion is related to the amount of LTA receptors	3. The degree of streptococcal adhesion is related to the amount of fibronectin expressed on cells.
4. Anti-LTA blocks adhesion.	4. Anti-fibronectin blocks adhesion
5. Antibiotics that induce a release of LTA cause a reduction in adhesion	5. Antibiotics that induce release of LTA cause a reduction in fibronectin binding to streptococci.
6. LTA competitively inhibits binding of fibronectin to streptococci.	6. Fibronectin binds to LTA
7. Deacylated LTA does not bind to or inhibit adhesion to epithelial cells.	7. Deacylated LTA does not bind to fibronectin.
8. Proteins that have fatty acid-binding sites block adhesion.	8. Fibronectin has fatty acid-binding sites.
9. LTA and anti-LTA passively protect mice against challenge streptococcal infection.	

reasons and also because it was also a difficult molecule to work with and one that could apparently not be manipulated genetically, we often hoped that a proteinaceous adhesin would be found. In the late 1970s, we were not necessarily convinced that LTA was the only adhesin, but we were convinced that M protein was not an adhesin, at least for human buccal epithelial cells. At the time, there was relatively little data in support of any other proteinaceous adhesins for GAS. Subsequently, Pietro Speziale and colleagues published a paper describing a GAS Fn-binding component that was thought to be a protein.[57] The data were very intriguing, but, curiously, they chose to study Fn binding to heat-treated (88°C for 20 minutes) organisms. We initiated a series of experiments on similarly heat-treated organisms and were somewhat disappointed to be led to the conclusion that Speziale and co-workers could be studying the binding of Fn to a cryptic receptor that was inaccessible on the surface of non-heated bacteria.[55] Additional studies by Nealon et al., showed that bound Fn that had been released from GAS by treatment of the organisms with penicillin treatment was precipitable with anti-LTA antibodies.[42] No other antibodies evoked by immunizing rabbits with intact GAS were capable of immunoprecipitating Fn. These data seemed to suggest quite strongly that the Fn-binding component of GAS was indeed LTA.

It was not until the early 1990s that there were truly convincing data published supporting the presence of an FBP in strains of GAS. In 1991, Suzanne Talay and her colleagues reported cloning the gene for a Fn-binding protein (FBP) which they called Sfb (for *s*treptococcal *F*n-*b*inding protein).[62] The next year, an isotype of Sfb, termed protein F, was described by Emanuel Hanski and Michael Caparon.[28] These and subsequent publications on protein F/Sfb[39,47,63,67] have made an enormous impact on the field. Several other distinct FBPs have also been found in the relatively few years since the two seminal publications. In fact, the principal reason so far for including some GAS components on the list of potential adhesins is only because they are surface molecules that contain motifs structurally similar to the Fn-binding motifs of protein F/Sfb[33,49] (or other known FBPs) and not necessarily because they have known effects on adhesion.

Due to the tremendous amount of work that has been done on GAS adhesion since 1972, we cannot undertake an exhaustive review in this brief chapter. Since the Sfb/protein F system will be presented elsewhere in this book, we will use the remainder of this article to present recent work we have done on two different adhesins. First, we will discuss some of our experiments concerning an FBP that serves as an adhesin for at least some host cells and some GAS strains. This protein, termed FBP54, is distinct from Sfb/protein F[15,17] . We will also present work we have done over the last few years to establish that M protein is indeed an important GAS adhesin for certain host surfaces. We will also review experiments we have performed to test the efficacy of two adhesins, LTA and M protein, in inhibiting colonization and preventing infection. Lastly, we will attempt to put into perspective some of the work from our laboratory and other laboratories as it relates to the concept that GAS utilize multiple adhesins in a stepwise fashion to become attached to host tissues and that the complement of the adhesins used varies among strains and that it can vary within a single strain, depending upon the physiological conditions within the different host niches.

2. A TALE OF TWO ADHESINS

Within the last several years, it has become increasingly clear that GAS utilize several different adhesins to attach to host tissues. In the first decade of GAS adhesion research, it appeared that each report on a potential adhesin had to take issue with reports of other adhesins, as if there could only be one true GAS adhesin, or only one truly important GAS adhesin. We now believe that there is probably significant merit in most of the adhesins that have been reported as well as the potential receptors. For instance, although we have never found an effect of sugars on streptococcal adhesion,[14] there have been too many reports in the literature to ignore the possibility that at least one adhesin/receptor pair responsible for GAS adhesion involves the saccharide moiety of a glycoprotein or glycolipid.[23,64,65,72,74] With our focus on FBP54 and M protein in the following paragraphs, we do not intend to imply that they are the only GAS adhesins or even the most important. There are a great many serotypes, a great many host tissue types and a vast array of physiological conditions to be explored before such a determination can be made with any degree of certainty for FBP54, M protein or for any of the other GAS adhesins thus far described.

2.1. The Role of FBP54 in GAS Adhesion

A gene library was constructed using chromosomal DNA from an M type 5 (Manfredo) group A streptococcus.[15] Clones were screened by SDS-PAGE for overexpression of proteins and blots of lysates containing foreign protein were probed by with biotinylated Fn and fibrinogen. One clone was found that bound both probes with apparent specificity, since other overexpressed proteins bound neither probe. One of the clones overexpressed a protein which only bound fibrinogen. This protein was subsequently identified as the M5 protein. The M3 protein has been found to bind Fn, but M5 protein does not. The recombinant protein that bound both Fn and fibrinogen was termed FBP54 due to its calculated molecular weight of 54-kDa and its ability to bind Fn. Antiserum specific for FBP54 produced in rabbits was found to react with 5 of 15 serotypes tested, including M1, M5, M6, M13 and M18. The low level of reactivity of some strains and the complete lack of reactivity of others could be due to antigenic differences, variation in levels of expression, culture conditions or to the absence of an *fbp*54 gene. At least one of the strains exhibiting little surface reactivity with anti-FBP54 serum, an M24 strain, does possess a copy of the gene for FBP54.

When presented with soluble or immobilized forms of Fn, some bacteria selectively react with only one form. In contrast to some strains of *E. coli* and *Streptococcus sanguis* which bind only to immobilized Fn,[40,56] GAS react with both soluble Fn and immobilized Fn.[11] This suggests that a Fn adhesin on GAS should react with both forms. Both soluble and immobilized forms of Fn reacted with FBP54, so its reaction is consistent with the reaction of GAS with Fn.

The protein sequence was deduced from the nucleotide sequence of the *fbp*54 gene.[15] It contains two repeat motifs, neither of which bear any striking homologies with other FBPs of gram positive bacteria. The motif within the amino-terminal region does exhibit some homology with the sequence of a *S. aureus* FBP. It does not exhibit the LPxTGE membrane anchor motif that is common to many, but not all, gram positive membrane proteins.

To determine whether the amino-terminus of FBP54 was exposed near the surface of GAS, we reacted GAS with immobilized antibodies specific for the amino-terminal fragment (residues 1-89).[17] The streptococci adhered avidly to this antibody, but not to a control antibody or BSA, indicating not only that FBP54 is present on the GAS surface, but also that its amino-terminal Fn-binding domain is exposed at the surface in such a manner that it can anchor the GAS to the substratum.

The Fn- and fibrinogen-binding domain(s) of FBP54 were localized by constructing truncated forms of FBP54.[15] Fusion proteins were engineered containing an N-terminal polyhistidine sequence for metal affinity chromatography purification. The constructs included the intact molecule (residues 1 to 474), an amino-terminal fragment (residues 1-89), and a carboxy-terminal fragment (residues 123-474). Both Fn and fibrinogen bound predominantly to the amino-terminal fragment, suggesting that the binding sites were in this region. The 28 kDa amino-terminal fragment of Fn is the primary domain that interacts with GAS, it is the most effective of the Fn domains as an inhibitor of GAS adhesion to human buccal epithelial cells and it was also found to be the primary domain that reacts with FBP54.[17]

The results described strongly suggested that FBP54 could be involved in adhesion of GAS to certain host cells. Subsequently, FBP54 was indeed found to block adhesion of GAS to buccal cells in a dose-dependent fashion.[17] It had essentially no effect, however, on the adhesion of GAS to HEp-2 tissue culture cells. The result with FBP54 and HEp-2 cells was not surprising, because, as detailed below, the primary adhesin for HEp-2 cells appears to be M protein.[14,16,74]

It is possible that FBP54 is expressed by GAS in culture, but not in the human host. Screening of antisera from ARF and AGN patients showed, however, significant levels of antibodies in a large number of the patient sera, although some reacted near background.[17] We do not yet know whether the significant variation in anti-FBP54 that was observed was due to absence of an FBP54 molecule from the infecting strain, antigenic variation of the molecule among strains or to variability of individuals in their response to this particular antigen.

In summary, we believe that FBP54 should be considered as one of the multiple adhesins utilized by GAS to adhere to certain host tissues. More work needs to be done to determine its prevalence among rheumatogenic as compared to non-rheumatogenic strains, its structural variation among strains and, eventually, its efficacy in actively or passively protecting mice from GAS challenge infection.

2.2. The Role of M Proteins in GAS Adhesion

The possibility that M protein was a GAS adhesin was the clear implication of the first two publications on GAS adhesion mechanisms in the early 1970s.[22,23] The fact that pep M proteins did not inhibit adhesion of GAS to buccal cells suggested that it was not the adhesin responsible for attachment.[43] In 1987-88, the Gibbons laboratory followed up their earlier reports by showing that while there was essentially no difference in the adhesion of natural M+ or M- variants of GAS strains to buccal cells, there were roughly 5-fold more M+ GAS than M- GAS that attached to pharyngeal epithelial cells.[64,65] Furthermore, recombinant M6 protein was shown to bind to both pharyngeal and buccal epithelial cells and it was also shown, in fact, to inhibit adhesion of GAS to buccal cells. Subsequently, Michael Caparon in June Scott's laboratory tested an isogenic pair of M6+ and M6- GAS for adhesion to buccal or tonsillar epithelial cells and reported that there were no differences between the strains.[9] There was a small increase in the number of mutant GAS bound than parent, but it was not significantly different. Interestingly, the M- strain bound significantly more Fn (80% increase) than did the M+ strain, suggesting that Fn ligand(s) on this strain were more accessible than on the parent strain. A basic issue that this highlights is that mutations do not necessarily have simple effects on the architecture of the bacterial surface. Otherwise, the elimination of M6 expression, a molecule that does not bind Fn, would have had no effect on Fn-binding by the organism. Regardless of any other issues, however, these results strongly suggested that M protein was not the primary adhesin for buccal or tonsillar epithelial cells.

Although we took a somewhat different approach, further experiments performed in our laboratory provided much the same conclusion as the earlier studies of Beachey and Ofek and colleagues and those of Caparon et al., that M protein was not the primary GAS adhesin for oropharyngeal epithelial cells. In our hands, LTA effectively inhibited an M type 5 GAS strain to buccal and pharyngeal epithelial cells and to HEp-2 tissue culture cells.[14] We also found that M protein also inhibited adhesion, but the inhibitory effects on adhesion to oropharyngeal epithelial cells appeared to be due to "buried" domains of the M molecule, regions that would not be expected to be accessible on an intact organism. Thus, the inhibition by recombinant M protein appeared, at least in our hands and with these particular cell types, to be artifactual. M-positive strains, on the other hand, did appear to bind much more effectively to HEp-2 tissue culture cells than did M-negative strains. Furthermore, although pep M protein failed to inhibit GAS adhesion to oropharyngeal epithelial cells, it effectively inhibited GAS adhesion to the tissue culture cells. Thus, M protein appeared to be an important GAS adhesin, depending upon the cell type being assayed. Wang and Stinson also reported evidence that M protein appeared to be an adhesin for HEp-2 cells, for an M type 6 GAS strain.[74]

The role of M protein as an adhesin for HEp-2 cells was confirmed by further experiments in our lab utilizing a genetically constructed M-negative strain and its parent.[15] The mutant was created by insertional inactivation of the *emm*24 gene with the Ω-interposon, followed by recombination into the M24 chromosome. The mutant strain, M24-Ω, did not react with antiserum to M24 nor did it survive in whole human blood, confirming the absence of M24 protein. The adhesion of the M+ parent to HEp-2 cells and to mouse oral epithelial cells was dramatically greater than the adhesion of the M24-Ω mutant, but there was no difference between the two in adhesion to human buccal cells. The parent was also dramatically more effective than the M24-Ω mutant in colonizing the oral cavity of mice.[15] We have also confirmed that M5 is an adhesin, because expressing the M5 protein in the M24-Ω background strain restores its ability to survive in whole human blood and to adhere to HEp-2 cells.[18] An M-negative derivative of an M18 strain, M18-Ω, appears also to fail to adhere to HEp-2 cells, suggesting that the M18 protein is also required for adhesion to HEp-2 cells. Whether or not the M18 protein can serve as an adhesin in the M24-Ω strain remains to be determined.

Recently, Nobuhiko Okada in Michael Caparon's laboratory, has confirmed our hypothesis that GAS can express multiple adhesins and that they can utilize these different adhesins to bind to different cells types.[47] The M+ parent strain JRS4, bound to histologic sections of human skin in the epidermal layer and also to cultured keratinocytes. Using isogenic mutants, they found that M protein appeared to confer the ability to adhere to keratinocytes, while protein F appeared to confer the ability to adhere to Langerhans' cells. Okada et al. have also shown that the receptor for M protein-mediated adhesion appears to be the membrane cofactor protein, CD46, and that it is the C-repeat region of M6 that serves as the CD46 binding domain.[48] Our preliminary data suggest that CD46 is also present on the surface of HEp-2 cells, so it will be interesting to determine whether it is also the receptor for M protein on these human cells. Our current speculation is that CD46 is not the receptor. Pep M protein effectively inhibits adhesion of the M5 Manfredo strain to HEp-2 cells,[14] but it lacks the C-repeat region thought to be required for interaction of M protein with CD46. Whether CD46, Fn or a Fn motif is expressed on Langerhans' cells was not determined. Interestingly, it appeared that perhaps the M-negative isogenic strain, JRS145, adhered in larger numbers to histologic sections of epidermis than did the parent strain, again raising the issue of what inaccessible or cryptic adhesins might be exposed in mutant strains. This possibility is a general one that could affect the work of any of us who produce and use mutant organisms.

3. PROTECTION OF MICE USING ADHESINS AND ANTI-ADHESIN ANTIBODIES

Thus far, the only adhesins that have been tested in animal models of infection are LTA and M protein. In our laboratory, we found that intranasal instillation of LTA, but not deacylated LTA, in mice blocked colonization and prevented death after intranasal challenge infection with group A streptococci.[20] Bacteria pretreated with rabbit antisera against LTA also failed to colonize or to infect mice after intranasal challenge. In vitro studies showed that LTA and M protein inhibited adherence of type 24 streptococci to mouse pharyngeal cells. Passive intranasal administration of purified type 24 M protein protected mice from death after challenge infection with type 24 streptococci, but had no significant effect on pharyngeal colonization. Similarly, the M24-Ω strain lacking M protein, failed to adhere to mouse epithelial cells in vitro and also failed to effectively colonize or kill mice.[16] These studies suggest that both LTA and M protein may mediate adherence of streptococci to mouse pharyngeal cells, thus differing somewhat from the results we obtained with human buccal and pharyngeal cells where M protein appeared not to be involved. It also shows that protection can be gained by pretreatment with these two adhesins or antibodies directed against them. Interestingly, the hyaluronate capsule, which is anti-adhesive in every instance where it has been studied, proved to be essential for colonization of mice by an M18 strain.[75] Thus, it appears that LTA, M protein and capsule are required for colonization of the oral cavity of mice by type 24 GAS. The requirement for capsule is unknown, but it could be due to its ability to prevent aggregation of GAS by saliva.[12] Using near-isogenic strains, Susan Hollingshead's group found M protein to be necessary for persistence of GAS in the pharynx of rats, but not for initial colonization.[30] Whether the differences in these studies is due to strain or host differences is not known. Of course, extrapolation of any these results to human disease must be done with caution. There are some potentially significant differences between intranasal challenge infection of mice and natural infection of humans. The number of streptococci required to consistently produce infection is high. Whatever differences may exist in the mechanisms of streptococcal infection in mice and humans, these results support the widely held concept that group A streptococcal adherence is an important first step in the process of infection.

4. MODELS AND PERSPECTIVES

Several years ago we proposed a model for GAS adhesion in which we attempted to find an explanation for the many apparently divergent publications that had appeared over the preceding 20 years.[29] In the four years since our model was proposed, essentially all of the data that have been published support our hypotheses. In essence, we proposed that GAS can express a variety of adhesins that are capable of reacting with different receptors. There

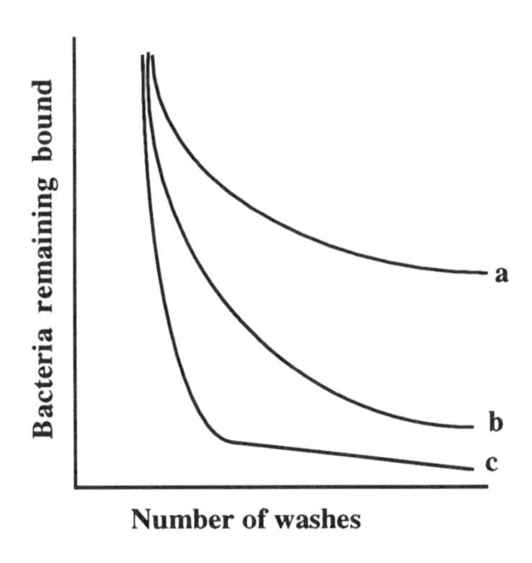

Figure 1. Theoretical analysis of the first and second steps in the firm adhesion of GAS to epithelial cells. According to our model, the numbers of bacteria that will remain bound to cells after successive washes will depend upon passing through distinct steps. Without the expression or exposure of the first-step adhesin (e.g. LTA), bacteria will be washed off rapidly (curve c). With the expression and exposure of the first-step adhesin, bacteria will bind significantly better, but their adhesion will be sufficiently loose, such that these organisms can eventually be washed away (curve b). Only adhesion that withstands the shearing forces of multiple washings would be considered firm adhesion (curve a). We believe that this would represent the completion of both the first and second steps of adhesion and that this adhesion is essentially irreversible.

is now little doubt that GAS utilize multiple mechanisms of adhesion. As examples, the studies from our laboratory regarding M protein and FBP54 as adhesins for attachment of the Manfredo strain of M type 5 GAS to HEp-2 and buccal epithelial cells, respectively, and those from Caparon's laboratory regarding M protein and protein F as adhesions of an M type 6 GAS to keratinocytes and Langerhans' cells, respectively, demonstrate this to be true. We believe that these data essentially support our model and that it will be found to be more generally true that the ability of a particular organism to adhere will be dependent upon, among other things, the organism's genotype (e.g. does it have a gene for adhesin X?), the physiologic conditions to which the organism exposed (e.g. is the gene expressed?) and the substratum (e.g. is the receptor present?). While these questions can be answered successfully using *in vitro* experiments, it is not easy to extrapolate to human disease. Perhaps some or all of the adhesins and receptors that have been studied are irrelevant to human disease. What is the relevant substratum for study of adhesive mechanisms at work in the pathogenesis of pharyngitis? Do GAS use the same molecular mechanisms to attach to isolated tonsillar epithelial cells and to tonsillar tissue *in situ* where they would be coated with salivary mucins? What are the effects of the secreted products of major or minor salivary glands that coat the oropharyngeal mucosa or the dietary components that are present? What are the effects of serum transudates or the secreted products of immune cells within the tonsils? The answers to these and many other questions have a major bearing on determining which are the most relevant adhesins.

Our model also proposes that the adhesion of GAS takes place in distinct kinetic steps.[29] In the first step, the bacteria must overcome the electrostatic repulsion separating its surface from that of the substratum. If this step does not occur, adhesion will not follow. We propose that LTA, or perhaps another hydrophobin, is likely to be the primary component responsible for passing through the first phase for most, if not all, GAS. Completion of the first step would provide for relatively weak, reversible adhesion and would not provide for tissue tropism. A second step must then occur in order to achieve firm adhesion. This step is dependent upon another tier of adhesins, such as protein F or M protein, that are capable of interacting with tissue-specific receptors, providing for less easily reversible or irreversible adhesion and for tissue tropism. It would be affected by environmental factors that regulate expression of second-step adhesins and on the availability of receptors. The regulation of M protein and protein F expression apparently leads, in many circumstances, to divergent expression of the two likely adhesins.[68] Even when these molecules are expressed at maximal levels, the effectiveness with which they confer adhesion would also be dependent upon the presence and accessibility of their receptors and potential inhibitory compounds.

It remains to be convincingly determined whether aspects of the model that invoke a need for sequential kinetic steps to obtain firm attachment, are true. The current availability of appropriate isogenic mutants should allow the model to be tested. The model would predict that if the first-step adhesin, purported to be LTA, is missing or masked, adhesion would not occur. Any bacteria bound to the substratum would be quickly and easily washed away (Figure 1, curve c). Although it has not been possible to create an LTA-negative mutant strain, this adhesin can be masked by utilizing growth conditions favoring expression of the hyaluronate capsule or it can be inhibited by soluble LTA. If the first-step adhesin was present, but the second-step adhesin was absent, adhesion would occur to a greater degree, but it too would also eventually be reversed (Figure 1, curve b). This can now be tested utilizing mutants lacking M protein, FBP54 or protein F/Sfb. Only if both first- and second-step adhesins were expressed and properly exposed at the streptococcal surface would firm adhesion that would resist the effects of washing occur (Figure 1, curve a).

In summary, there is a large body of data regarding candidate adhesins for GAS and receptors. Some of the data would appear, on the surface, to be in conflict. We are of the opinion that it is only the interpretations that are in conflict. We also believe that when all of the data are analyzed, most of the theories regarding which GAS components can serve as adhesins will be found to be correct. There is, however, a paucity of data regarding the efficacy of adhesins in a vaccine. A great many basic and important questions remain to be answered regarding GAS adhesins, adhesive motifs, prevalence of adhesin genes among clinically important strains, regulation of adhesin gene expression and of other genes that directly or indirectly affect adhesin function and regarding relevant host receptors. However, there is already a sufficient amount of information to warrant testing a larger number of GAS adhesins for their ability to protect, actively or passively, against challenge GAS infections.

ACKNOWLEDGMENTS

The authors would like to thank Y. Li for exceptional technical assistance and Dr. James B. Dale and Itzhak Ofek for insightful contributions to the experiments and publications described that have originated from our laboratory over the last several years.

REFERENCES

1. Abraham, S.N., E.H. Beachey and W.A. Simpson. 1983. Adherence of *Streptococcus pyogenes*, *Escherichia coli*, and *Pseudomonas aeruginosa* to fibronectin-coated and uncoated epithelial cells. Infect. Immun. 41: 1261-1268.
2. Alkan, M.L., and E.H. Beachey. 1978. Excretion of lipoteichoic acid by group A streptococci. Influence of penicillin on excretion and loss of ability to adhere to human oral mucosal cells. J. Clin. Invest. 61: 671-677.
3. Beachey, E.H. and I. Ofek. 1976. Epithelial cell binding of group A streptococci by lipoteichoic acid on fimbriae denuded of M protein. J. Exp. Med. 143: 759-771.
4. Beachey, E.H., J.B. Dale, W.A. Simpson, J.D. Evans, K.W. Knox, I. Ofek and A.J. Wicken. 1979. Erythrocyte binding properties of streptococcal lipoteichoic acids. Infect. Immun. 23: 618-625.
5. Beachey, E.H. 1981. Bacterial adherence: Adhesin-receptor interactions mediating the attachment of bacteria to mucosal surfaces. J. Inf. Dis. 143: 325-345.
6. Bisno, A.L. 1980. The concept of rheumatogenic and nonrheumatogenic group A streptococci. In Streptococcal Diseases and the Immune Response. Reed, S.E. and J.B. Zabriskie (Eds.), Academic Press, New York, pp. 789-803.
7. Bisno, A.L. 1991. Group A streptococcal infections and acute rheumatic fever. New Engl. J. Med. 325: 783-793.
8. Botta, G. 1981. Surface components in adhesion of group A streptococci to pharyngeal epithelial cells. Curr. Microbiol. 6: 101-104.
9. Caparon, M.G., D.S. Stephens, A. Olsén and J.R. Scott. 1991. Role of M protein in adherence of group A streptococci. Infect. Immun. 59: 1811-1817.
9a. Courtney, H.S., I. Ofek, W.A. Simpson and E.H. Beachey. 1981. Characterization of lipoteichoic acid binding to polymorphonuclear leukocytes of human blood. Infect. Immun. 32: 625-631.
10. Courtney, H.S., W.A. Simpson and E.H. Beachey. 1983. Binding of streptococcal lipoteichoic acid to fatty acid-binding sites on human plasma fibronectin. J. Bacteriol. 153: 763-770.
11. Courtney, H.S., I. Ofek, W.A. Simpson, D.L. Hasty and E.H. Beachey. 1986. Binding of *Streptococcus pyogenes* to soluble and insoluble fibronectin. Infect. Immun. 53: 454-459.
12. Courtney, H.S. and D.L. Hasty. 1991. Aggregation of group A streptococci by human saliva and effect of saliva on streptococcal adherence to host cells. Infect. Immun. 59: 1661-1666.
13. Courtney, H.S., D.L. Hasty, J.B. Dale and T.P. Poirier. 1992. A 29-kilodalton fibronectin binding protein of group A streptococci. Curr. Microbiol. 25: 245-250.
14. Courtney, H.S., C. von Hunolstein, J.B. Dale, M.S. Bronze, E.H. Beachey and D.L. Hasty. 1992. Lipoteichoic acid and M protein: dual adhesins of group A streptococci. Microb. Path. 12: 199-208.
15. Courtney, H.S., Y. Li, J.B. Dale and D.L. Hasty. 1994. Cloning, sequencing, and expression of a fibronectin/fibrinogen-binding protein from group A streptococci. Infect. Immun. 62: 3937-3946.

16. Courtney, H.S., M.S. Bronze, J.B. Dale and D.L. Hasty. 1994. Analysis of the role of M24 protein in streptococcal adhesion and colonization by use of _-interposon mutagenesis. Infect. Immun. 62: 4868-4873.
17. Courtney, H. S., J. B. Dale and D. L. Hasty. 1996. Differential effects of the streptococcal fibronectin-binding protein, FBP54, on adhesion of group A streptococci to human buccal cells and HEp-2 tissue culture cells. Manuscript submitted.
18. Courtney, H.S., S.Y. Liu, J.B. Dale and D.L. Hasty. 1996. Conversion of M serotype 24 *Streptococcus pyogenes* to M serotypes 1, 5 and 18: Effect on resistance to phagocytosis and adhesion to host cells, manuscript submitted.
19. Dale, J.B. and E.H. Beachey. 1985. Multiple heart-cross-reactive epitopes of streptococcal M proteins. J. Exp. Med. 161: 113-122.
20. Dale, J.B., R.W. Baird, H.S. Courtney, D.L. Hasty and M.S. Bronze. 1994. Passive protection of mice against group A streptococcal pharyngeal infection by lipoteichoic acid. J. Infect. Dis. 169: 319-323.
21. Denny, F.W. 1994. A 45-year perspective on the streptococcus and rheumatic fever. the Edward H. Kass lecture in infectious disease history. Clin. Infect. Dis 19: 1110.
22. Ellen, R.P. and R.J. Gibbons. 1972. M protein-associated adherence of *Streptococcus pyogenes* to epithelial surfaces: Prerequisite for virulence. Infect. Immun. 5: 826-830.
23. Ellen, R.P., and R.J. Gibbons. 1974. Parameters affecting the adherence and tissue tropisms of *Streptococcus pyogenes*. Infect. Immun. 9: 85-91.
24. Gerlach, D., C. Schalén, Z. Tigyi, B. Nilsson, A. Forsgren and A.S. Naidu. 1994. Identification of a novel lectin in *Streptococcus pyogenes* and its possible role in bacterial adherence to pharyngeal cells. Curr. Microbiol. 28: 331-338.
25. Gibbons, R.J. and I. Dankers. 1983. Association of food lectins with human oroepithelial cells in vivo. Arch. Oral Biol. 28: 561-566.
26. Gordis, L. 1985. The virtual disappearance of rheumatic fever in the United States: lessons in the rise and fall of disease. T. Duckett Jones Memorial Lecture. Circulation 72: 1155.
27. Grabovskaya, K.B., A.A. Totoljan, M. Ryc, J. Havlicek, L.R. Burova and R. Bicova. 1980. Adherence of group A streptococci to epithelial cells in tissue culture. Zbl. Bakt. Hygl. 247: 303-314.
28. Hanski, E. and M.G. Caparon. 1992. Protein F, a fibronectin-binding protein, is an adhesin of the group A streptococcus, *Streptococcus pyogenes*. Proc. Natl. Acad. Sci. USA 89: 6172-6176.
29. Hasty, D.L., I. Ofek, H.S. Courtney and R.J. Doyle. 1992. Multiple adhesins of streptococci. Infect. Immun. 60: 2147-2152.
30. Hollingshead, S.K., J.W. Simecka and S.M Michalek. 1993. Role of M protein in pharyngeal colonization by group A streptococci in rats. Infect. Immun. 61: 2277-2283.
31. Kass, E.H. 1971. Infectious diseases and social change. J. Infect. Dis. 123: 100.
32. Kolenbrander, P. and J. London. 1993. Adhere today, here tomorrow: Oral bacterial adherence. J. Bacteriol. 175: 3247-3252.
33. Kreikemeyer, B., S.R. Talay and G.S. Chhatwal. 1995. Characterization of a novel fibronectin-binding surface protein in group A streptococci. Mol. Microbiol. 17: 137-145.
34. Krisher, K. and M.W. Cunningham. 1985. Myosin: a link between streptococci and heart. Science 227: 413-415.
35. Kuusela, P., T. Vartio, M. Vuento and E.B. Myhre. 1984. Binding sites for streptococci and staphylococci in fibronectin. Infect. Immun. 45: 433-436.
36. Lancefield, R.C. 1948. Differentiation of group A streptococci with a common R antigen into three serologic types with special reference to the bactericidal test. J. Exp. Med. 106: 525.
37. Lancefield, R.C. 1959. Persistence of type-specific antibodies in man following infection with group A streptococci. J. Exp. Med. 110: 271.
38. Lancefield, R.C. 1962. Current knowledge of the type specific M antigens of group A streptococci. J. Immunol. 89: 307.
39. Lee, J.-Y. and M. Caparon. 1996. An oxygen-induced but protein F-independent fibronectin-binding pathway in *Streptococcus pyogenes*. Infect. Immun. 64: 413-421.
39a. Leon, O. and C. Panos. 1990. *Streptococcus pyogenes* clinical isolates and lipoteichoic acid. Infect. Immun. 58: 3779-3787.
40. Lowrance, J.H., D.L. Hasty and W.A. Simpson. 1988. Adherence of *Streptococcus sanguis* to conformationally specific determinants in fibronectin. Infect. Immun. 56: 2279-2285.
41. Natanson, S., S. Sela, A. Moses, J. Musser, M. Caparon and E. Hanski. 1995. Distribution of fibronectin-binding proteins among group A streptococci of different M types. J. Infect. Dis. 171: 871-878.
42. Nealon, T.J., E.H. Beachey, H.S. Courtney and W.A. Simpson. 1986. Release of fibronectin-lipoteichoic acid complexes from group A streptococci with penicillin. Infect. Immun. 51: 529-535.

43. Ofek, I., E.H. Beachey, W. Jefferson and G.L. Campbell. 1975. Cell Membrane-binding properties of group A streptococcal lipoteichoic acid. J. Exp. Med. 141: 990-1003.
44. Ofek, I., E.H. Beachey, F. Eyal, and J.C. Morrison. 1977. Postnatal development of binding of streptococci and lipoteichoic acid by oral mucosal cells of humans. J. Infect. Dis. 135: 267-274.
45. Ofek, I., W.A. Simpson and E.H. Beachey. 1982. Formation of molecular complexes between a structurally defined M protein and acylated or deacylated lipoteichoic acid of *Streptococcus pyogenes*. J. Bacteriol. 149: 426-433.
47. Okada, N., A.P. Pentland, P. Falk and M.G. Caparon. 1994. M protein and protein F act as important determinants of cell-specific tropism of *Streptococcus pyogenes* in skin tissue. J. Clin. Invest. 94: 965-977.
48. Okada, N., M.K. Liszewski, J.P. Atkinson and M.G. Caparon. 1995. Membrane cofactor protein (CD46) is a keratinocyte receptor for the M protein of the group A streptococcus. Proc. Natl. Acad. Sci. USA 92: 2489-2493.
49. Pancholi, V. and V.A. Fischetti. 1992. A major surface protein on group A streptococci is a glyceraldehyde-3-phosphate-dehydrogenase with multiple binding activity. J. Exp. Med. 176: 415-426.
50. Perez-Casal, J., N. Okada, M.G. Caparon and J.R. Scott. 1995. Role of the conserved C-repeat region of the M protein of *Streptococcus pyogenes*. Mol. Microbiol. 15: 907-916.
51. Sanford, B.A., V.E. Davison and M.A. Ramsay. 1982. Fibrinogen-mediated adherence of group A *Streptococcus* to influenza A virus-infected cell cultures.
52. Simpson, W.A., I. Ofek, C. Sarasohn, J.C. Morrison and E.H. Beachey. 1980. Characteristics of the binding of streptococcal lipoteichoic acid to human oral epithelial cells. J. Infect. Dis. 141: 457-462.
53. Simpson, W.A., I. Ofek and E.H. Beachey. 1980. Fatty acid binding sites of serum albumin as membrane receptor analogs for streptococcal lipoteichoic acid. Infect. Immun. 29: 119-122.
54. Simpson, W.A. and E.H. Beachey. 1983. Adherence of group A streptococci to fibronectin on oral epithelial cells. Infect. Immun. 39: 275-279.
55. Simpson, W.A., T.J. Nealon, D.L. Hasty, H.S. Courtney and E.H. Beachey. 1988. Binding of fibronectin to *Streptococcus pyogenes*: Heat induction of fibronectin receptors. In Mucosal Immunity and Infections at Mucosal Surfaces, (W. Strober, M.E. Lamm, J.R. McGhee and S.P. James, eds). Oxford University Press, NY, pp. 349-354.
56. Sokurenko, E.V., H.S. Courtney, S.N. Abraham, P. Klemm and D.L. Hasty. 1992. Functional heterogeneity of type 1 fimbriae of *Escherichia coli*. Infect. Immun. 60: 4709-4719.
57. Speziale, P., M. Höök, L.M. Switalski and T. Wadström. 1984. Fibronectin binding to a *Streptococcus pyogenes* strain. J. Bacteriol. 157: 420-427.
58. Stevens, D.L., M.H. Tanner, J. Winship, et al. 1989. Severe group A streptococcal infections associated with a toxic shock-like syndrome and scarlet fever toxin A. N. Engl. J. Med. 321: 1.
59. Stollerman, G.H. 1975. Rheumatic fever and streptococcal infection. Grune and Stratton (New York).
60. Stollerman, G.H. 1988. The return of rheumatic fever. Hosp. Pract. 23: 100.
61. Switalski, L.M., P. Speziale, M. Höök, T. Wadström and R. Timpl. 1983. Binding of *Streptococcus pyogenes* to laminin. J. Biol. Chem. 259: 3734-3738.
62. Talay, S.R., E. Ehrenfeld, G.S. Chhatwal and K.N. Timmis. 1991. Expression of the fibronectin-binding components of *Streptococcus pyogenes* in *Escherichia coli* demonstrates that they are proteins. Mol. Microbiol. 5: 1727-1734.
63. Talay, S. R., P. Valentin-Weigand, P.G. Jerlström, K.N. Timmis and G.S. Chhatwal. 1992. Fibronectin-binding protein of *Streptococcus pyogenes*: Sequence of the binding domain involved in adherence of streptococci to epithelial cells. Infect. Immun. 60: 3837-3844.
64. Tylewska, S.K. and R.J. Gibbons. 1987. Application of Percoll density gradients in studies of the adhesion of *Streptococcus pyogenes* to human epithelial cells. Curr. Microbiol. 16: 129-135.
65. Tylewska, S.K., V.A. Fischetti and R.J. Gibbons. 1988. Binding selectivity of *Streptococcus pyogenes* and M-protein to epithelial cells differs from that of lipoteichoic acid. Curr. Microbiol. 16: 209-216.
66. Valentin-Wiegand, P., J. Grulich-Henn, G.S. Chhatwal, G. Müller-Berghaus, H. Blobel and K.T. Preissner. 1988. Mediation of adherence of streptococci to human endothelial cells by complement S protein (vitronectin). Infect. Immun. 56: 2851-2855.
67. Valentin-Weigand, P., S.R. Talay, K.N. Timmis and G.S. Chhatwal. 1993. Identification of a fibronectin-binding protein as adhesin of *Streptococcus pyogenes*. Zbl. Bakt. 278: 238-245.
68. VanHeyningen, T., G. Fogg, D. Yates, E. Hanski and M. Caparon. 1993. Adherence and fibronectin binding are environmentally regulated in group A streptococci. Mol. Microbiol. 9: 1213-1222.
69. Veasey, L.G., S.E. Wiedmeir, G.S. Orsmond, et al. 1987. Resurgence of acute rheumatic fever in the intermountain area of the United States. N. Engl. J. Med. 316: 421.
70. Veasey, L.G., L.Y. Tani and H.R. Hill. 1994. Persistence of acute rheumatic fever in the intermountain area of the United States. J. Pediatr. 124: 9.

71. Visai, L., S. Bozzini, G. Raucci, A. Toniolo and P. Speziale. 1995. Isolation and characterization of a novel collagen-binding protein from *Streptococcus pyogenes* strain 6414. J. Biol. Chem. 270: 347-353.
72. Wadström, T. and S. Tylewska. 1982. Glycoconjugates as possible receptors for *Streptococcus pyogenes*. Curr. Microbiol. 7: 343-346.
73. Wagner, B., K-H. Schmidt, M. Wagner and T. Wadström. 1988. Localization and characterization of fibronectin-binding to group A streptococci. An electron microscopic study using protein-gold-complexes. Zbl. Bakt. Hyg. A 269: 479-491.
74. Wang, J. and M. Stinson. 1994. M protein mediates streptococcal adhesion to HEp-2 cells. Infect. Immun. 62: 442-448.
75. Wessells, M.R. and M.S. Bronze. 1994. Critical role of the group A streptococcal capsule in pharyngeal colonization and infection in mice. Proc. Natl. Acad. Sci. USA 91: 12238-12242.

11

ADHESINS OF *STAPHYLOCOCCUS AUREUS* THAT BIND LEWIS[a] ANTIGEN

Relationship to Sudden Infant Death Syndrome

C. C. Blackwell, A. T. Saadi, S. D. Essery, M. W. Raza, A. A. Zorgani, O. R. Elahmer, A. H. Alkout, V. S. James, D. A. C. MacKenzie, D. M. Weir, and A. Busuttil

Department of Medical Microbiology and Forensic Medicine Unit
University of Edinburgh
Edinburgh, Scotland

1. WHAT IS SUDDEN INFANT DEATH SYNDROME?

Sudden Infant Death Syndrome (SIDS) is defined as " the sudden death of any infant or young child which is unexpected by history, and in which a thorough postmortem examination fails to demonstrate an adequate cause of death" [Beckwith, 1969]. Since 1990, there has been a steady reduction in the numbers of SIDS in Britain [Court, 1995; Scottish Cot Death Trust, personal communication]; however, SIDS is still the major cause of post perinatal mortality during the first year of life. Petechiae in the lungs and thymus, liquid heart blood and empty bladder are common findings at autopsy [Berry, 1992]. While there is little evidence that could explain why the infant died, there are common findings that suggest immune or inflammatory reactions have been elicited before death (Table 1).

2. PARALLELS BETWEEN RISK FACTORS FOR SIDS AND FOR RESPIRATORY INFECTION

Epidemiological factors associated with susceptibility to respiratory infections are similar to those associated with SIDS. Infants are vulnerable to infectious agents during the age range in which most SIDS cases occur (2-4 months) due to the decline in passively acquired maternal antibodies and the lack of active antibody production by the infant in response to natural infection or immunization. In most countries, SIDS cases occur more frequently during winter months when there are more respiratory tract infections. Maternal smoking is associated with SIDS, and among studies on exposure to cigarette smoke and infection, the strongest associations are usually with smoking habits of the mother [Pershagen, 1986]. Bottle feeding is a risk factor for SIDS [Ford *et al.*, 1993] and breast-feeding contributes to protection against respiratory and gastrointestinal

Toward Anti-Adhesion Therapy for Microbial Diseases, edited by Kahane and Ofek
Plenum Press, New York, 1996

Table 1. Evidence for inflammatory/immune responses in SIDS infants

Respiratory tract –	inflammatory infiltrates [Howat *et al.*, 1994]
	ì numbers of IgM cells in trachea [Stoltenberg *et al.*, 1992]
	pulmonary edema [Berry, 1992]
	mast cell degranulation [Harrison *et al.*, 1993]
Digestive tract –	↑ numbers of IgA cells in duodenum [Stoltenberg *et al.*, 1992]
Nervous system –	interferon α in brain [Howatson, 1992]
	ì level of interleukin 6 in cerebral spinal fluid [Vege *et al.*, 1995]
Blood –	↓ levels of IgG to endotoxin core [Oppenheim *et al.*, 1994]
	ì levels of IgM to endotoxin core [Oppenheim *et al.*, 1994]
	ì levels of mast cell tryptase [Holgate *et al.*,1994; Platt *et al.*, 1994]
	ì levels of acute phase reactants [Amberg *et al.*, 1993]
	ì cross-linked fibrin degradation products [Goldwater *et al.*, 1990]

infections. Lower socioeconomic conditions associated with SIDS in some studies could reflect the higher proportion of women who smoke and the lower proportion of those who breast feed in these groups [Gibson, 1992; Gilbert *et al.*, 1995]. While it is postulated that infectious agents are involved in some cases of SIDS, no definitive link between specific viruses or bacteria has been established [Fleming, 1992; Blackwell *et al.*, 1994]. By definition, deaths due to invasive bacterial disease are not SIDS; consequently, the search for an infectious aetiology in SIDS has concentrated on identification of viruses or toxigenic bacteria in these infants.

3. TOXIGENIC BACTERIA AND SIDS

It has been suggested that common bacterial toxins are involved in some of these infant deaths [Morris *et al.*, 1987]. Toxigenic strains of *Staphylococcus aureus, Escherichia coli* and *Clostirdium perfringens* have been isolated from a high proportion of SIDS infants and toxins from these bacteria identified in serum, faeces and tissues [Telford *et al.*, 1989; Murrell *et al.*, 1993; Bettiol *et al.*, 1994; Lindsay *et al.*, 1993].

Our research has concentrated on the role of *S. aureus* strains producing pyrogenic toxins, superantigens which can elicit severe inflammatory responses [Bohach *et al.*, 1990; Schlievert, 1995]. These bacteria have been isolated from the nasopharynx of SIDS cases [Telford *et al.*, 1989] and staphylococcal enterotoxin C_1 (SEC) and the toxic shock syndrome toxin (TSST-1) identified by immunohistochemical methods in tissues of many of these infants [Malam *et al.*, 1992; Newbould *et al.*, 1989]. In an Australian study in which identification of toxins from SIDS infants was assessed by season, the staphylococcal toxins showed the winter peak but the clostridial toxins did not [Murrell *et al.*, 1993]. An enzyme linked immunosorbent assay (ELISA) recently developed in our laboratory for detection of these toxins in tissue homogenates has identified TSST-1 and/or SEC_1 in 6/8 SIDS infants examined. Toxigenic staphylococci were isolated from the 6 toxin-positive infants but not the 2 toxin-negative ones [unpublished observations].

4. THE AGE DISTRIBUTION OF SIDS CASES

One of the most consistent epidemiological findings in studies of SIDS world-wide is the peak age range during which the majority of these deaths occur, 2-4 months. In addition to being a period in which the infant is particularly vulnerable to infection, a number of important developmental and physiological changes are occurring [Lodemore *et al.*, 1992; Peterson *et al.*, in press]. During the 2-4 month age range there is also a peak in expression of the Lewis^a antigen on red cells of infants (80-90%) [Issit, 1986]. This is thought to be due to the fucosyltransferase coded for by the Lewis gene (*Le*) being more efficient during this period than the fucosyl transferase coded for by the secretor gene (*Se*) in infants who possess both. If the *Se* transferase adds its fucose to the terminal sugar of the type 1 precursor chain first to form H type 1, the *Le* transferase can add fucose to the subterminal sugar of the chain resulting in Lewis[b]. If the fucosyltransferase encoded by the *Le* gene adds its fucose to the subterminal sugar first, the Lewis[a] antigen formed cannot be used as a substrate by the *Se* transferase. We identified Lewis[a] antigen in respiratory secretions of 63/89 (71%) SIDS infants [Blackwell *et al.*, 1992].

Another set of observations that implicated staphylococci was a study of the nasopharyngeal flora of Swedish infants during the first year of life. *S. aureus* was the predominant isolate from the youngest age group. Among infants in the 1-3 month age range, staphylococci were isolated from 39% but this fell to 11% among the 4-7 month agen range and to 4% for those examined between 8-18 months [Aniansson *et al.*, 1992].

5. OBJECTIVES OF RESEARCH PROGRAMME

Our original hypothesis was that there are adhesins on staphylococci that bind Lewis[a] and this might contribute to the colonisation pattern observed in infants. Other risk factors associated with SIDS, maternal smoking, virus infection, and bottle feeding, might influence adherence of staphylococci to epithelial cells and these have been examined with reference to expression of the Lewis[a] antigen on cells and in human milk. The objectives of our research programme were: 1) to determine if Lewis[a] was a receptor for toxigenic strains of staphylococci; 2) to determine if an adhesin could be isolated by affinity purification with synthetic Lewis[a]; 3) to assess the effect of smoking and virus infection on adherence of staphylococci to epithelial cells and expression of the Lewis antigens; 4) to assess the effect of human milk on inhibition of binding of staphylococci to epithelial cells; 5) to compare the isolation rate of *S. aureus* among healthy infants and SIDS infants over a 3-year period.

6. BACTERIAL BINDING STUDIES

In these studies, bacterial binding and binding of monoclonal antibodies to blood group antigens and other cell surface markers to epithelial cells were assessed by flow cytometry. Results were expressed as binding index calculated as the product of the percentage of cells with fluorescence above the untreated control sample multiplied by the mean fluorescence on the linear scale [Saadi *et al.*, 1993].

Initially, we compared adherence of staphylococcal strains to buccal epithelial cells from secretors and non-secretors obtained from adults. Secretors express Lewis[b] predominantly and non-secretors can express only Lewis[a]. Three of the 8 toxigenic strains tested bound in greater numbers to cells of the non-secretors. This suggested that if there was an adhesin that bound to Lewis[a], it was not present on all the strains tested.

Binding of a strain of *S. aureus* (NCTC 10655) that adhered in greater numbers to buccal epithelial cells of non-secretors was significantly inhibited by pre-treatment of the cells with monoclonal antibodies to Lewis[a] ($P < 0.001$, 95% CI 55-81); however, similar inhibitory effects were observed with a non-toxigenic strain (NCTC 8532 ($P<0.001$, 95% CI 43-72)) which had not shown any difference in binding to cells of secretors and non-secretors. While the binding indices for anti-Lewis[x] (CD15) to epithelial cells of 6 donors were significantly lower than those for Lewis[a] ($P < 0.001$, 95% CI 1-13) (Figure 1), inhibition of binding of staphylococci was significantly greater for cells treated with anti-Lewis[x] compared to those treated with anti-Lewis[a] ($P<0.01$) (Figure 2). Binding of the bacteria was significantly correlated only with the amount of

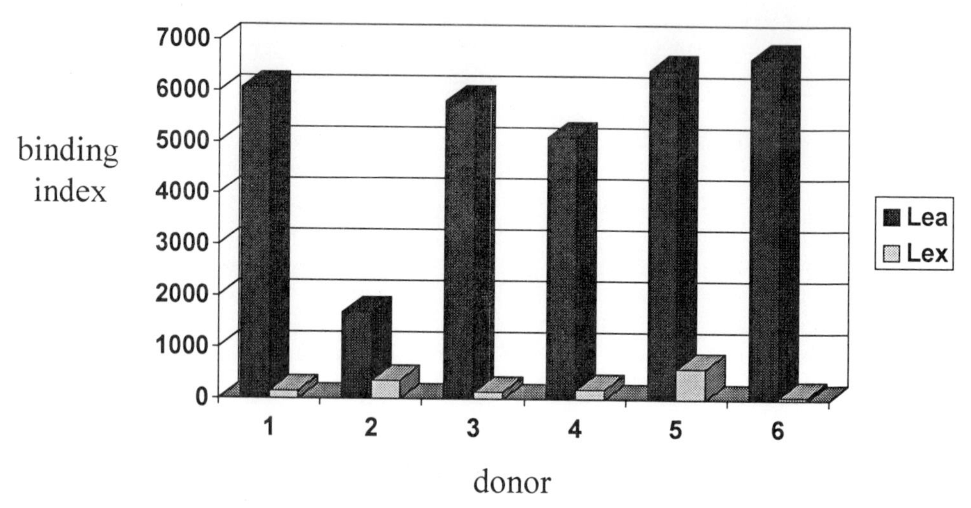

Figure 1. Binding of monoclonal anti-Lewis[x] and anti-Lewis[a] to epithelial cells.

Lewis[a] on the cells of each individual donor reflected in the amount of the monoclonal anti-Lewis[a] bound to the cells; toxigenic strain ($P < 0.001$); non-toxigenic strain ($P < 0.001$). Binding of the staphylococcal strains was not associated with the amount of anti-precursor type 1 or anti-Lewis[x] [Saadi *et al.*, 1993].

While we had predicted that in adults, the levels of Lewis[a] would be significantly higher among non-secretors, re-examination of cells from our donors found that some secretors bound as much anti-Lewis[a] on their cells as non-secretors while others bound very small amounts, equivalent to those observed for individuals who lack the Lewis gene [Saadi *et al.*, 1993].

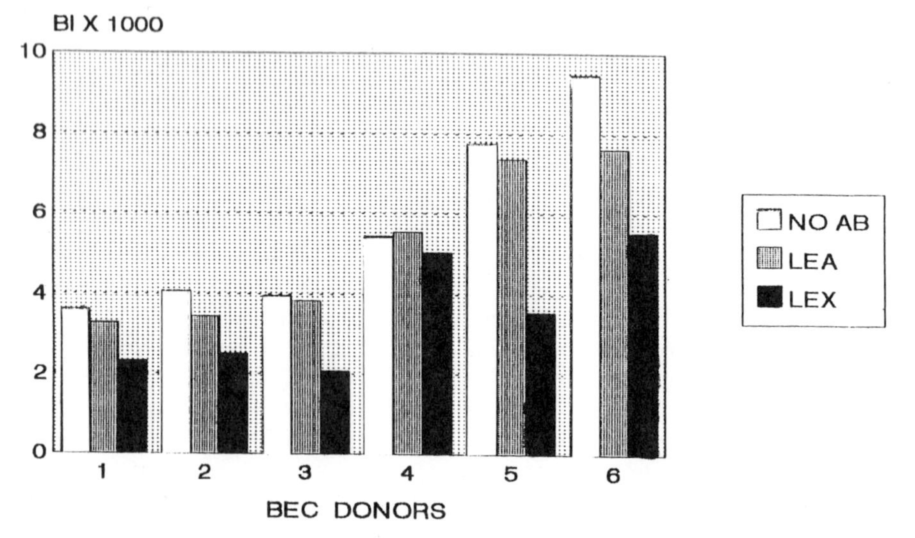

Figure 2. Inhibition of binding of *S. aureus* NCTC 10655 to cells pretreated with anti-Lewis[a] (LEa) or anti-Lewis[x] (LEx).

7. AGGLUTINATION OF BACTERIA BY ANTI-IDIOTYPIC ANTIBODIES PRODUCED AGAINST MONOCLONAL ANTI-LEWIS[a]

Studies in which binding of bacteria is inhibited by pre-treatment of cells with antibodies can always be criticised in that the inhibition might due to stearic hindrance of an epitope adjacent to that to which the monoclonal antibody binds. An anti-idiotypic antibody was produced by immunising mice with the monoclonal anti-Lewis[a] used in the inhibition of binding experiments and a co-agglutination reagent prepared to screen staphylococcal strains for surface antigens that bind Lewis[a] [Essery *et al.*, 1994].

Each of the 7 strains of staphylococci tested, non-toxigenic as well as the toxigenic ones, were agglutinated with the anti-idiotypic reagent as were 8/13 isolates of type b *Haemophilus influenzae* and 10/18 meningococci tested. An isolate of *Candida albicans* expressing an adhesin that binds fucose was also agglutinated by the reagent but a strain with an adhesin that binds N-acetylgalactosamine was not [Essery *et al.*, 1994]. This suggested that the adhesin is present on all the isolates of staphylococci and that our initial results for binding of staphylococci to cells of secretors and non-secretors was due to the high levels of Lewis[a] present on cells of some of our secretor donors.

8. ISOLATION OF AN ADHESIN THAT BINDS TO SYNTHETIC LEWIS[a]

Cell membrane preparations from *S. aureus* NCTC 10655 were extracted by the method of Sharp and Poxton [1988], applied to Synsorb® beads conjugated with synthetic Lewis[a] and the bound material eluted with dilute ammonia [Saadi *et al.*, 1994]. A 67 kDa protein was isolated from cell membrane preparations of *S. aureus* NCTC 10655 by affinity adsorption with synthetic Lewis[a]. Pre-treatment of buccal epithelial cells with the protein reduced binding of the staphylococcal strain to a greater extent than the material not bound by the Synsorb® beads [Saadi *et al.*, 1994a].

9. INHIBITION OF BACTERIAL BINDING WITH HUMAN MILK

The mean binding index of staphylococci to buccal epithelial cells was reduced by 91.7% following pre-treatment of the bacteria with pooled human milk (95% CI 53.7-98.5). Preliminary studies with the IgA fraction or the glycoprotein fraction obtained by affinity adsorption with anti-Lewis[a] also appear to reduce binding of the bacteria. In addition, in 4 experiments each of these preparations reduced the ability of the TSST-1 to elicit nitric oxide from human monocytes (Figure 3).

10. DO OTHER RISK FACTORS FOR SIDS ENHANCE BACTERIAL BINDING?

In 24 studies between 1963 and 1994, maternal smoking was identified as an important risk factor for SIDS [Mitchell, 1995], and in the Scottish population we have studied, 80% of the mothers of SIDS infants were smokers [Scottish Cot Death Trust, personal communication]. Infants exposed to cigarette smoke metabolize nicotine to

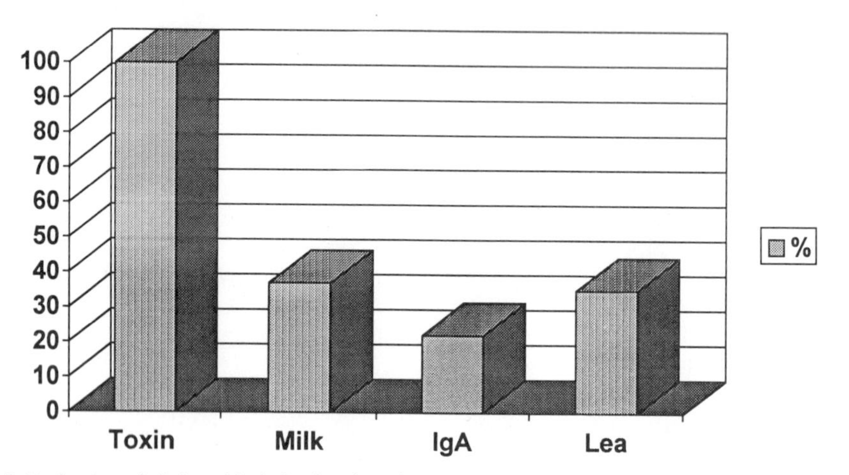

Figure 3. Reduction of nitric oxide induction from human monocytes by TSST-1 treated with human milk, IgA or glycoproteins.

cotinine, and there was a significant correlation between the level of cotinine in the saliva of 89 infants and the number of cigarettes smoked in the household per day ($r = 0.71$, $P < 0.001$). In studies with buccal epithelial cells from 8 pairs of smokers and non-smokers, the mean binding index of staphylococci to cells from smokers was 156% that of non-smokers ($P < 0.05$, 95% CI 105-223) [Saadi *et al.*, 1994b]. Bacterial binding to epithelial cells of smokers is not a non-specific phenomenon; some species exhibit enhanced binding to epitheial cells of smokers, e.g., *Neisseria lactamica* but not *Neisseria meningitidis* [El Ahmer *et al.*, 1994]. Comparison of the expression of cell surface antigens of smokers and non-smokers found that the increase in bacterial binding was not associated with increased expression of Lewis^a, Lewis^b, H type 2, CD14 or CD18 on the cells of smokers or expression of fibrinogen or fibronectin, other host cell antigens to which staphylococci bind [El Ahmer *et al.*, 1994; Alkout *et al.*, 1996].

Many SIDS infants had a mild respiratory infection during the week prior to death. We have demonstrated *in vitro* that an epithelial cell line (HEp-2) infected with respiratory syncytial virus (RSV) binds significantly more staphylococci than uninfected cells [Saadi *et al.*, 1993]. This enhanced binding is not mediated through increased expression of Lewis^a as this cell line does not produce the antigen [Raza, 1992].

11. ARE THERE SIMILAR INTERACTIONS BETWEEN HOST CELLS AND OTHER BACTERIA IMPLICATED IN SIDS?

Epidemiological studies in Britain and Scandinavia suggested that some cases of SIDS might be asymptomatic whooping cough [Nicol and Gardner, 1988; Lindgren *et al.*, 1994]. The pertussis toxin produced by *Bordetella pertussis* acts as an adhesin for attachment of these bacteria to Lewis^a and Lewis^x on human monocytes [van t'Wout *et al.*, 1992]. In studies similar to that for staphylococci, we have demonstrated inhibition of binding of two isolates of *B. pertussis* to epithelial cells treated with anti-Lewis^a or anti-Lewis^x. For strain 8002 (Fimbriate, type 1,2) treatment of epithelial cells with anti-Lewis^a significantly reduced bacterial binding ($t=-2.93$, 95% CI 86-99, $P < 0.05$) as did treatment with anti-Lewis^x ($t = -4.32$, 95% CI 45-82, $P < 0.01$). A similar pattern was found with strain 250815 (non-fimbriate, type 1,3); however, there was no significant association between binding of the bacteria and expression of Lewis[a] as we had observed with staphylococci. Treatment of the bacteria with pooled human milk also reduced the mean binding index of pertussis to epithelial cells by 64% (95% CI 26.3-82.4).

Enhanced binding of these bacteria to RSV-infected cells and to cells from 15 pairs of smokers was demonstrated [Saadi *et al.*, 1994; submitted for publication].

12. NASOPHARYNGEAL FLORA OF HEALTHY INFANTS AND SIDS INFANTS

In our longitudinal survey of healthy infants and their mothers, *S. aureus* was the predominant isolate among infants screened during the 6-12 week age range, 56% of 252 infants had staphylococci in their nose and/or throat compared with *H. influenzae* (29%), *Streptococcus pyogenes* (18%) or *N. lactamica* (6%). This pattern changed in the next periods during which the infants were tested, 4-6 months, 7-9 months and 10-13 months; *H. influenzae* became the dominant isolate, 44%, 49% and 47% respectively compared with *S. aureus*, 39%, 34% and 19% respectively. For staphylococci and haemophilus, there was a significant correlation between isolation of each of these species from both mother and baby at each visit ($P < 0.05$ - $P < 0.001$). There was not, however, any association between isolation of these bacteria from infants and maternal smoking, symptoms of viral infection or breast feeding compared with bottle feeding. Although the infants and their mothers were sampled 4 times during the year, only 1 isolate of *B. pertussis* was obtained and this was from one of the mothers.

In Britain, the majority of SIDS deaths now occur in infants under the age of 3 months [Court, 1995]. Respiratory secretions, nose and throat swabs of SIDS infants were cultured for pathogenic bacteria during the same period as our survey of healthy infants. Among 22 SIDS infants who died before the age of 3 months, 19 (86.4%) had *S. aureus* in either nasal swabs or bronchial secretions. This is a significantly higher proportion than that among healthy infants in the same age range (56%) ($X^2 = 6.16$, $P = 0.012$). *H. influenzae* was the second most common isolate among healthy infants, but the proportion of SIDS infants from whom these bacteria were isolated was only 26%.

13. IMMUNIZATION AND SIDS

In the early 1980's it was suggested that SIDS was precipitated by immunization with the diphtheria, pertussis and tetanus (DPT) vaccine [Baraf *et al.*, 1983]. Subsequent epidemiological studies in the United States indicated that the risk of SIDS was greater among unimmunized infants or those immunized late [Hofman *et al.*, 1987; Walker *et al.*, 1987]. In Britain, the decline in SIDS has been attributed mainly to the campaign to discourage the prone sleeping position which began in 1990 in England and Wales. There was, however, another major change in health care practices that affected all infants started a year earlier; the immunization schedule was changed in October 1990 to begin at 2 months rather than 3 months.

The pertussis toxin binds to Lewis[a] and Lewis[x] on monocytes, and we demonstrated that binding of staphylococcal enterotoxin b (SEB) to human monocytes could be inhibited by pre-treatment with anti-Lewis[a] [Essery *et al.*, 1994b]. By ELISA, we demonstrated that antibodies elicited by immunization of a rabbit with the current DPT vaccine elicited antibodies that bound to the pertussis toxin and also to 4 of the staphylococcal toxins, SEA,

SEC , SEB and TSST [Essery *et al.*, 1994c]. In addition, the immune rabbit serum, but not normal rabbit serum, could partially neutralise the ability of the staphylococcal toxins to elicit nitric oxide from human monocytes [Blackwell et al., 1995; Blackwell *et al.*, submitted for publication]. If DPT immunization induced antibodies in infants that reduce the effects of these toxins, this might contribute to the reduced risk of SIDS associated with immunization in epidemiological studies. We predicted there would be a shift in the age distribution with a higher proportion of these deaths now occurring in the unimmunized group < 2 months of age. In southeast Scotland, the total numbers of SIDS cases has decreased from 1990, but the proportion of cases < 2 months of age has increased significantly from 11% in 1988 to 56% in 1994 (X^2 =14.37, $P < 0.001$) [Blackwell *et al.*, 1995; Blackwell *et al.*, submitted for publication].

By ELISA studies, we have found significant correlations between levels of IgG to the DPT vaccine and IgG to pertussis toxin, SEC and TSST in sera of infants in the age range 1 week - 5 months.

14. CONCLUSIONS

Our studies indicate that Lewis[a] is an epithelial cell receptor for staphylococci which might explain the isolation pattern of bacterial flora observed among infants by our group and in studies in Sweden [Aniansson *et al.*, 1992] (Figure 4). Neither virus infection nor exposure to cigarette smoke enhanced expression of the Lewis[a] antigen or any of the other antigens examined. While viral infection, exposure to cigarette smoke or method of feeding might affect density of colonization, it did not affect frequency of colonization in our epidemiological studies. Other work by our group indicates that exposure to cigarette smoke or virus infection might enhance induction of inflammatory responses by both pyrogenic toxins and endotoxin. This rather than effects on colonisation by toxigenic bacteria might be the means by which they act as risk factors for SIDS.

Although asymptomatic infection with *B. pertussis* has been suggested to be a cause of SIDS and the same factors associated with enhanced binding of staphylococci enhanced binding of pertussis, our epidemiological studies of healthy infants and SIDS cases do not

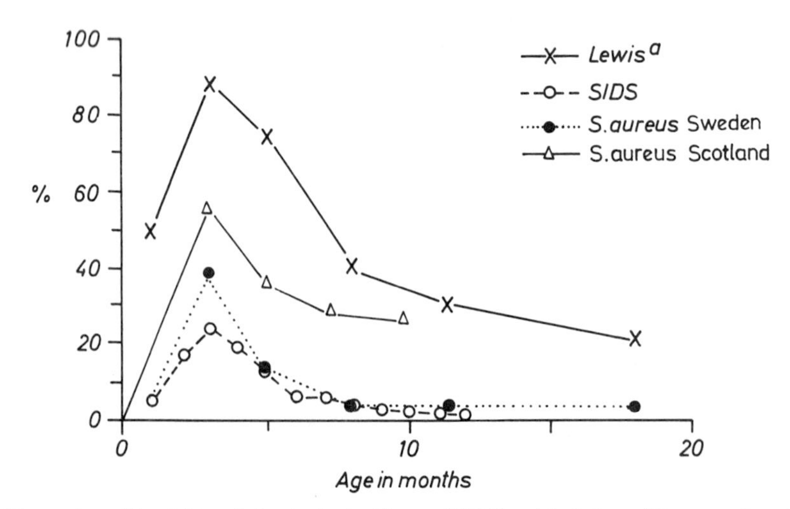

Figure 4. Expression of Lewis[a] in relation to the incidence of SIDS and isolation of *S. aureus* from infants in Sweden and Scotland.

support this hypothesis. In Britain, the majority of SIDS deaths (60%) now occur in infants under the age of 3 months and *S. aureus* was isolate from 86.4% of SIDS infants in our study. Identification of TSST-1 and/or SEC_1 in SIDS infants indicates these toxins might play a role in some of these unexpected deaths.

We have proposed the hypothesis that some cases of SIDS are due to pathophysiological responses elicited by combinations of microbial infections, and possibly cigarette smoke, which occur during a developmental stage when the infants hormonal responses cannot "damp down" the effects of inflammatory mediators [Blackwell *et al.*, 1995a,b,].. Further investigations of the components of the DPT vaccine that elicit antibodies cross-reactive with toxins or adhesins of species implicated in SIDS might provide the basis for improving passive or active immunity to toxigenic bacteria during the age range in which infants are at increased risk to the inflammatory responses elicited by bacterial exotoxins or endotoxins.

ACKNOWLEDGMENTS

This work was supported by the Scottish Cot Death Trust and Babes in Arms.

REFERENCES

Alkout AH, Blackwell CC, Weir DM, Luman W, Palmer K. Adhesin of *Helicobacter pylori* that binds to H type 2 and Lewis blood groupa: an explanation of increased susceptibility of blood group O and non-secretors to peptic ulcers. In Toward Anti-Adhesin Therapy of Microbial Diseases Ed. I. Kahane and I. Ofek Plenum

Amberg R, Furanol R, Pelz K, Goebel U, Pollak S. Acute phase reactions in cases of SIDS and infants suffering from infection. 13 Meeting International Assocociation of Forensic Sciences, Dusseldorof, Dermany 1993. A260.

Aniansson G, Alm B, Andersson B, Larsson P, Nyleu O, Peterson H, Ringer P, Svanborg M, Svanborg C. Nasopharyngeal colonization during the first year of life. J Infect Dis 165 (suppl): s38-s42.

Anonymous. 38. Cot deaths. BMJ 1995; 310: 7-10.

Baraff LJ, Ablon WJ, Weiss RC. Possible temporal association between diphtheria-tetanus toxoid-pertussis vaccination and sudden infant death syndrome. Pediatr Infect Dis 1983; 2: 7-11.

Beckwith JB Discussion of terminology and definition of the sudden infant death syndrome. in Sudden Infant Death Syndrome: Proceedings of the Second International Conference on the Causes of Sudden Deaths in Infants (Bergman, A.B., Beckwith, J.B. and Ray, CG., eds.), University of Washington Press, 1970 pp. 14-22.

Berry PJ. Pathological findings in SIDS. J Clin Pathol 1992; 45 (suppl): 11-16.

Bettiol SS, Radcliff FJ, Hunt AL, Goldsmid JM.. Bacterial flora of Tasmanian SIDS infants with special reference to pathogenic strains of *Escherichia coli*. Epidemioll Infect 1994; 112: 275-284.

Blackwell CC, Saadi AT, Raza MW, Stewart J, Weir DM Susceptibility to infection in relation to sudden infant death syndrome. J Clin Pathol 1992; 45 (Supplement): 20-24.

Blackwell CC, Weir DM, Busuttil A. Infectious agents, the inflammatory responses of infants and sudden infants death syndrome (SIDS). Mol Med Today 1995; 1: 72-78.

Blackwell CC, Weir DM, Busuttil A, Saadi AT, Essery SD, Raza MW, Zorgani AA, James VS, Mackenzie DAC. 1995. Infectious agents and SIDS: a new concept involving interactions between microorganisms the immune system and developmental stage of infants. In Sudden Infant Death Syndrome, New Trends in the Nineties. Ed. T.O. Rognum. Scandinavian University Press PP. 189-198.

Blackwell CC, Weir D M, Busuttil A, Saadi AT, Essery SD, Raza MW, James VS Mackenzie DAC. The role of infectious agents in sudden infant death syndrome. FEMS Immunol Med Microbiol 1994; 9: 91-100.

Bohach GA, Fast DJ, Nelson RD, Schlievert PM. Staphylococcal and streptococcal pyrogenic toxins involved in toxic shock syndrome and related illnesses. Crit Rev Microbiol 1990; 17: 251-272.

Court C. Cot deaths: Britain: Incidence reduced by two thirds in five years. Br Med J 1995; 310: 7-8.

El-Ahmer OR, Mackenzie DAC, James VS, Blackwell CC, Raza MW, Saadi AT, Elton RA, Ogilvie MM, Weir DM. Exposure to cigarette smoke and colonization by Neisseria species. Neisseria 94 (eds. JS Evans et al.) England: SCC. pp 391-392.

Essery, S.D., Blackwell, C.C., Weir, D.M. and Busuttil A. 1994. Antigenic cross reactivity of bacterial toxins. Third SIDS International Congress.p 121.

Essery S.D., Saadi, A.T., Twite, S.J., Weir, D.M. Blackwell C.C. and Busuttil A 1994. Lewis antigen expression on human monocytes and binding of pyrogenic toxins. Agents and Actions 41: 108-110.

Essery SD Weir, D.M., James, V.S., Saadi, A.T., Blackwell, C.C., Tzanakai G, Busuttil, A. Detection of microbial surface antigens that bind Lewis[a] blood group antigen. FEMS Immunol Med Microbiol 1994; 9: 15-22.

Fleming K. Upper respiratory inflammation and detection of viral nucleic acid. J Clin Pathol 1992; 45 (suppl): 17-19.

Ford RP, Taylor BJ, Mitchell EA, Enright SA, Stewart AW, Becroft DM, Scragg R, Hassall IB, Barry DM, Ellen EM *et al.*. Breastfeeding and the risk of sudden infant death syndrome. Int J Epidemiol 1993; 22: 885-890.

Gibson AAM. Current epidemiology of SIDS. J Clin Pathol 1992; 45 (suppl) : 7-10.

Gilbert RE , Rudd PT, Berry PJ, Fleming PJ, Hall E, White DG, Oreffo VO, James R, Evans JA Bottle feeding and the sudden infant death syndrome. Br Med J 1995; 310: 88-90.

Goldwater PN Williams V, Bourne AJ, Byard RW. Sudden infant death syndrome: a possible clue to causation. Med J Austral. 1990; 153: 59-60.

Harrison M, Carson C, Gillan JE.. Mast cell degranulation suggests non-immune anaphylaxis as a cause of death in SIDS, an electron microscopic study. Third European Congress, European Society for the Study and Prevention of Infant Deaths 1993: 34.

Hoffman HS, Hunter JC, Damus K et al. Diphtheria-tetanus-pertussis immunization and sudden infant death: results of the National Institute of Child Health and Human Development Co-operative Epidemiological Study of Sudden Infant Death Syndrome Risk Factors.

Pediatrics 1987; 79: 598-611.

Holgate ST, Walters, Walls AF, Lawrence S, Shell DJ, Variend S, Fleming PJ, Berry PJ, Gilbert RE, Robinson C. The anaphylaxis hypothesis of sudden infant death syndrome (SIDS): mast cell degranulation in cot death revealed by elevated concentrations of tryptase in serum. Clin Exp Allergy 1994; 24: 115-123

Howat WJ, Moore IE, Judd M, Roche WR. Pulmonary immunopathology of sudden infant death syndrome. Lancet 1994; 343: 1390-1392.

Howatson AG. Viral infection and alpha interferon in SIDS. J Clin Pathol 1992; 45 (suppl):25-28.

Issit PD. Applied blood group serology. 3rd edition. Miami: Montogemery, 1986.

Lindsay JA, Mach AS, Wilkinson MA, Martin M, Wallace MF, Keller AM, Wojciechowski LM. *Clostridium perfringens* type a cytotoxic-enterotoxin(s) as triggers for death in the sudden infant death syndrome: development of a toxico-infection hypothesis. Curr Microbiol 1993; 27: 51-59.

Lodemore MR, Peterson SA, Wailoo MP. Factors affecting the development of night-time temperature rhythms. Arch Dis Child 1992; 67: 1259-1261.

MacKenzie DAC, James V.S, Elton RA, Zorgani AA, Blackwell CC, Weir DM, Busuttil A., Gibson AAM. Toxigenic bacteria and SIDS: nasopharyngeal flora in the first year of life. Fourth SIDS International, 1996.

Malam JE, Carrick GF, Telford DR, Morris JA. Staphylococcal toxins and sudden infant death syndrome. J Clin Pathol 1992; 45: 716-721.

Mitchell EA. Smoking: the next major and modifiable risk factor. In Sudden Infant Death Syndrome. New Trends in the Nineties. (Ed. T.O. Rognum) Scandinavian University Press, Oslo, 1995 pp. 114-118.

Morris JA, Haran D, Smith A. Common bacterial toxins are a possible cause of the sudden infant death syndrome. Med Hypotheses 1987;22: 211-222.

Murrell WG, Stewart BJ, O'Neill C, Siarakas S, Kariks S. Enterotoxigenic bacteria in the sudden infant death syndrome. J Med Microbiol 1993; 39: 114-127.

Newbould MJ, Malam J, McIllmurray JM, Morris JA, Telford DR, Barson AJ. Immunohistological localisation of staphylococcal toxic shock syndrome toxin (TSST-1) in sudden infant death syndrome. J. Clin. Pathol 1989; 42: 935-939.

Nichol A and Gardner A. Whooping cough and unrecognized post-perinatal mortality. Arch Dis Child 1988; 63: 41-47.

Oppenheim BA, Barclay GR, Morris J, Know F, Barson A, Drucker DB, Crawley BA, Morris JA. Antibodies to endotoxin core in sudden infant death syndrome. Arch Dis Child 1994; 70: 95-98.

Pershagen G. Review of epidemiology in relation to passive smoking. Arch Toxicol 1986; 9 (suppl 9): 63-73.

Peterson S Wailoo M. Relationships between the development of physiological systems in infancy. In. Wailoo, M (ed.) Proceedings of the Babes in Arms Symposium of Developmental Physiology in Relation to SIDS. Beaconsfield: Chiltern Publishing Ltd., (in press).

Platt MS, Yunginger JW, Sekula-Perlman A, Irani AM, Smialek J, Mirchandani HG, Schwartz LB. Involvement of mast cells in sudden infant death syndrome. J Allergy Clin Immunol 1994; 94: 250-256.

Raza MW. Viral infections as predisposing factors for bacterial meningitis. PhD Thesis University of Edinburgh 1992.

Saadi AT, Blackwell CC, Raza MW, James VS, Stewart J, Elton RA, Weir DM. Factors enhancing adherence of toxigenic *Staphylococcus aureus* to epithelial cells and their possible role in sudden infant death syndrome. Epidemiol Infect 1993; 110: 507-517.

Saadi AT, Weir DM, Poxton IR, Stewart J, Essery SD, Raza MW, Blackwell CC, Busuttil A . Isolation of an adhesin from *Staphylococcus aureus* that binds Lewis[a] blood group antigen and its relevance to sudden infant death syndrome. FEMS Immunol Med Microbiol 1994; 8: 315-320.

Saadi AT, Blackwell, C.C., Essery, S.D., Raza, M.W., Weir, D.M., Elton, R.A., Busuttil, A., Keeling J.W.Developmental and environmental factors that enhance binding of *Bordetella pertussis* to human epithelial cells in relation to sudden infant death syndrome. (submitted for publication)

Saadi, A.T., Raza, M.W., Blackwell, C.C., Weir, D.M. and Busuttil A. 1994. Binding of toxigenic bacteria to epithelial cells of smokers and non-smokers. Third SIDS International Congress. p. 52.

Schlievert, PM. The role of superantigens in human disease. Curr Opin Infect Dis. 1995; 8: 170-174.

Sharp J, Poxton IR. The cell wall proteins of Clostridium difficile.FEMS Microbiol Lett 1988; 55: 99-104.

Stoltenberg L, Saugstad OD, Rognum TO. Sudden infant death syndrome victime show local immunoglobulin M response in tracheal wall and immunoglobulin A response in duodenal mucosa. Pediatr Res 1992; 31: 372-375.

Telford DR, Morris JA, Hughes P, Conway AR, Lee S, Barson AJ, Drucker DB. The nasopharyngeal bacterial flora in sudden infant death syndrome. J Infect 1989; 18: 125-130.

van t'Wout J, Burnette WN, Mar VL, Rozdzinski E, Wright SD, Tuomanen E. Role of carbohydrate recognition domains of pertussis toxin in adherence of Bordetella pertussis to human macrophages. Infect Immun 1992; 60: 3303-3308.

Vege A, Rognum TO, Scott H, Aasen AO, Saugstad OD. SIDS cases jave increased levels of interleukin-6 in cerebral spinal fluid. Acta Paediatr 1995; 84: 193-196.

Walker AM, Jick H, Perera DR et al. Diphtheria-tetanus-pertussis immunization and sudden infant death syndrome. Am J Public Health 1987; 77: 945-51.

ALTERNATIVE INHIBITORS OF MYCOPLASMA ADHERENCE

Itzhak Kahane[1], Abed Athamna[1], David Yogev,[1] and Mordechai R. Kramer[2]

[1] Department of Membrane and Ultrastructure Research
The Hebrew University-Hadassah Medical School
[2] Pulmonary Institute
Hadassah University Hospital
Jerusalem, Israel 91120

1. INTRODUCTION

Mycoplasma is the trivial name of the class *Mollicutes*, a group of minute wall-less bacteria comprising now more than 140 species. They are characterized by their small genome size; the smallest of bacteria. It comprises about 500 - 1000 predicted coding sequences (genes) as compared to, for example, 1727 and about 4000 in *Hæmphilus influenza* and *Escherichia coli* respectively. The complete genome of the smallest mycoplasma, *Mycoplasma genitalium* which is a human pathogen , was elucidated very recently by Fraser *et al.*. It is 580 kb long and comprises only 470 predicted coding sequences. By evolution, the mycoplasmas belong to the gram-positive bacteria with low G+C genome. They underwent genome reduction and with it, also lost much of their biosynthetic machinery. The organisms grow in broth, but need a complex medium. It is therefore not surprising that most mycoplasmas are parasites, and many are pathogens of a vast variety of hosts, including humans. In addition, they easily contaminate the rich culture media and then adhere to the cells grown there and therefore are considered to be pests of cells in culture.

2. MYCOPLASMA PATHOGENICITY

Many of the mycoplasmas are parasites of the epithelial lining of the respiratory and the urogenital tracts, and in humans and animals they primarily cause diseases of these tracts. The usual antibiotic treatment to combat mycoplasmal infection is erythromycin or tetracycline or their newer derivatives. However, the efficient therapy is now less easy reached due to the increasing numbers of mycoplasmal infections caused by the drug resistant strains, *e.g.*, the *Ureaplasma urealyticum* Tet M-resistant strains, as well as *Ureaplasma* and other mycoplasmas erythromycin-resistant strains.

Toward Anti-Adhesion Therapy for Microbial Diseases, edited by Kahane and Ofek
Plenum Press, New York, 1996

The list of mycolasmas that are known or suspected to be involved in infections in human has expanded quite extensively in the recent years (Table 1). These include the "classic" 1. *Mycoplasma pneumoniae* - which is primarily the cause of primary-atypical cold-agglutinin-positive pneumonia (Brunner, 1993), but may also be involved in infections of other organs as well; 2. *Mycoplasma hominis* which is involved in infections of the urogenital tract, but is quite commonly a cause of bacteremia after birth and may cause arthritis, and 3. *U. urealyticum*. This species is comprised of many strains which actually, by most systematic criteria , deserve to be split into two species. Next to *Chlamydia, U. urealyticum* is most probably the second most common microorganism involved in STD. In addition, *U. urealyticum* was demonstrated to be involved in respiratory tract infections in newborns. This is primarily because of the high incidence that these bacteria may be present in the lower genital tract of women. When they are giving birth, the organisms are infecting the respiratory tract of newborns while passing in the birth canal and may become a major threat to newborns of low birth weight who suffer from respiratory distress syndrome.

Altogether *U. urealyticum* is associated with a number of clinical conditions. These include also prostatitis , a segment of NGU , infertility, sterility and spontaneous abortions. The data about several of these are still ambiguous and more detailed controlled studies are needed. The data also suggest , but is not conclusive, that some strains are more pathogenic and involved more frequently in several of these infections. Less is known about the pathogenicity of the "newer" mycoplasma species (Table 1).

M. genitalium is considered a genital tract pathogen. However, very seldom was isolated from this site. It can be traced in samples from this organ, but also from the respiratory tract, by *in vitro* nucleic acid amplification (e.g. PCR). The data about *Mycoplasma fermentans* (incognitus strain) and *Mycoplasma penetrans* is still meager. However, it indicate that they may cause acute systemic, life threatening disease. On top of it is the incrimination of at least several mycoplasmas to be an initial factor in setting the stage for HIV induction (Montagnier, 1993). These finding were supported by the *in vitro* experiments of our group which indicate that these mycoplasmas as well as others including *U. urealyticum* can regulate HIV-LTR (Nir Paz *et al.*, 1995).

The development of mycoplasmal diseases at the initial stages are primarily due to adherence of the organisms to the host cell membrane and induction of oxidative stress, both by inhibition of the host cell catalase and superoxide dismutase (Almagor *et al.*, 1984; Kahane, 1992). This in turn, causes subtle, but severe enough damage, to the cell membrane followed by leakage of nutrients that ooze to the outer surface of the membrane at the site of settlement. In this way the mycoplasmas are provided with the needed supplements to further their settlement and colonization and deepen their deleterious effects on the host.

Table 1. Mycoplasma: Human pathogens

"Classic"
M. pneumoniae
M. hominis
Ureaplasma urealyticum
"Newcomers"
M. genitalium
M. fermentans
M. penetrans
M. pirum

3. ADHERENCE AND ADHESINS

Lacking a cell wall, the adherence of these mycoplasmas to cells seemed to present the questions related to the topic in a more focused way and at the onset of the studies when the concept was that the mycoplasma presented a very primitive, almost a fossil-bacteria, the idea was that simple microorganisms will have simple solutions. The initial quantitative studies with the well-recognized human and poultry pathogens, *M. pneumoniae* and *M. gallisepticum*, respectively, indicated that mycoplasma adhere via sialic acid residues (Thomas & Bitensky, 1966). At that time it was also reported that these mycoplasmas have sialidase and we have toyed with the idea that this enzyme will start the process that mycoplasmas can be inducers of fusion in a process similar to that present in several viruses (Kahane & Loyter, unpublished results). We could not provide evidence for this, but also could not support the observations of others of the presence of sialidases in mycoplasmas.

The early findings that mycoplasmas adhere to host cell sialoglyconjugates made them quite unique, but also loners, not part of the common "Mannose Sensitive / Resistant" Club. It took awhile for microbiologists to find out that microorganisms do adhere to a large variety of ligands and many bind to sialic acids which are the most common ultimate sugar of glycoconjugates.

With the rapid increase in the number of known mycoplasma species the studies of the various adhering mycoplasmas are still incomplete. Therefore, we will summarize several principles that are already visible and add several points of interest.

As it stands now:

1. Many adhering mycoplasma species bind specifically to sialic acid residues of sialoglycoconjugate receptors on the host cell membranes.
2. The linkage of sialic acid to these mycoplasmas is not uniform.
3. At least *M. pneumoniae* recognizes an asialoglycoconjugate receptor on fibroblasts.
4. Several mycoplasmas adhere to sulfatides of glycolipids.
5. Many mycoplasmas can adhere to more than one type of receptors on the host (e.g. sialoglycoconjugate and sulfatides).

The major conclusion from these is that studies of a mycoplasma should be conducted on various receptors and on various host cells as the data with the latter may vary too. The data presented in Table 2 in regard to the adherence of *U. urealyticum* is an example which supports several of these points. Several other mycoplasmas adhere to sialoglycoconjugates of the host cell receptors. These include *M. genitalium, M. synoviae, M. gallisepticum.*

Our recent study on the adherence of *M. pneumoniae* to human alveolar macrophages was presented during this seminar (Athamna, Kramer and Kahane, 1996a). We found that *M. pneumoniae* adherence is inhibited by dextran-sulfate, but not by dextran or several monosaccharides, showing that sulfated components on the macrophage surface act as receptors for *M. pneumoniae* binding. In addition sialylated compounds, such as fetuin and α1-acid glycoprotein, were found to be potent inhibitors of the attachment, also indicating the role of sialic acid residues in recognition and attachment of *M. pneumoniae* to human alveolar macrophages.

So far, the most detailed experiments on the nature of the sialoreceptor were conducted using *M. pneumoniae.* This was based using various ligands and inhibitors including purified glycoproteins, glycophorins, laminin, fetuin, and human chorionic gonadotropin and were reveiwed recently by Kahane (1995).

An example of studies of mycoplasma specificity to sialic acid are those of Feizi's group (Loveless & Feizi, 1989). Based on restoration of binding to neuraminidase-treated

Table 2. Effects of sialoglycoproteins, asialoglycophorin and dextran sulfate adherence of *Ureaplasma urealyticum* to human erythrocytes

Treatment[a]	% Inhibition[b]
Glycophorin	52 ± 20
Asiaglycophorin	0
α1-Acid glycoprotein	29 ± 10
Fetuin	8 ± 6
Bovine serum albumin	4 ± 4
Dextran sulfate	53 ± 10
Dextran	16 ± 6
Buffer (control)	0

[a] α1-Acid glycoprotein, asiaglycophorin, fetuin, glycophorin dextran sulfate and dextran (M_w ~500,000) (0.1 mg/ml), were added to the suspension of *U. urealyticum* (serotype 8) prior to the erythrocytes in adherence experiments. In control experiments, bovine serum albumin or buffer was added.

[b] Results are expressed as percentage inhibition compared with controls with buffer (5.5 to 7% adherence). Each value is the mean ± standard error of eight experiments.

[c] Adapted from Saada *et al.*, 1991, Adherence of *Ureaplasma urealyticum* to human erythrocytes, Infect. Immun. 59:467-469.

erythrocytes using CMP-sialic and purified sialytransferases, adhesion of the erythrocytes on surface-grown sheet cultures of *M. pneumoniae* specifically recognized sialic acid linked α(2-3) to *N*-acetyllactosamine sequences.

The detailed studies of adhesins of mycoplasmas were carried primarily on the human pathogens *M. pneumoniae* and *M. genitalium* and on the poultry mycoplasma *M. galllisepticum*. In the first two the genes that are involved in the adherence process have been cloned and analyzed. The major adhesin P1 and MgPa of *M. pneumoniae* and *M. genitalium* respectively are part of an operon. Moreover, other proteins are involved in the adherence process. P1 is a protein of 1560 amino acids and spans the membrane several times as reviewed recently by Kahane (1995). The studies of Jacobs' group have indicated that three segments of the protein are involed in the binding to sialic acid but still we lack the detailed molecular understanding of the interaction (Gerstenecker and Jacobs, 1993).

It turned out that the mycoplasmas take advantage of their specific adherence to ligands, but they are not "sitting ducks" for the macrophage to feed on. They can change the repertoire of proteins exposed on the outer surface by a genetically controlled mode (Yogev *et al.*, 1991). We were interested to test whether these include also the adhesins or other components involved in the adherence process. This was found to be the case and we reported our recent findings during this seminar (Athamna *et al.*, 1996b). In essence it was found that hemadsorption (HA) of *M. gallisepticum* involved several membrane surface proteins undergoing high-frequency phase variation. The most interesting observation was the identification of colonies exhibiting, within a colony, sectorial regions of hemadsorbing to red blood cells (RBC). Antiserum prepared against an hemadsorbing clone (HA+) was used in Western immunoblot analysis to identify antigenic differences between HA+ and HA- phenotypes. Several proteins (P80, P50, P48, and P16) were found in the HA+, but not in HA-, clones. This HA+ antigenic profile was further monitored throughout sequentially subcloned HA population of *M. gallisepticum* (HA+ → HA- → HA+). The appearance of these additional proteins was consistent with the HA+ phenotype. Experiments in which HA+ proteins were bound to RBC revealed that P80 and P50 were selectively bound.

Altogether the data show that hemadsorption of *M. gallisepticum* involves switching of variable surface membrane proteins. We now learn its initial principles, but do not know how it is controlled. Thus, there is actually a cross-talk, or a war, between the host and the invader. This also has been taken into account when planning the anti-adhesion therapy.

4. CONCLUDING REMARKS

A therapy strategy should include the following:

1. Identifying the best competing ligand to block the invader from adherence to its natural ligand on the host cell target.
2. Since many times there are various ligands we should design a cocktail of competing ligands at the correct ratio.
3. We should add the conditions for the OFF position for induction of the adherence plus variation.

By combining these parameters, we design an approach for a newer generation of treatment.

REFERENCES

Athamna, A. , M. R. Kramer and I. Kahane 1996a. Adherence of *Mycoplasma Pneumoniae* to human alveolar macrophages, In: Toward Anti-Adhesion Therapy of Microbial Diseases (Kahane, I. and I. Ofek, eds) Plenum Press, New York and London (p. 266)

Athamna, A. , R. Rosengarten, S. Levisohn, I. Kahane and D. Yogev . 1996b. Hemadsorption of *Mycoplasma Gallisepticum* involves switching of variable surface membrane proteins. In: Toward Anti-Adhesion Therapy of Microbial Diseases (Kahane, I. and I. Ofek, eds) Plenum Press, New York and London (p. 265)

Almagor, M., I. Kahane and S. Yatsiv. 1984. The role of superoxide anion in host cell injury induced by *Mycoplamsa pneumoniae* infection: Study in normal and trisomy 21 cells. J. Clin. Investigation, 73:842-847

Brunner, H. 1993. Mycoplasma Infections of Man: Respiratory and Male Genital Tract Diseases, In: Rapid Diagnosis Of Mycoplasmas (I. Kahane and A. Adoni, eds) pp.39-56, Plenum Press, New York and London

Fraser, C. M. *et al.* 1995. The minimal gene complement of *Mycoplama genitalium.*Science. 270:397-398

Gerstenecker, B. and Jacobs, E. 1993. Development Of A Capture-Elisa For The Specific Detection of *Mycoplasma Pneumoniae* In Patients' Material, In: Rapid Diagnosis Of Mycoplasmas (I. Kahane and A. Adoni, eds) pp.195-205, Plenum Press, New York and London

Kahane, I. 1992. Frontiers in mycoplasma pathogenicity. World J. Microbiol. Biotech. 8:50-51

Kahane, I. 1995. Adhesions of mycoplasmas, In: Methods in Enzymology: Adhesion of Microbial Pathogens (R. J. Doyle and I. Ofek, eds) vol.253, pp.367-373. Academic Press,Inc .Orlando,FL.

Krivan, H. C., Olson, L.D., Barile, M. F. , Ginsburg, V. and Roberts, D. D. 1989. Adhesion of *Mycoplasma pneumoniae* to sulfated glycolipids and inhibition by dextran sulfate. J. Biol. Chem. 264:9283-9288

Loveless, R. W. and Feizi, T. 1989. Sialo-oligosaccharide receptors for *Mycoplasma pneumoniae* and related oligosaccharide of poly-N-acetyl-lactosamine series are polarised at the cilia and apical-microvillar domains of the ciliated cells in human bronchial epithelium. Infect. Immun. 57:1285-1289

Montagnier, L. 1993. HIV cofactors and AIDS. Proceedings of IX International Congress on AIDS and STD.

Nir-Paz, R., Israel, S., Honigman, A. and Kahane, I. 1995. Mycoplasmas regulate HIV-LTR-dependent gene expression. FEMS Microbial. Lett. 128:63-68

Thomas, L. and Bitensky, M. W. 1966. Studies of PPLO infection. IV. The neurotoxicity of intact mycoplasma, and their production of toxin *in vivo* and *in vitro*. J. Exp. Med. 124:1089-1098.

Yogev, D., Rosengarten,R., Watson-McKown, R. and Wise, K. S. 1991. Molecular basis of *Mycoplasma* surface antigenic variation: a novel set of divergent genes undergo spontaneous mutations of periodic coding regions and 5′regulatory sequences. EMBO J. 10:4069-4079.

13

STUDIES ON THE MOLECULAR MECHANISMS OF MENINGOCOCCAL INTERACTIONS WITH HUMAN CELLS

Towards Anti-Adhesion Measures for the Control of Meningococcal Disease

Mumtaz Virji

University of Oxford Department of Paediatrics
John Radcliffe Hospital
Oxford, OX3 9DU, United Kingdom

INTRODUCTION

The meningitis-causing organism, *Neisseria meningitidis*, is a human pathogen with exquisite affinity for the nasopharyngeal tissue of its host where it may exist as a harmless commensal in up to 30% of healthy individuals (DeVoe, 1982). However, in some cases, predisposing factors which are poorly understood, make the host susceptible to further invasion by these bacteria. In such cases, meningococci breach the epithelial barrier of the nasopharynx and disseminate throughout the body via the blood stream. Blood stream invasion is often associated with endothelial necrosis and intra-vascular coagulation which result in skin lesions common in disseminated meningococcal infections. Further tissue infiltration including that of the central nervous system may ensue in untreated cases, with serious outcomes including death. Factors that are implicated in bacterial virulence include polymeric proteins, pili, which are composed of repeating subunits (pilins). Pili extend beyond the capsule which is invariably present on the surface of bacteria isolated from blood or cerebrospinal fluid (csf) (DeVoe and Gilchrist, 1975). Two structural classes of pili occur in *N. meningitidis*. Class I pili are similar to gonococcal pili whereas Class II pili produce pilins of smaller Mr (Diaz *et al.*, 1984, Perry *et al.*,1988, Virji *et al.*, 1989). Both classes of pili occur in clinical isolates but are mutually exclusive within a given strain. Pili undergo phase and antigenic variation as a result of inter- and intra-genomic recombinational events (reviewed Saunders *et al.*, 1994). In addition, two outer membrane proteins, Opa and Opc, may also increase the potential of meningococci to interact with human cells. These proteins are classified as Class 5 proteins or opacity proteins (since they impart opacity to colonies of bacteria expressing the proteins). Opa proteins are a family of antigenically variable proteins and occur in *N. meningitidis* as well as in the closely related organism *N. gonor-*

rhoeae (Cannon, 1994). In meningococci, 3-4 *opa* gene loci code for related proteins with conserved, semivariable and hypervariable domains. Opc protein is largely invariant and is expressed in all epidemic and highly pathogenic strains other than ET37 complex of serogroup C meningococci and gonococci (Olyhoek *et al.*, 1991, Wang *et al.*, 1993).

In our investigations on the molecular basis of meningococcal pathogenesis, we have studied the roles of pili and the opacity proteins in bacterial interactions with host cells. The studies have focused on the mechanisms of bacterial-ligand / host-receptor interactions. A number of observations made during these studies have suggested that structural features that modulate bacterial adhesion via particular adhesins (such as pili) may also alter the severity of cellular toxicity induced by meningococci. The identification of these features is of importance in understanding meningococcal pathogenesis and ultimately for molecular targeting for intervention during the course of meningococcal infection. Also, since adhesion is the first step to colonisation which precedes dissemination, anti-adhesion measures (vaccination or other) are likely to be the best means for the prevention of colonisation which may ultimately lead to eradication of meningococcal disease. This article provides an overview of our recent studies on adhesive and invasive mechanisms of *N. meningitidis*.

IMPORTANCE OF PILI IN HOST SPECIFICITY AND MULTIPLE CELLULAR TARGETING

Meningococcal pili have been implicated in mediating epithelial interactions (Stephens, 1989). However, since blood and csf isolates are piliated and pili are lost rapidly on non-selective subculture (Virji *et al.*, 1995a), pili are selected for *in vivo* and may play an important role during dissemination. Pili were shown to mediate haemagglutination (Trust *et al.*, 1983). Our studies using numerous disease isolates as well as defined derivatives of a single strain demonstrated that pili are also important for endothelial adhesion and that they are essential adhesins in capsulate bacteria and also in acapsulate bacteria with sialylated lipopolysaccharide (LPS) (Virji *et al.*, 1991a, 1992a, 1993a, 1995b). Meningococcal pili mediate adhesion to human epithelial and endothelial cells but not to human phagocytic cells or cells originating from other animals.

STRUCTURE / FUNCTION RELATIONSHIPS OF MENINGOCOCCAL PILI, PILUS-ASSOCIATED ADHESINS

Studies using adhesion variants (derived by single colony isolation with or without prior selection on host cells) implied that structural variations of pilins affect epithelial interactions significantly but have a lesser effect on endothelial interactions (Virji *et al.*, 1992a, 1993a). Thus the pilin subunit may contain a human cellular binding domain or, at least, has influence on adhesion if mediated by an accessory protein. The influence of pilin structure on meningococcal as well as gonococcal adhesion to host cells has been recently confirmed by others (Nassif *et al.*, 1993, Jonsson *et al.*, 1994). Other factors may also affect pilus-dependent interactions of the pathogenic *Neisseriae*. A pilus-associated protein, PilC has been implicated in cellular adhesion and in biogenesis in both meningococci and gonococci (Jonsson *et al.*, *1991,* Rudel *et al.*, *1992,* Nassif *et al.*, 1994, Virji *et al.*, 1995a). Figure 1 shows adhesion of two clonal isolates of strain C311 that produce distinct pilin subunits but have identical expression of PilC and other major surface structures (Virji *et al.*, 1993a).

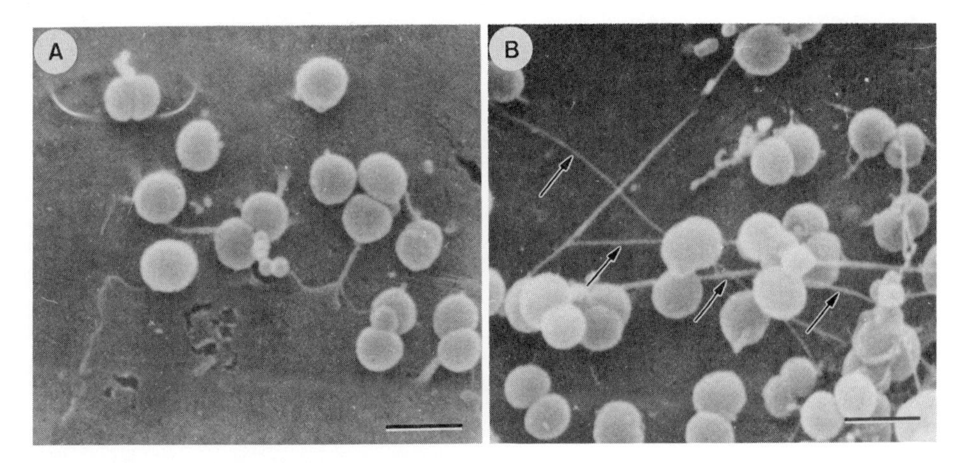

Figure 1. Scanning electron micrograph (SEM) showing differences in pilus-mediated adhesion to human umbilical vein endothelial cells: a) parental piliated phenotype of strain C311 which elaborates individual non-bundling pili. Bacteria show moderate adhesion: 50-100 bacteria per cell; b) a derivative of strain C311 with bundle-forming pili easily visible by SEM (arrows). Bacteria show increased adhesion: 200-300 bacteria per cell. Adhesion differences of these variants were more marked in the case of epithelial cells (5-10 and 100-150 respectively, Virji et al, 1993a). Pilins of these variants exhibited differences in migration on SDS-PAGE (apparent Mr differences: 1500-2000). However, molecular weights of these pilins deduced from their DNA sequences were identical. Subsequent studies have shown that variant pilins of strain C311 contain some common and some distinct post translational modifications. (Bars represent 1.0 μm.).

Meningococcal Pili Are Post-Translationally Modified: Identification of Glycosyl Substituents of Pilins

During the studies on the molecular basis for adhesion variation (Fig. 1), we observed that pilins of identical molecular weight, as deduced from DNA sequence, migrated to different positions on SDS-PAGE. Using biotin hydrazide under the conditions that labelled glycans, as well as chemical deglycosylation, the presence of carbohydrates on pilins of strain C311 was shown (Virji *et al.*, 1993a). In further studies, using a combination of molecular biology and biochemical methods, we have shown that variant pilins of strain C311 contain a common trisaccharide substitution. The precise structure of this O-linked trisaccharide was elucidated by the use purified pilins derived from parental strains and their galactose epimerase (*galE*) mutants. Pilins were digested using trypsin and peptides were separated by reverse phase HPLC. Purified peptides were subjected to Fast Atom Bombardment- and Electrospray- mass spectrometry. Glycan structures were derived by sugar and linkage analysis and by Electron Impact mass spectrometry (Stimson *et al.*, 1995). Anomeric stereochemistry has been determined recently by sequential treatment of the trisaccharide by α and β galactosidases (Virji *et al.*, 1996). The structure of the trisaccharide (digalactosyl, diacetamidotrideoxyhexose) is presented in Figure 2.

Prokaryotes are capable of producing a wide variety of sugar structures and among the rarest of these are diamino sugars, the first of which to be isolated was 4-acetamido-2-amino-2,4,6-trideoxy-D-glucose (N-Acetylbacillosamine) (Sharon and Jeanloz, 1960, Zehavi and Sharon, 1973). Since then, diacetamidotrideoxyhexoses have been found to be constituents of polysaccharides (e.g. *E. coli* K48 capsule, Whittaker *et al.*, 1994) but such sugars have not been previously found as constituents of glycoproteins. The presence of such

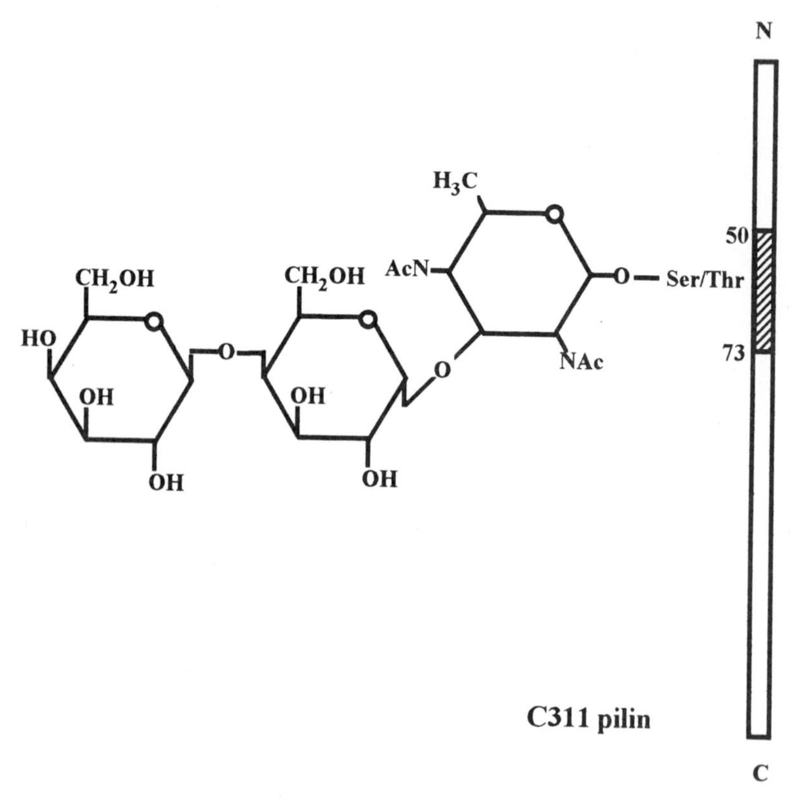

Figure 2. Proposed structure of the trisaccharide covalently attached to the tryptic peptide ^{50}Y-K^{73} derived from pilins of strain C311. The stereochemistry of the 2,4-diacetamido-2,4,6-trideoxyhexose is not known.

unusual structures on pili suggests that they may be functionally important but their role(s) remain to be shown.

Multiple Modifications of Meningococcal Pili

Further recent studies have shown that at a distinct site, meningococcal pilins contain, perhaps, a unique substitution, α-glycerophosphate which is linked to serine at position 93 (Stimson *et al.*, 1996). Such a modification has not been reported for proteins and suggests a special role for this moiety. Over and above these common modifications, meningococcal variant pili contain distinct substitutions - the existence of which is known from mass spectrometry studies (Stimson *et al.*, 1995) but the structure is not known at present.

Cytotoxicity and Meningococcal Pili

Neisseria meningitidis and *Haemophilus influenzae*, both human respiratory pathogens, are able to cause invasive disease in their human host. From the sequence of events noted in patients, it appears that in both cases bacteria enter the systemic circulation from the oropharynx and disseminate by the haematogenous route to other sites throughout the body. However, whilst *N. meningitidis* causes endothelial necrosis and skin lesions that characterise meningococcal disease, *H. influenzae* is only occasionally associated with haemorrhagic skin lesions. In order to elucidate the mechanisms of endothelial necrosis, we developed an *in vitro* assay using human umbilical vein endothelial cells (Virji *et al.*, 1991a, 1991b) and observed that meningococcus, but not haemophilus, causes cytopathic damage to cultured human endothelial cells reflecting the situation *in vivo* (Virji *et al.*, 1991b). We also observed that piliated *N. meningitidis* are more toxic for endothelial cells than non-piliated bacteria (Virji *et al.*, 1991a). However, in both piliated and pilus-deficient meningococcal phenotypes, LPS is the major toxic factor (Dunn *et al.*, 1995). These recent studies have also shown that the level of *N. meningitidis* toxicity increases with the level of pilus mediated adhesion but Opc-mediated adhesion does not enhance toxicity. These studies imply that pili may deliver independent deleterious signal to endothelial cells over that delivered by meningococcal LPS (Dunn *et al.*, 1995).

Interactions Mediated by the Outer Membrane Protein Opc: Molecular Mechanisms and Identification of Host Cell Receptors

In studies using variants of a serogroup A strain C751, expressing distinct opacity proteins, the potential of the outer membrane protein Opc to mediate cellular invasion was discovered (Virji *et al*, 1992b, Fig. 3). In addition, the protein was only functional in bacterial phenotypes that were asialylated (non-capsulate bacteria with LPS lacking sialic acid residues: a phenotype that is often isolated from the nasopharynx, Cartwright *et al,* 1987). Further studies have identified distinct mechanisms of bacterial interactions at the apical and basolateral surfaces of polarised cells (Virji *et al,*, 1994a). Interactions of Opc-expressing bacteria at the apical surfaces of polarised cells were totally dependent on the presence of serum in the incubation medium. Inhibition studies using peptides with arginine-glycine-aspartic acid (RGD) motif and monoclonal antibodies to RGD-recognising integrins (Ruoslahti and Pierschbacher, 1987), suggested a 'bridge' mechanism of Opc-mediated interactions at the apical surface of human endothelial cells. This involves a molecular

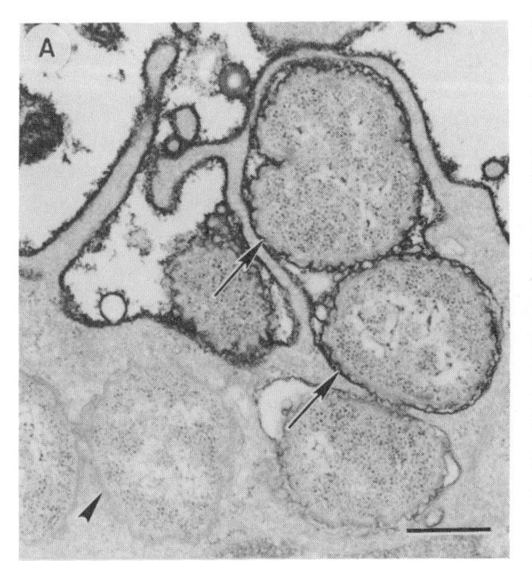

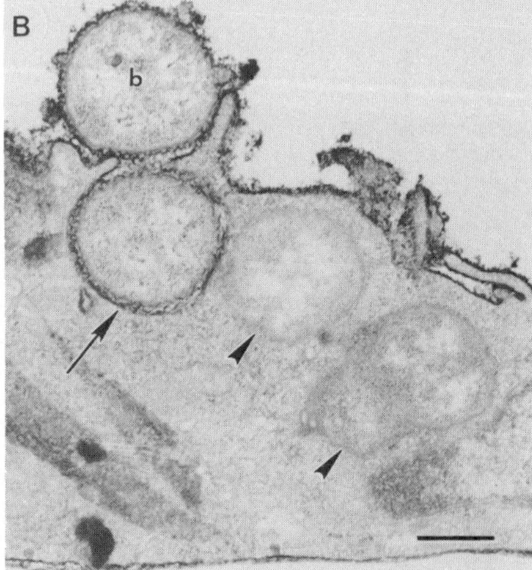

Figure 3. Transmission EM showing invasion of human epithelial (a) and endothelial (b) cells by Opc-expressing asialylated variants of meningococci. Ruthenium red staining shows extracellular bacteria (b) and the glycocalyx of partly enclosed vacuoles (arrows) but not of truely intracellularly located bacteria or vacuolar membranes surrounding them (arrowheads). (Bars represent 0.5 μm.)

complex of Opc and serum-derived ligands that in turn bind to the integrin $\alpha v\beta 3$ and perhaps also to $\alpha 5\beta 1$ (Fig. 4). Purified preparations of vitronectin and fibronectin can replace serum and mediate RGD-dependent cellular invasion. These studies have also shown that intact cytoskeletal function and receptor clustering are additionally required for Opc-mediated invasion (Virji *et al,* 1994a).

Interactions of Opc with Extracellular Matrix

Opc-expressing meningococci can adhere to multiple extracellular matrix (ECM) components (Virji *et al,* 1994b) and serum proteins giving the organism the potential to interact with several different integrins by bridging via their respective ligands (see above). Using immobilised matrigel or purified human proteins (fibronectin, vitronectin, laminin, collagens, thrombospondin (TSP) and von Willebrand factor (vWF), Opc was shown to increase adherence of bacteria to several but not all proteins (2-10 fold increase) compared to Opc-deficient bacteria. Minimal or no interactions were observed with TSP or vWF. This could be an effective strategy for the organism for cellular invasion and for adherence to substrata of damaged mucosa as well as for penetration of deeper tissues after cellular invasion.

PHENOTYPIC REQUIREMENTS OF OPC-DEPENDENT INVASION AND INTERPLAY BETWEEN SURFACE VIRULENCE FACTORS

The interplay between four surface-expressed virulence factors (pili, Opc, capsule and LPS) in host cell adhesion and invasion was examined using derivatives of a serogroup B strain, MC58, created by mutation (capsule, Opc) and selection of variants. To examine the role of Opc and of additional expression of pili, bacteria lacking the expression of Opa proteins were used. The effect of different LPS structures was examined in variants express-

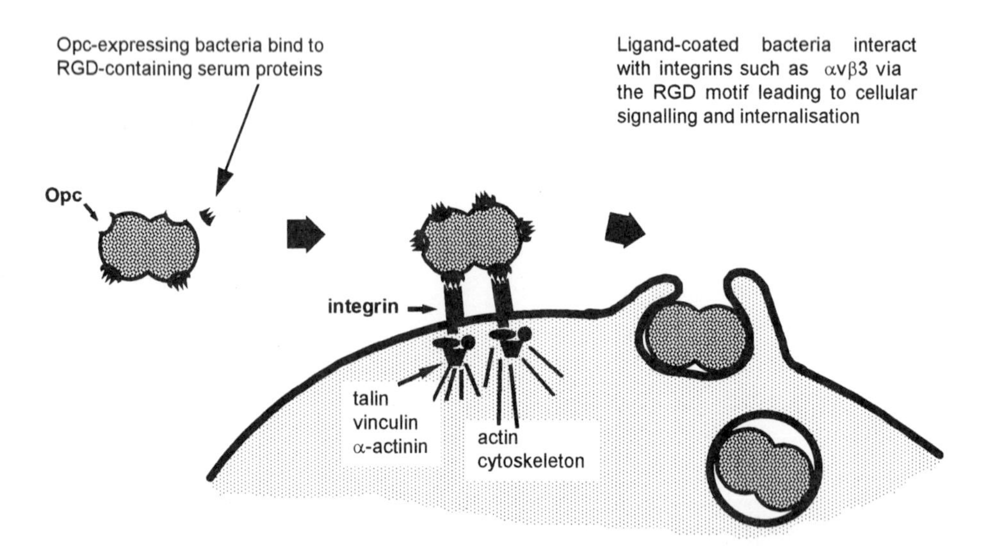

Figure 4. Mechanism of meningococcal entry into human endothelial cells.

ing either sialylated (L3 immunotype) or truncated non-sialylated (L8 immunotype) LPS. Studies showed that a) pili were essential for meningococcal interactions with host cells in both capsulate and acapsulate bacteria with the sialylated L3 LPS immunotype; b) the Opc-mediated invasion of host cells by piliated and non-piliated bacteria was observed only in acapsulate organisms with L8 LPS immunotype and c) expression of pili in Opc-expressing bacteria resulted in increased invasion (Virji *et al,* 1995b).

These studies have confirmed the inhibitory effects of surface sialic acids on Opc function. It is envisaged that Opc (basic in nature) needs down-modulation of negatively charged surface structures to be functionally active. It is also possible that surface sialic acids sterically hinder outer membrane proteins. The studies also demonstrated the potentiating role of pili in cellular invasion. Thus pili appear to modulate multiple properties of meningococci.

Investigations on the mechanisms of cellular invasion in strain MC58 have also confirmed the previous observations on variants of a serogroup A strain (above). In summary, Opc mediates cellular invasion in distinct meningococcal strains via a sequence of molecular events resulting in tri-molecular complexes at the apical surface of host cells. The expression of Opc appears to enable bacteria to utilise normal signal transduction mechanisms of host cells by attaching to accessory molecules derived from serum that adhere to their cognate receptors (Fig. 4).

The Role of Opa Proteins

Three different Opa proteins are produced by meningococcal strain C751. OpaB and OpaD increase interactions of bacteria with endothelial and epithelial cells whereas OpaA is ineffective. Opa proteins that increase cellular adhesion also mediate cellular invasion (Virji *et al,* 1993b). Mechanisms of cellular interactions mediated by Opc and Opa proteins are distinct. Preliminary studies show that Opa proteins may also interact with some matrix proteins (Virji *et al,* 1994b).

Interactions of Meningococci with Monocytes and Polymorphs: the Role of Surface Virulence Factors (Capsule, Lps, Pili, Opa and Opc)

Although disseminated strains are invariably capsulate, nasopharyngeal isolates may be deficient in capsule expression. Moreover, since up to 16% of the cells present in the nasal mucosa are monocytes, we have investigated the interactions of distinct phenotypic variants of *N. meningitidis* with both human monocytes and polymorphs. These studies have shown that capsulate bacteria resist phagocytosis in the absence of opsonins but acapsulate bacteria are internalised and opacity proteins mediate bacterial uptake (McNeil *et al,* 1994a). Pili of distinct structural make-up or the pilus-associated protein PilC were ineffective in mediating interactions with phagocytic cells. Opc protein was the most effective protein in mediating monocyte interactions but failed to increase neutrophil interactions (McNeil *et al,* 1994b). The molecular mechanisms of interactions of Opa and Opc with polymorphs and monocytes appear to be distinct (McNeil and Virji 1996).

Concluding Comments

The above studies have identified several important aspects of outer membrane structure-dependent cellular interactions and also molecular details of some of the interactions of *N. meningitidis* with human target cells. Until recently, few studies had reported glycosylation of prokaryotic proteins. However, it is becoming clear that surface structures of pathogens may, in some cases, be glycosylated. For example, glycans have been detected on mycobacterial 19 kDa protein (Garbe *et al,* 1993) and in *Pseudomonas aeruginosa* pilin (Castric, 1995). However, neither the structures nor the functions of these glycans are known. Our studies on *N. meningitidis* pili have identified unusual glycans and other unique substituents such as glycerophosphate. In addition, recent studies using *galE* mutants of distinct strains and antisera against Galβ1-4Gal suggest that glycans are present on numerous clinical isolates and on both Class I and Class II pili. Also, the absence of digalactose structure leads to modulation of bacterial adhesion in several variants (Virji *et al,* unpublished).

Successful nasopharyngeal colonisation by meningococci may require not only pilus- but also Opa/Opc-mediated interactions. It is conceivable that pili help in initial colonisation but longer term persistence in the nasopharynx requires the bacteria to be internalised by the nasopharyngeal epithelial cells. Internalisation may help evade host immune responses. Nasopharyngeal isolates often exhibit phenotypic characteristics that permit outer membrane invasins (Opa and Opc) to be functional. Elucidation of the mechanisms of adhesion and invasion via the outer membrane proteins are also essential in our attempts to prevent colonisation. Our studies have shown that Opc requires serum-derived factors to interact with host cells and that ECM is also targeted by Opc-expressing bacteria.

Further studies are needed to elucidate the key components of the multi-faceted interactions of meningococci with its human host and to identify likely candidates for use in anti-adhesion strategy aimed at prevention of meningococcal colonisation of its unique human niche.

ACKNOWLEDGEMENTS

I would like to acknowledge my colleagues and collaborators who have contributed to the above studies on *Neisseria meningitidis*, particularly Professors E. Richard Moxon, Jon Saunders and Anne Dell. I would also like to thank Dr. David J.P. Ferguson for his help with electron microscopy.

REFERENCES

Cannon, J.G. (1994) In Neisseria 94. Genetics and function of Opa proteins: Progress and unanswered questions. Proceedings of the Ninth International Pathogenic Neisseria Conference. pp. 70-72. Editors J.S. Evans, S.E. Yost, M.C.J. Maiden and I.M. Feavers.

Cartwright, K.A.V., Stuart, J.M., Jones, D.M. and Noah, N.D. (1987) The Stonehouse survey: nasopharyngeal carriage of meningococcal and Neisseria lactamica. Epidem Inf 99: 591-601.

Castric, P. (1995) PilO, a gene required for glycosylation of *Pseudomonas aeruginosa* 1244 pilin. Microbiology 141: 1247-1254.

DeVoe, I.W. (1982) The Meningococcus and Mechanisms of Pathogenicity. Microb Reviews 46: 162-190.

DeVoe, I.W., and Gilchrist, J.E. (1975) Pili on meningococci from primary cultures of nasopharyngeal carriers and cerebrospinal fluid of patients with acute disease. J Exp Med 141: 297-305.

Diaz, J.-L., Virji, M. & Heckels, J.E. (1984). Structural and antigenic differences between two types of meningococcal pili. FEMS Microb Lett 21, 181-184.

Dunn, K.L.R., Virji, M. and Moxon, E.R. (1995) Investigations into the molecular basis of meningococcal toxicity for human endothelial and epithelial cells: the synergistic effect of LPS and pili. Microb Pathogen 18: 81-96.

Garbe, T., Harris, D., Vordermeier, M., Lathigra, R., Ivanyi, J. and Young, D. (1993) Expression of the *Mycobacterium tuberculosis* 19-kilodalton antigen in *Mycobacterium smegmatis* - immunological analysis and evidence of glycosylation. Infect Immun 61: 260-267.

Jonsson, A.-B., Nyberg, G. and Normark, S. (1991) Phase variation of gonococcal pili by frameshift mutation in *pilC*, a novel gene for pilus assembly. EMBO J 10: 477-488.

Jonsson, A. -B., Iver, D., Falk. P., Pepose, J. and Normark, S. (1994) Sequence changes in the pilus subunit lead to tropism variation of *Neisseria gonorrhoeae* to human tissue. Mol Microbiol 13: 403-416.

McNeil, G., Virji, M. and Moxon, E.R. (1994a) Interactions of *Neisseria meningitidis* with human monocytes. Microb Pathogen 16, 153-163.

McNeil, G., Virji, M. and Moxon, E.R. (1994b). Interactions of *Neisseria meningitidis* expressing Opc and Opa proteins with human phagocytes. In Neisseria 94. Proceedings of the Ninth International Pathogenic Neisseria Conference. pp. 265-267. Editors J.S. Evans, S.E. Yost, M.C.J. Maiden and I.M. Feavers.

McNeil, G. and Virji, M. (1996). Phenotypic variants of meningococci and their potential in phagocytic interactions: the influence of opacity proteins, pili, PilC and surface sialic acids (submitted to Microb Pathogen).

Nassif, X., Lowry, J. Stenberg, P., O'Gaora, P., Ganji, A. and So, M. (1993) Antigenic variation of pilin regulates adhesion of *Neisseria meningitidis* to human epithelial cells. Mol Microbiol 8: 719-725.

Nassif, X., Beretti, J-L., Lowy, J., Stenberg, P., O'Gaora, P., Pfeifer, J., Normark, S., and So, M. (1994) Roles of pilin and PilC in adhesion of *Neisseria meningitidis* to human epithelial and endothelial cells. Proc Nat Acad Sci USA 91: 3769-3773.

Olyhoek, A.J.M., Sarkari, J., Bopp, M., Morelli, G. and Achtman, M. (1991) Cloning and expression in *Escherichia coli* of *opc*, the gene for an unusual Class 5 outer membrane protein from Neisseria meningitidis. Microb Pathogen 11: 249-257.

Perry, A.C.F., Nicolson, I.J. and Saunders, J.R. (1988) *Neisseria meningitidis* C114 contains silent, truncated pilin genes that are homologous to *Neisseria gonorrhoeae pil* sequences J Bacteriol 170: 1691-1697.

Rudel, T., van Putten, J.P.M., Gibbs, C.P., Haas, R. and Meyer, T.F. (1992) Interaction of two variable proteins (pilE and pilC) required for pilus-mediated adherence of *Neisseria gonorrhoeae* to human epithelial cells. Mol Microbiol 6: 3439-3450.

Ruoslahti, E., and Pierschbacher, M.D. (1987) New Perspectives in Cell Adhesion: RGD and Integrins. Science 238: 491-497.

Saunders, J.R., O'Sullivan, H., Wakeman, J., Sims, G., Hart, C.A., Virji, M., Heckels, J.E., Winstanley, C., Morgan, J.A.W. & Pickup, R.W. (1994) Flagella and pili as antigenically variable structures on the bacterial surface. In Microbial Cell Envelopes: Interactions and Biofilms, pp. 33-42, Edited by L.B. Quesnel, P. Gilbert and P.S. Handley, Blackwell Scientific Publications.

Sharon, N. and Jeanloz, R.W. (1960) The diaminohexose component of a polysaccharide isolated from *Bacillus subtilis*. J Biol Chem 235: 1-5.

Stephens, D.S. (1989) Gonococcal and meningococcal pathogenesis as defined by human cell, cell culture and organ culture assays. Clin Microbiol Rev 2: S104-S111.

Stimson, E., Virji, M., Makepeace, K., Dell, A., Morris, H.R., Payne, G., Saunders, J.R., Jennings, M.P., Barker, S., Panico, M., Blench, I. and Moxon, E.R. (1995) Meningococcal pilin: a glycoprotein substituted with digalactosyl 2,4-diacetamido-2,4,6-trideoxyhexose. Mol Microbiol 17(6): 1201-1214.

Stimson, E., Virji, M., Barker, S., Panico, M., Blench, I. Saunders, J.R., Payne, G., Moxon, E.R., Dell, A. and Morris, H.R. (1996) Discovery of a novel protein modification: α-glycerophosphate is a substituent of meningococcal pilin. (Biochem J. in press)

Trust, T.J., Gillespie, R.M., Bhatti, A.R. and White, L.A. (1983) Differences in the adhesive properties of *Neisssseria meningitidis* for human buccal epithilial cells and erythrocytes. Infect Immun 41: 106-113.

Virji, M., Heckels, J.E., Potts, W.J., Hart, C.A. and Saunders, J.R. (1989). Identification of the epitopes recognised by monoclonal antibodies SM1 and SM2 which react with all pili of Neisseria gonorrhoeae but which differentiate between two structural classes of pili expressed by Neisseria meningitidis and the distribution of their encoding sequences in the genomes of Neisseria spp. J Gen Microbiol 135: 3239-3251.

Virji, M., Kayhty, H., Ferguson, D.J.P., Alexandrescu, C., Heckels, J.E. and Moxon, E.R. (1991a) The role of pili in the interactions of pathogenic Neisseria with cultured human endothelial cells. Mol Microbiol 5: 1831-1841.

Virji, M., Kayhty, H., Ferguson, D.J.P., Alexandrescu, C., and Moxon, E.R. (1991b) Interactions of Haemophilus influenzae with cultured human endothelial cells. Microb Pathogen 10: 231-245.

Virji, M., Alexandrescu, C., Ferguson, D.J.P., Saunders, J.R. and Moxon, E.R. (1992a) Variations in the expression of pili: the effect on adherence of Neisseria meningitidis to human epithelial and endothelial cells. Mol Microbiol 6: 1271-1279.

Virji, M., Makepeace, K., Ferguson, D.J.P., Achtman, M., Sarkari, J. and Moxon, E.R. (1992b) Expression of the Opc protein correlates with invasion of epithelial and endothelial cells by Neisseria meningitidis. Mol Microbiol 6: 2785-2795.

Virji, M., Saunders, J.R., Sims, G., Makepeace, K., Maskell, D. and Ferguson, J.P. (1993a) Pilus-facilitated adherence of Neisseria meningitidis to human epithelial and endothelial cells: modulation of adherence phenotype occurs concurrently with changes in amino acid sequence and the glycosylation status of pilin. Mol Microbiol 10: 1013-28.

Virji, M., Makepeace, K., Ferguson, D.J.P., Achtman, M. & Moxon, E.R. (1993b). Meningococcal Opa and Opc proteins: their role in colonisation and invasion of human epithelial and endothelial cells. Mol Microbiol 10(3), 499-510.

Virji, M., Makepeace, K. & Moxon, E.R. (1994a). Distinct mechanisms of interaction of Opc-expressing meningococci at apical and basolateral surfaces of human endothelial cells; the role of integrins in apical interactions. Mol Microbiol 14(1), 173-184.

Virji, M., Makepeace, K. & Moxon, E.R. (1994b). In Neisseria 94. Meningococcal outer-membrane Opc mediates interactions with multiple extracellular matrix components. Proceedings of the Ninth International Pathogenic Neisseria Conference. pp. 263-264. Editors J.S. Evans, S.E. Yost, M.C.J. Maiden and I.M. Feavers.

Virji, M. Makepeace, K., Peak, I., Payne, G., Saunders, J.R., Ferguson, D.J.P. and Moxon, E.R. (1995a) Functional implications of the expression of PilC proteins in meningococci. Mol Microbiol 16(6): 1087-1097.

Virji, M., Makepeace, K., Peak, I.R.A., Ferguson, D.J.P., Jennings, M.P. and Moxon, E.R. (1995b) Opc- and pilus- dependent interactions of meningococci with human endothelial cells: molecular mechanisms and modulation by surface polysaccharides. Mol Microbiol, 18 (4): 741-754.

Virji, M., Stimson, E., Makepeace, K., Dell, A., Morris, H.R., Payne, G., Saunders, J.R. and Moxon, E.R. (1996) Post-translational modifications of meningococcal pili: identification of a common trisaccharide substitution on variant pilins of strain C311. In Microbial Pathogenesis and Immune Response II. Annals of The New York Academy of Sciences. (in press).

Wang, J.F., Caugant, D.A., Morelli, G., Koumare, B. and Achtman, M. (1993) Antigenic and epidemiologic properties of the ET-37 complex of Neisseria meningitidis. J Infect 167(6):1320-1329

Whittaker, D.V., Parolis, L.A.S. and Parolis, H. (1994) *Escherichia coli* K48 capsular polysaccharide: a glycan containing a novel diacetamido sugar. Carbohydr Res 256: 289-301.

Zehavi, U. and Sharon, N. (1973) Structural studies of 4-acetamido-2-amino-2,4,6-trideoxy-D-glucose (N-Acetylbacillosamine), the N-acetyl-diamino sugar of *Bacillus licheniformis*. J. Biol. Chem. 248: 433-438.

14

LIPOPOLYSACCHARIDE'S ROLE IN THE ASSOCIATION OF *SALMONELLA* CELLS TO THE MOUSE INTESTINE STUDIED BY RIBOSOMAL *IN SITU* HYBRIDIZATION

K. A. Krogfelt,[1] T. R. Licht,[1,2] and S. Molin[2]

[1] Department of Gastrointestinal Infections
Statens Seruminstitut
DK-2300 Copenhagen S, Denmark
[2] Department of Microbiology
Technical University of Denmark
DK-2800 Lyngby, Denmark

BACTERIAL ADHESION

The majority of microbes most probably exist in nature in close association with particular surfaces. The adhesive properties of microorganisms were first recognized at the beginning of this century. Since then it has been shown that bacterial adhesion is important in plant and animal hosts, pathogenesis, medical devices, aquatic and soil ecosystems, biodegradation, and industrial processes.

In human and animal pathogenesis bacterial adhesion plays an important role in diseases such as urinary and gastrointestinal infections, respiratory diseases, wound infections, septicemia, infections at the site of artificial joints and other implants, dental plaque and dental cavities; and numerous other infections caused by bacteria. The disease provoking ability of pathogenic microorganisms can be assigned to four major properties of the organism, including the capacity to a) enter the host, b) to resist or not stimulate host defenses c) to acquire essential nutrients for growth and multiplication in/on the host tissues, and finally d) to damage the host. These properties can be ascribed to specific virulence factors, and these often turn out to be surface components of the organism (Krogfelt 1991).

One very important property of the bacteria is their ability to adhere and subsequently colonize cell surfaces of the host. A clear correlation between adhesion and the development of infectious diseases has been demonstrated (rev. Ofek and Beachey 1980). Nevertheless, it also seems to be an important aspect for commensals of the normal flora that are harmless, or even helpful, in their usual habitat but which may become significant agents of disease if and when they establish growth in other habitats. It is important to point out that disease is

Toward Anti-Adhesion Therapy for Microbial Diseases, edited by Kahane and Ofek
Plenum Press, New York, 1996

not caused by adhesion as such, but by other virulence factors produced by the microorganism.

Mechanisms by which bacteria maintain close proximity to the host's cell surface can be roughly categorized as *association*, *adhesion* and *invasion*; according to the degree of intimacy between bacterial and epithelial surfaces (Krogfelt 1993).

The above mechanisms have been studied in detail by numerous research groups mainly *in vitro* on cultured cell lines and/or solid surfaces, e.g. coated with different compounds and *ex vivo* on epithelial cells and/or biopsies. Studies on bacterial adhesion *in vivo* have been very difficult to conduct. Here we present a tool for studying adhesion and colonization *in vivo*.

BACTERIAL ADHESION AND GROWTH IN THE LARGE INTESTINE

The large intestine is the most heavily colonized part of the gastrointestinal tract of mammals. At least 500 different bacterial species are thought to be present at any time in the healthy human intestinal tract and up to 10^{12} bacteria are found in every gram of feces (Borrielo, 1986). In this complex ecosystem the microorganisms co-exist in a fine balance. They must grow slowly enough not to overgrow the host, but fast enough not to be flushed out by the host's intestinal activities, e.g. peristaltic movements, fluid flow etc. (Finegold et al 1983).

Most studies performed on the bacterial flora of the gut has concentrated on the analysis of fecal specimens. Much less information is found on the flora of the cecum or that associated with the intestinal mucosa. *In situ* investigations of the growth physiology and adhesion of intestinal bacteria are therefore of great interest.

By using *in situ* hybridization we have estimated the growth rates of *Escherichia coli* and *S. typhimurium* to 30-80 min in the large intestine and localized specifically the strains in mouse intestinal sections (Poulsen et al 1994 and 1995; Licht et al 1996).

IN SITU HYBRIDIZATION: THE METHOD

Sectioning and Fixation

Tissue specimens from the mouse large and small intestine were prepared either by embedding in tissue glue and freezing in liquid nitrogen for cryostat sections, or by embedding in paraffin for microtome sections. Cryostat sections were fixed in 3% paraformaldehyde immediately after cryostat cutting. Tissue for microtome sections was fixed in formalin prior to embedding.

The sections on slides were air dried and stored at 4°C until they were either stained with Alcian blue PAS and Meyers Hematoxylin, or used for *in situ* rRNA-hybridization.

Bacterial cells growing in laboratory cultures and bacterial cell smears from the ceca of colonized mice were fixed in 3% paraformaldehyde. Fixed cells were stored in storage buffer (Poulsen et al 1994). Hybridizations were carried out as described by Poulsen 1994. Briefly, the specimens on slides were hybridized at 37°C in a hybridization solution containing various concentrations of formamide, depending on the melting point of the probe and on the desired specificity. The slides were then washed twice in washing solutions with decreasing specificity, rinsed quickly in distilled water and air dried.

Probes

Specific probes to *S. typhimurium* (Licht et al 1996) and *E. coli* (Poulsen et al 1993) were designed in order to perform *in situ* hybridization of bacteria in the mouse gut. The specificity of the probes was tested by use of the CHECK-PROBE program (Larsen et al 1993).

Probe EUB338 (Stahl et al 1991) specific to the eubacterial domain was used to visualize the total bacterial population in the intestine of streptomycin-treated mice. The probes were labeled with various fluorocromes, i.e. fluorescein, Lissamin Rhodamine B or CY3 as previously described (Poulsen et al 1994).

Different fluorochromes were assessed for labeling the probes in order to obtain the lowest background binding to the tissue. The use of hydrophobic fluorochromes sometimes resulted in high non-specific binding of the probe to hydrophobic compartments of the mouse epithelial cells. When hydrophilic fluorochromes were tested, a higher signal to noise ratio was obtained. In order to overcome the inherent fluorescence of epithelial cells and material trapped in the mucosal layer, we used narrow band by-pass filters for the emission.

Microscopy and Image Analysis

The hybridizations were visualized as described in detail by Poulsen et al 1994. An Axioplan epi-fluorescence microscope was equipped with filter sets depending on the excitation and emission wave lengths of the fluorocromes.

A slow scan CCD (Charged Coupled Device) camera was used for capturing digitalized images. The camera was operated at -40°C. The integration time for the CCD camera varied between 500 msec and 3-6 sec depending on the intensity of the hybridized probe. Image analysis was performed in order to determine the amount of fluorescence per bacterial cell volume. This ratio corresponds to the ribosomal concentration in individual bacterial cells. Bacterial growth rates can then be estimated from the cellular RNA concentrations, since RNA content is dependent on the growth rate (Poulsen et al 1993, Schaechter et al 1958, Neidhart and Magasanik 1960). The applied software has previously been described (Poulsen et al 1994).

THE STREPTOMYCIN-TREATED MOUSE MODEL

Treatment of conventional mice with streptomycin-containing drinking water removes the indigenous facultative flora, and thereby allows orally introduced streptomycin-resistant strains of *S. typhimurium* SL5319 (Franklin et al 1990) or *E. coli* (Krogfelt et al 1993) to colonize.

This animal model has been studied in great detail and was therefore chosen for the present work (Franklin et al 1990; Krivan et al 1992; Myhal et al 1982 and Wadolkowsky et al 1988).

LOCALIZATION OF *S. TYPHIMURIUM* STRAIN AND ITS LPS-DEFICIENT MUTANT IN THE MOUSE LARGE INTESTINE

Bacterial membrane-associated lipopolysaccharide (LPS) plays a major role in host-pathogen interaction and is generally accepted as an important antigen (Käthy 1985). The importance of LPS in colonization and interaction with intestinal mucosa has also been recognized: For *E. coli*, LPS appears to be a major bacterial adhesin for specific binding to one or more of the glycoprotein receptors present in mouse colonic mucus (Cohen et al 1985). Furthermore, it was observed that LPS is necessary for *Klebsiella pneumoniae* to colonize the gut of germfree chicken (Camprubi 1993) and that LPS enhances persistence of *Salmonella* in the avian intestinal tract (Craven 1994).

The role of LPS in the ability of *S. typhimurium* to colonize the large intestine of streptomycin-treated mice has been studied by McCormick et al 1988, Nevola et al 1985 and 1986. It was reported that the avirulent, streptomycin-resistant *S. typhimurium* SL5316 containing wild-type LPS is a better colonizer of the intestine of streptomycin-treated mice than LPS-deficient mutants of this strain (Nevola et al 1985). In the present study, we used the hybridization method to study the role of LPS in colonization of the mouse intestine, monitoring *in situ* the adhesion and growth characteristics of *Salmonella typhimurium* strain SL5319 which is smooth, and SL5325, which carries an *rfaJ* mutation and is consequently rough.

To investigate the *in vivo* location in the mouse intestinal mucus of the two avirulent *Salmonella* strains, 6 mice were inoculated with 3.0×10^{10} colony forming units of either *S. typhimurium* SL5319 (smooth) or *S. typhimurium* SL5325 (rough). The total number of *Salmonella* per gram of cecum ranged from 10^{10} to 10^{11} in all cases. Histological sections of the cecum and colon were prepared and hybridized.

The total bacterial population in these sections was visualized by the fluorescein labeled (green), eubacterial probe. The *S. typhimurium* cells were visualized by the specific CY3 labeled (red) probe.

In figure 1 the association of the smooth *S. typhimurium* SL5319 cells to cecal epithelial cells is clearly observed. When mice were colonized with the rough strain *S. typhimurium* SL5325 all bacterial cells were found situated in the outer layer of the intestinal mucosa towards the lumen (Licht et al 1996).

The two *S. typhimurium* strains were found to differ markedly in their binding to the host epithelium *in vivo*. In the hybridized cecal and colonic cross-sections, almost no rough cells were found adhering to the epithelial cells while many of the smooth cells found in

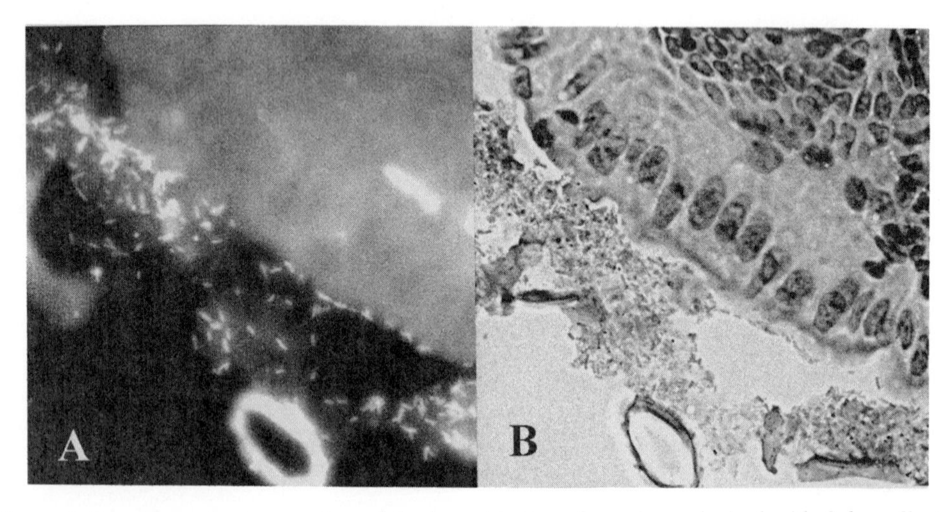

Figure 1. 5 μm thin microtome section of the large intestine of a mouse colonized with *Salmonella typhimurium* SL5319. A) *S. typhimurium* cells are visualized by hybridization with a specific to *Salmonella* fluorescent labeled probe. B) The same section is stained with hematoxylin shown by phase contrast microscopy.

mucus were binding to the cecal and colonic epithelium. Nevertheless, it has previously been shown *in vitro* that the efficiency of binding to MDCK epithelial cell lines (Licht et al 1996) was ten times higher for SL5325 (rough) than for SL5319 (smooth). Furthermore, it was seen that the LPS-deficient mutant bound to preparations of mouse epithelial cells at levels 3 to 5 times higher than the wild-type LPS strain (Nevola et al 1986).

Therefore, it seems highly unlikely that the LPS-deficient *S. typhimurium* cells would not bind to the epithelium *in vivo* if they were able to get in contact with the intestinal wall. Based on these observations we suggest that the difference in binding to the epithelium *in vivo* observed for the two strains is caused by the reduced ability of the rough strain to penetrate the mucus layer. *In vitro* assays have shown that the rough *Salmonella* strain gets trapped in the mucus, while its smooth counterpart does not.

CONCLUDING REMARKS

Application of *in situ* Hybridization in the Gut

Studies on the adhesion of bacterial cells have usually been performed with pure cultures growing in well defined liquid laboratory media under well controlled conditions. In natural environments bacterial growth is influenced by a number of factors that are often unknown. The complexity of the natural ecosystems makes it difficult to mimic the conditions *in vitro*. Therefore, it is important to develop methods by which bacterial growth physiology can be studied *in situ*.

It is clear that *in situ* rRNA hybridization can be applied in the complex environment of the intestine, and reveal useful information about bacterial distribution and bacterial binding sites in the gut. In addition, the concentration of ribosomal RNA in a given bacterial cell as measured by *in situ* hybridization can provide further information about the physiological state of the microorganism.

The correlation between adhesion and pathogenesis has led to the idea of finding means by which adhesion can be prevented. It is therefore important to study bacterial adhesion *in vivo* for developing relevant antiadhesive drugs.

REFERENCES

Borriello, S. P. 1986. Microbial flora of the gastrointestinal tract, p. 2-16. in M. J. Hill (ed.), *Microbial metabolism in the digestive tract.* CRC Press, Inc. Roca Raton, FL.

Camprubi, S., S. Merino, J. F. Guillot, and J. M. Tomás. 1993. The role of the O-antigen lipopolysaccharide on the colonization of in vivo of the germfree chicken gut by *Klebsiella pneumoniae*. Microb. Pathog. 14:433-440.

Cohen, P. S., J. C. Arruda, T. J. Williams, and D. C. Laux. 1985. Adhesion of a human fecal *Escherichia coli* strain to mouse colonic mucus. Infect. Immun. 48:139-145.

Craven S.E. 1994. Altered colonising ability for the cecae of broiler chicks by lipopolysaccharide-deficient mutants of Salmonella typhimurium. Avian Diseas. 38:401-408.

Finegold, S. M., V. L. Sutter and G. E. Mathisen. 1983. Normal indigenous intestinal flora, p. 3-31. in D. J. Hentges (ed.), *Human intestinal microflora in health and disease.* Academic Press, New York.

Franklin, D. P., D. C. Laux, T. J. Williams, M. C. Falk and P. S. Cohen. 1990. Growth of *Salmonella typhimurium* SL5319 and *Escherichia coli* F-18 in mouse cecal mucus: role of peptides and iron. *FEMS Microbiol. Ecol.* 74, p. 229-240.

Käthy, H. 1985. Antibody response to bacterial surface components, p. 96-97. *In* T. K. Korhonen (ed.), E.A. Dawes and P. H. Mäkelä, Enterobacterial surface antigens, FEMS symposion No.25, Elsevier, Amsterdam, New York, Oxford.

Krivan, H. C., D. P. Franklin, W. Wang, D. C. Laux and P. S. Cohen. 1992. Phosphatidylserine found in intestinal mucus serves as a sole source of carbon and nitrogen for salmonellae and *Escherichia coli*. *Infect. Immun.* 60, p. 3943-3946.

Krogfelt K.A. 1991. Bacterial Adhesion: Genetics, biogenesis, and role in pathogenesis of fimbrial adhesins of *Escherichia coli* Rev. Infect. Diseas., 13:721-735,

Krogfelt K. A. 1993. Cellular Adhesion. *In*: McGraw-Hill Yearbook of Science and Technology 1994. McGraw-Hill, Inc. pp 68-70.

Krogfelt, K. A., L. K. Poulsen and S. Molin. 1993. Identification of coccoid *Escherichia coli* BJ4 cells in the large intestine of streptomycin-treated mice. *Infect. Immun.* 61, p. 5029-5034.

Larsen, N., Olsen, G. J., B. L. Maidak, M. J. McCaughey, R. Overbeek, T. J. Macke, T. L. Marsh and C. R. Woese. 1993. The ribosomal RNA database project. *Nuc. Acid Res.* 21, p. 3021-3023.

Licht T.R., K.A. Krogfelt, Cohen P.S., Poulsen L.K. Urbace J.and S. Molin. 1996. Role of lipopolysaccharide in colonization of the mouse intestine by *Salmonella typhimurium* studied by *in situ* hybridization Infect. Immun. submitted.

McCormick, B. A., B. A. D. Stocker, D. C. Laux, and P. S. Cohen. 1988. Roles of motility, chemotaxis, and penetration through and growth in intestinal mucus in the ability of an avirulent strain of *Salmonella typhimurium* to colonize the large intestine of streptomycin-treated mice. Infect. Immun. 56:2209-2217.

Myhal, M. L., D. C. Laux and P. S. Cohen. 1982. Relative colonizing abilities of human fecal and K12 strains of *Escherichia coli* in the large intestines of streptomycin-treated mice. *Eur. J. Clin. Microbiol.* 1, p. 186-192.

Nevola, J. J., B. A. D. Stocker, D. C. Laux, and P. S. Cohen. 1985. Colonization of the mouse intestine by an avirulent *Salmonella typhimurium* strain and its lipopolysaccharide-defective mutants. Infect. Immun. 50:152-159.

Nevola, J. J., D. C. Laux, and P. S. Cohen. 1987. *In vivo* colonization of the mouse large intestine and *in vitro* penetration of intestinal mucus by an avirulent smooth strain of *Salmonella typhimurium* and its Lipopolysaccharide-deficient mutant. Infect. Immun. 55:2884-2890.

Neidhardt, F. C. and B. Magasanik. 1960. Studies on the role of ribonucleic acid in the growth of bacteria. *Biochem. Biophys. Acta* 42, p. 99-116.

Ofek I. and E.H. Beachey. 1980. General concepts and principles of bacterial adherence in animals and man. In: Bacterial Adherence Ed.: E.H. Beachey, Chapman & Hall, p.1-30

Poulsen, L. K., G. Ballard and D. A. Stahl. 1993. Use of rRNA fluorescence in situ hybridization for measuring the activity of single cells in young and established biofilms. *Appl. Environ. Microbiol.* 59, p. 1354-1360.

Poulsen L. K., F. Lan, C. S. Kristensen, P. Hobolth, S. Molin and K.A. Krogfelt. 1994. Spatial distribution of *Escherichia coli* in the mouse large intestine inferred from rRNA *in situ* hybridization. *Infect. Immun.* 62, p. 5191-5194.

Poulsen L. K., T. R. Licht, C. Rang, K.A. Krogfelt and S. Molin, 1995, The physiological state of *E. coli* BJ4 growing in the large intestine of streptomycin-treated mice. *J. Bact.* 177:5840-5845

Schaechter, M., O. Maaløe and N. O. Kjeldgaard, 1958, Dependency on medium and temperature of cell size and chemical composition during balanced growth of *Salmonella typhimurium J. Gen. Microbiol.* 19, p. 592-606.

Stahl, D. A. and R. I. Amann. 1991. Development and application of nucleic acid probes, p. 205-248. in E. Stackebrandt and M. Goodfellow (ed.), *Nucleic acid techniques in bacterial systematics*. John Wiley and Sons, New York.

Wadolkowski, E. A., D. C. Laux and P. S. Cohen. 1988. Colonization of streptomycin-treated mouse large intestine by a human fecal *Escherichia coli* strain: Role of growth in mucus. *Infect. Immun.* 56, p. 1030-1035.

15

INTERACTIONS OF BACTERIAL ADHESINS WITH THE EXTRACELLULAR MATRIX

Åsa Ljungh and Torkel Wadström

Department of Medical Microbiology
Lund University
Sölvegatan 23
S-223 62 Lund, Sweden

1. INTRODUCTION

The role of microbial adhesion to epithelia in mucosa-associated infections has been recognized for a long time. The discovery by Kuusela in 1978[23] that *Staphylococcus aureus* strains specifically bind the extracellular matrix (ECM) and plasma glycoprotein fibronectin, Fn, launched a broad field of research on (i) how microbes, i.e. bacteria, parasites, fungi and viruses, interact with ECM components, and (ii) the role of interaction with these subepithelial structures in the pathogenesis of various infectious diseases[30]. We will here describe some of these interactions, and emphasize some possible targets for anti-adhesion therapy.

2. EXTRACELLULAR MATRIX

The ECM comprises glycoproteins and glycosaminoglycans (GAG:s) which form a network through a number of specific interactions between different ECM components. The composition of ECM differs in various organs but dominating structures are Fn, collagens (Cn) type I-XV, laminin (Ln), and GAG:s like heparan sulphate, chondroitin sulfate, and others[21]. Of these, Ln , type IV Cn and HS are major constituents of basement membranes. Binding of type IV Cn to Ln is mediated by entactin (nidogen). Vitronectin (Vn, S-protein) is found in vessel walls, aging dermis, and the ECM of certain tissues, as well as in plasma. More recently, thrombospondins (TSP:s), major glycoproteins of platelets and developing tissues, have been recognized as ECM proteins, synthesized during cell proliferation[1]. Bone sialoprotein II (BSP II) is an ECM of developing tissue but only found in osseous tissue in adults. Plasminogen (Plg) is a key component of the fibrinolytic system, present in plasma and subepithelial and subendothelial basement membranes. It can be converted to a serine proteinase by a number of activators of bacterial or eukaryotic origin[31]. The ECM serves a structural function but also profoundly affects eukaryotic cell adhesion, differentiation, migration and proliferation. Several ECM components are present in plasma and other body

Toward Anti-Adhesion Therapy for Microbial Diseases, edited by Kahane and Ofek
Plenum Press, New York, 1996

fluids. Fn and Vn have e.g. been defined as constituents of bile[72]. The conformation of some ECM components, like Fn and Vn, differs significantly when the glycoprotein is immobilized in tissues or on surfaces, or appear in soluble form. Reports that microbial adhesins like P fimbriae of *Escherichia coli* and Yad A of *Yersinia enterocolitica* bind immobilized but not soluble Fn may reflect that these adhesins recognize a domain which is exposed only when Fn is immobilized in tissues[51,69].

How does the ECM become exposed to bacteria? One obvious mechanism is trauma disrupting the skin or epithelial integrity. However, viral infections of mucosal surfaces cause exposure of ECM, and bacterial infections likewise, through the action of proteolytic enzymes and toxins (Fig 1). Furthermore, the normal shedding of epithelia (18-24 hrs for gastric epithelium) provides the epithelial surface with ECM components. Finally, pathogenic bacteria may gain access to various body fluids. Bacterial cells which express binding of ECM components may then coat themselves with these which may be an important mechanism to evade the host immune defense mechanisms[18].

3. ENVIRONMENTAL CONTROL

Expression of binding of various ECM components by various bacteria is influenced by culture conditions, and may appear at different stages during growth. *In vitro* growth under nutrient poor conditions, like on the skin, promotes expression of binding of Fn, Vn, Cn, Ln and heparan sulphate by *S.aureus* [28]. *Helicobacter pylori* express binding of ECM components during the stationary phase of growth, and both coccoidal and spiral forms of the organism express binding (T Wadström, unpublished observation). Some factors which influence expression of surface proteins are pH, redox potential, divalent cations and the availability of iron[3,36]. When studying the role of interactions with ECM components it is important to grow the bacteria under *in vivo*-like conditions.

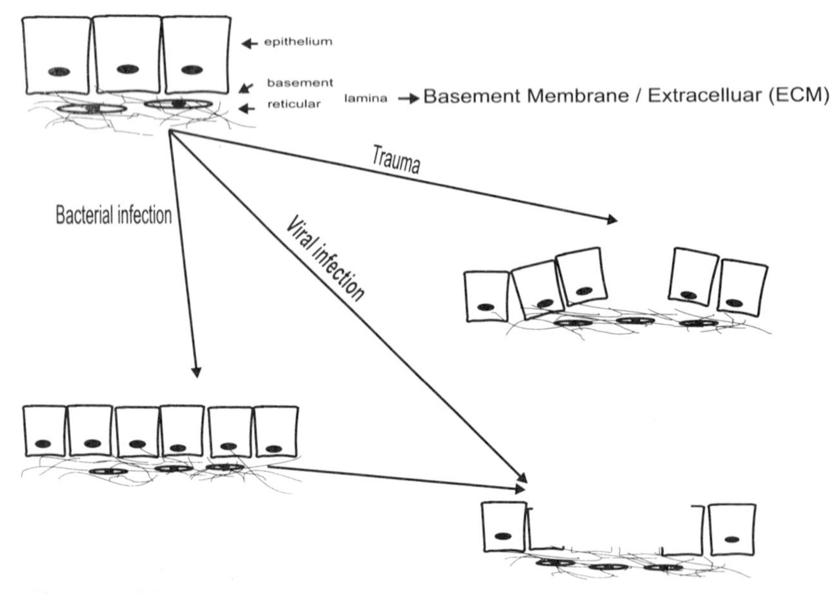

Figure 1. The extracellular matrix components may become exposed by trauma, by viral infection, and by bacterial infection through the activity by toxins and lytic enzymes.

4. INTERACTIONS OF *HELICOBACTER PYLORI* WITH ECM

H. pylori is the causative agent of acute and chronic gastritis, a main determinant for gastric and duodenal ulcer, and strongly associated with gastric cancer[65]. Whether the infection will remain as a superficial infection, or proceed into subepithelial tissues and ECM is likely to depend on strain characteristics as well as on the host immune response. The low toxicity of *H. pylori* LPS with resulting low stimulation of inflammatory mediators[68] may be one important factor in the development of chronic gastritis.

A number of putative adhesins mediating the intial adhesion to gastric mucus and epithelium have been described such as sialic acid-specific hemagglutinins[9,16,26], other hemagglutinins and factors recognizing the Lewis b blood group substance[66]. *H. pylori* was earlier shown to bind ECM components - Cn type IV, Vn, Plg, Ln and heparan sulphate [5,47,59,61].

4.1. Heparan Sulphate Binding

The binding of heparan sulphate and heparin is mediated by specific proteins (with molecular weights between 14 and 65 kDa) on the cell surface of *H. pylori* strains[17]. Binding is optimal at pH 4-5, and is inhibited by sulphated polysaccharides, like heparin, chondroitin and dextran sulphate, fucoidan, pentosan sulphate, and carrageenans but not by carboxylated or nonsulphated compounds, i.e. hyaluronic acid. It is almost abolished when 0.5 M NaCl is added, indicating the involvement of mainly ionic interactions[17]. Heparan sulphate binding to *H. pylori* cells is of high affinity (Fig 2)[5], and does not affect hemagglutinating ability[17]. The earlier reported binding by *H.pylori* strains to sulphatides in the gastric mucus[54] is likely to be mediated by these heparan sulphate binding proteins.

4.2. Laminin Binding

H. pylori, particularly hemagglutinating strains, were earlier shown to bind the basement membrane protein Ln[61]. Further studies showed surface exposed lipopolysaccharide, LPS, to mediate Ln binding., confirming studies by Slomiany and coworkers that *H. pylori* LPS could inhibit binding of Ln to its receptor on epithelial cells[55,62]. The Ln binding by hemagglutinating strains differs from that of non-hemagglutinating strains in that a phosphorylated structure in the core oligosaccharide mediates binding of hemagglutinating strains whereas a conserved nonphosphorylated part of the core oligosaccharide mediates that of nonhemagglutinating strains[62]. The binding domain for LPS on the Ln molecule has not been identified. The earlier proposed lectin-like binding of Ln by *H. pylori*[61] has been shown to be mediated by a 25 kDa protein recognizing sialyllactose on the Ln molecule[62].

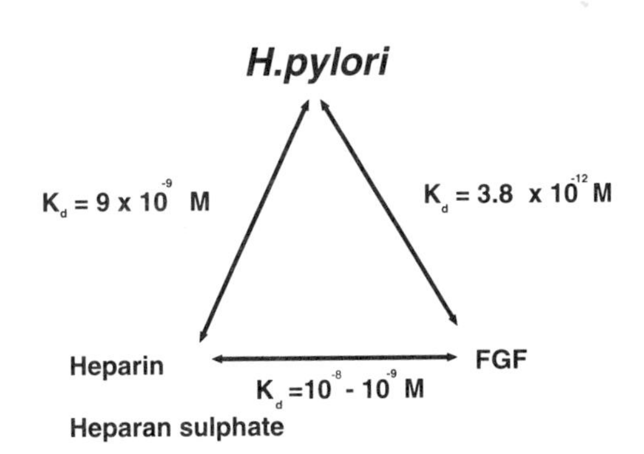

Figure 2. Affinities of heparin-dependant growth factors (FGF) for *H. pylori*, for heparin and heparan sulphate.

Using isomers of sialyllactose it was further shown that *H. pylori* Ln binding protein binds to NeuAc(2-3)Lac and not to NeuAc(2-6)Lac (Moran AP et al., submitted for publication). The 25 kDa protein is present on hemagglutinating as well as of non-hemagglutinating strains. Interestingly, non-hemagglutinating strains bind Ln with a higher affinity (K_d =4.1 pM) than hemagglutinating strains (K_d=8.5 pm) but more binding sites are available on the surface of hemagglutinating strains, thereby explaining the higher extent of Ln binding by these strains. It is now proposed that the initial binding of Ln by *H. pylori* strains is mediated by LPS, and that this is followed by the specific lectin binding (Moran AP et al., submitted for publication). Ln binding by *H. pylori* is unlikely to play a role in the initial adherence mechanisms. However, once adherent to gastric epithelium Ln binding exerted by *H. pylori* LPS may inhibit the binding between Ln and its cell receptor on the gastric epithelium[55,66]. Furthermore, 3-sialyllactose residues on Ln serve an important function in cellular adhesion. The lectin binding to this molecule may further disrupt epithelial cell -basement membrane interactions and create the gastric epithelial leakiness typical of *H. pylori* infections[66].

5. OTHER LECTIN-LIKE INTERACTIONS WITH ECM

Ln is a highly glycosylated molecule. Hence, it is not surprising that other microbes also have been shown to bind Ln by lectin interactions, like the 160 kDa hemagglutinin of *S. saprophyticus*, recognizing N-acetyllactoseamine[13], S-fimbriae of *E. coli* recognizing sialyl (α2-3) galactose, and the Fim H of Type 1 fimbriae mannose[22,64]. Interestingly, also Lactobacillus strains, which form part of the normal intestinal flora, were shown to bind Cn, and this was proposed as a means of colonization of the gut epithelium[2,58]. Glycosidase treatment of Cn abolished binding indicating that Lactobacillus strain recognize carbohydrate structures on the Cn molecule[2].

6. INTERACTIONS OF STAPHYLOCOCCI WITH ECM

Since the first description of Fn binding by *S. aureus*[23] specific binding of Cn, BSPII, TSP, Vn, Plg, Ln and heparin by *S. aureus* as well as by coagulase-negative staphylococci (CNS) has been reported[30,67]. Most information is available regarding Fn and Cn binding which will be summarized below.

6.1. Fibronectin Binding by Staphylococci

Fn binding by *S. aureus* is mediated by two surface exposed proteins with MW of 110 kDa , named FNBP-A and FNBP-B. The genes for these have been cloned and sequenced[19]. Significant homology is found between FNBP-A and B as well as with FNBP of *Streptococcus pyogenes,* with one of the FNBP:s of *S. dysgalactiae*[39] (Fig 3), and to a lesser extent with the PapE of P fimbriae which also mediates binding to Fn[69]. Most bacterial strains bind to the N-terminal domain of the Fn molecule. However, the presence of a second binding domain in the C-terminal part has been clearly demonstrated[49]. Fn binding by Grampositive organisms has been the target for vaccines as new prophylaxis against bovine mastitis. Fusion proteins encompassing the *S. aureus* FNBP-A gave significant protection in a mouse model as determined by reduced recovery of challenge organisms and histopathological mastitis[33]. Vaccination with FNBP-A combined with staphylococcal α-toxoid did not improve the protection. Vaccine trials in cows are now conducted using FNBP-A with iscomes or purified Quil A (batch 21) with promising results (Cambridge Biotech., Woster, Mass. USA and Virbact, Nice, France). Studies are in progress to elucidate if

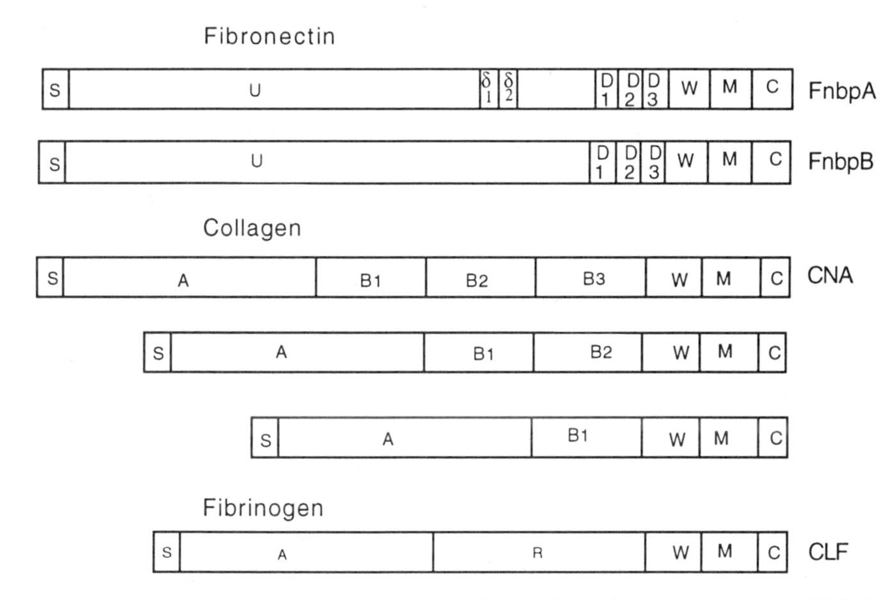

Figure 3. Cloned ECM binding molecules of *S. aureus* - fibronectin binding protein A and B, FNBP-A and B, collagen binding protein, CNBP; fibrinogen binding (clumping factor)[35]. W = cell wall spanning domain; M = membrane spanning domain (hydrophobic); C = carboxy terminus; S = signal sequence; U or A = unique nonrepetitive sequence; δ = upstream repeat sequence; D, B or R = repeated domain.

antibodies to FNBP-A actually block Fn binding, or if they stimulate opsonization[48]. Another potential application for vaccination with FNBP:s is endocarditis where strains of *S. aureus* as well as of *S. sanguis* with reduced or lacking Fn binding showed significant reduced ability to induce endocarditis in rat and rabbit models[25,32]. On the other hand, isogenic strains of *S. aureus* differing in Fn binding, did not differ in their ability to induce endocarditis in a rat model, showing the multifactorial nature of bacterial adhesion in the pathogenesis of endocarditis[11]. We have earlier shown that administration of Fn-substituted Sepharose beads to wounds infected with *S. aureus* resulted in reduced healing time and reduced bacterial counts. The effect was potentiated if protease inhibitors was administered concomitant with the Fn-beads[29]. These data suggest that Fn-substituted matrices could be used also as a prophylaxis for wound infections, e.g. in surgery.

6.2. Collagen Binding by Staphylococci

Cn binding is commonly expressed by staphylococcal strains. In contrast to Fn binding, only one gene encoding Cn binding has been identified[41]. Phenotypically, however, strains can express Cn binding proteins with MW:s of 133, 110 and 87 kDa. This has been interpreted as the presence of 1-3 B repeats, possibly as a result of gene duplication (Fig 3). However, the B repeats are not involved in binding to the Cn molecule, and consequently strains expressing CNBP:s with different MW:s do not differ in their Cn binding ability[57]. Cn is the major constituent of cartilage. The role of Cn binding by *S. aureus* in arthritis was elucidated using polystyrene beads coated with purified CNBP. Beads coated with CNBP adhered to cartilage wheras beads coated with FNBP did not[57]. These *in vitro* studies were followed up in a mouse model of arthritis where *S. aureus* expressing Cn binding induced severe arthritis whereas its isogenic mutant, lacking Cn binding did not[40].

7. EFFECTS OF BACTERIAL BINDING OF ECM COMPONENTS

While bacterial binding of any ECM molecule immobilized in tissue may induce conformational changes causing intracellular signalling with different biological effects, biological effects of binding of heparan sulphate and plasminogen will be discussed.

7.1. Effects of Bacterial Binding of Heparan Sulphate and Heparin

The important wound pathogens, *S. aureus* and *S. pyogenes* of the skin and *H. pylori* of gastric mucosa, have all been shown to bind HS and heparin[5,27]. They also bind heparin dependent growth factors, basic fibroblast growth factor, bFGF, acidic FGF, aFGF, and platelet derived growth factor (PDGF) to varying extents[4,38]. For *H. pylori* cells the binding of heparan sulphate was of much higher affinity ($K_d = 9 \times 10^{-9}$ M) than the binding of heparin to FGF (Kd = 10^{-8} -10^{-9} M) but with lower affinity than the binding of aFGF by *H. pylori* (K_d 3.8×10^{-12}M) (Fig 2)[4]. The affinity of *S. aureus* and group A,C,G streptococci cells for heparan sulphate was lower, about K_d 10^{-5} M[27].

7.2. Effects of Bacterial Binding of Plasminogen

Specific binding of plasminogen was first described for *Y. pestis*[56]. Since then specific binding has been described for *S. aureus* and group A streptococci[24], S-fimbriae and curly fimbriae of *E. coli*, *Salmonella enterica*[53] and *H. pylori*[47]. In group A streptococci a 42 kDa protein, related to M protein[7,8] mediates binding of Plg. Surface-bound Plg is activated by streptokinase[7]. With the other bacterial species the mechanisms of binding of Plg have not been elucidated. However, surface-bound Plg is commonly converted to plasmin which is insensitive to tissue plasmin inactivators, and provide the bacterial cell with a potent proteolytic determinant which has been proposed to enhance tissue invasion[20,31].

8. IMPLICATIONS FOR THERAPY

Targeting therapy towards microbial interactions with ECM components implies attempts to prevent deeper, tissue invasive infections and could result in minimized effect on innocuous bacterial colonization than therapy directed towards initial adhesion by bacteria to mucosal surfaces. With the increasing prevalence of antibiotic resistant strains of various bacterial species targeting therapy towards ECM binding is a promising alternative regime. Our recent observation that heparin as well as a number of polysulphated agents (suramin, fucoidan, pentosan polyphosphate and aromatic dyes like Evans blue and Congo red) inhibited binding of Fn, Vn, TSP, and of lactoferrin by *S. aureus* strains[37], provide alternatives to antibiotic for treatment of *S. aureus* induced wounds and superficial infections. *In vivo* heparin binding by bacteria may result in lack of activation of heparin-dependant growth factors and reduced availability of these factors. This may implicate that heparin and various growth factors should form part of complimentary therapy of wound infections of the skin. Likewise, these agents should be considered as complimentary or alternative treatment of ulcers of the gastrointestinal tract, i.e. *H. pylori*-induced ulcers of the upper intestine, and ulcers in the large bowel in ulcerative colitis[6,12,52,60,66].

9. BIOMATERIAL-ASSOCIATED INFECTIONS

Clinical applications of biomaterials (polymers, metals and cerams) for permanent or temporary use in the human body is a rapidly expanding field. Beside adverse tissue reactions, thrombus formation and complement activation, infections constitute a serious complication[67]. The incidence of infections is higher in tissue-penetrating devices, and indeed low-virulent microorganisms of the normal skin flora, like CNS strains, are dominating etiological agents[67]. Within seconds after introduction in the body host proteins and GAG:s adsorb to the biomaterial surface. The adsorption depends on characteristics of the biomaterial, such as surface hydrophobicity, charge, ultrastructure and rigidity but also on the composition of the host fluid and if the biomaterial is exposed to the host fluid under static or perfusion conditions[10,71]. With time the surface exposure of proteins varies, partly due to postadsorptional changes of configuration of molecules like Fn and Vn, and in part due to formation of layers of proteins because of protein-protein interactions between ECM and plasma proteins[10,21]. This implies that the exposure of potentially bacteria binding domains also varies with time.

9.1. Pathogenesis of Biomaterial-Associated Infections

The first step in the pathogenesis of biomaterial-associated infections is that bacteria gain access to the device during the surgical implantation, or at a later stage, and binds to adsorbed protein or GAG molecules by receptor-specific or e.g. hydrophobic interaction. Once adherent, the bacteria turn down their metabolism, grow slowly as microcolonies, and start to produce the extracellular polysaccharide, glycocalyx or slime, which serves a protective function towards host defense mechanisms and antibiotics, and also provides the bacteria with nutrients[34,46]. As a consequence, biomaterial-associated infections are in general impossible to treat with antibiotics, and the device has to be extirpated[14]. We are thus left with preventive measures to block adhesion and colonization of bacteria on these surfaces. Antibiotics and disinfectants have been tried but none has gained widespread use [50]. The use of antibiotics carries the risk of inducing antibiotic resistance, and of allergic reactions.

9.2. Fibronectin Mediating Bacterial Adhesion to Biomaterial Surfaces

Fn and TSP, as well as fibrinogen have been proposed to mediate adhesion of bacteria to implanted devices[15,63]. Indeed, CNS strains express binding of Fn and TSP, and in general bind the immobilized form of ECM proteins to a higher extent than the soluble form [44]. This could imply that they recognize a domain on the ECM molecule which is preferentially exposed on the immobilized form. So far, however, no other binding domain than the N-terminal of Fn has been identified for CNS strains. We have earlier shown that Fn and fibrinogen are exposed on the surface of central venous catheters extirpated from patients[73], and that preincubation of catheters, extirpated from the biliary duct of rats, with antibodies to Fn reduced subsequent adhesion of *E. coli*[70] In a similar study on staphylococcal adhesion to polymers in the presence of cerebrospinal fluid, however, anti-Fn antibodies did not reduce adhesion of any CNS strain (Lundberg F et al., submitted for publication).

9.3. Vitronectin Mediating Bacterial Adhesion to Biomaterial Surfaces

Vn, another adhesive ECM protein, has been shown preferentially to adsorb to rigid surfaces, and to adsorb to a proportionally higher extent than other proteins[10]. We have recently confirmed this in a study on polymers exposed to cerebrospinal fluid, and shown that preincubation of polymers with anti-Vn antibodies reduced adhesion of two CNS strains (Fig.4). Furthermore, preincubation of bacterial cells with Vn reduced subsequent adhesion of 11 out of 19 strains (Lundberg F et al., submitted for publication). Bacterial binding of Vn is of particular interest in association with biomaterial infections since this "consumption" of Vn inhibits an important down-regulatory mechanism of the complement system, and may result in enhanced complement activation, a serious complication to e.g. the use of extracorporeal circulation[45].

Studies on bacterial adhesion to polymer surfaces show that CNS strains vary in their ability to adhere to polymer surfaces both quantitatively and qualitatively[4243,71]. Whereas protease treatment significantly reduced adhesion of several strains, adhesion of other strains is not affected[43,71]. This may reflect that proteinaceous as well as carbohydrate moieties are involved in the binding, and emphasizes the heterogeneity of these organisms. The way proteins adsorb determines if they expose domains which bind and/or activate components of the coagulation and complement systems, eukaryotic cell binding, and induce an inflammatory reaction. Microbial adhesion to surfaces may interfere with these biological systems, and with binding and activation of different growth factors[38], by virtue of specific binding to active components, or by steric hindrance of exposure of a specific domain.

9.4. Future Perspectives

Treatment of established biomaterial-associated infections usually fails. Hence, the most promising perspective is to create biomaterial surfaces which are less prone to become colonized and infected. Before anti-bacterial surface coatings can be optimized for applica-

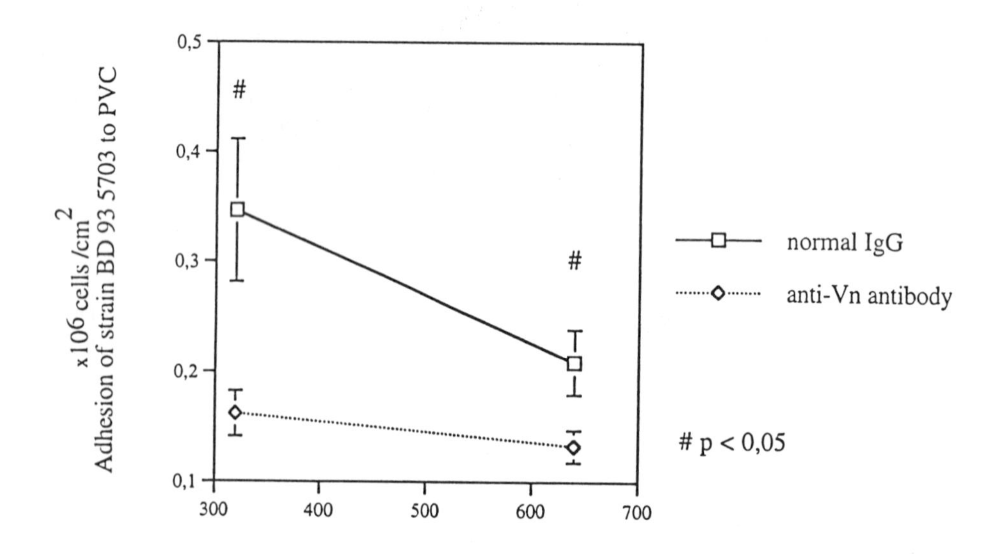

Figure 4. Adhesion of *S. epidermidis* to polyvinyl chloride preperfused with cerebrospinal fluid with 10% human plasma (24 hrs), and preincubated with antibodies to human Vn. Adhesion was significantly inhibited by anti-Vn antibodies but not by normal IgG ($p < 0.05$).

tion in various body compartment the most common binding mechanisms have to be defined. It is also evident that we have to tailor-make surfaces for application in different compartments since the protein composition of these differs, and also the demands for e.g. eukaryotic cell adhesion. Since any intervention with protein domains to block bacterial adhesion on biomaterial surfaces may interfere with important biological functions it is of vital importance that such surfaces are subjected to detailed studies with respect to complement and platelet activation, inflammatory response and tissue compatibility before introduction in the clinics.

ACKNOWLEDGMENTS

Studies by the authors were supported by grants from the Swedish Medical Research Council (16x-04723 and 11229), the Board for Technical Development (NUTEK), and the Faculty of Medicine, Lund University, Lund, Sweden.

REFERENCES

1. Adams, J. C. and J. Lawler. 1993. The thrombospondin family. Curr. Biol. 3:188-190.
2. Aleljung, P., W. Shen, B. Rozalska, U. Hellman, Å. Ljungh, and T. Wadström. 1994. Purification of collagen binding proteins of *Lactobacillus reuteri*. Curr. Microbiol. 28:231-236.
3. Arvidson, S., L. Janzon and S. Löfdahl. 1990. The role of the δ-lysin gene (*hld*) in the *agr*-dependent regulation of exoprotein synthesis in *Staphylococcus aureus*. p. 421-433. In R. P. Novick. (ed.), Molecular Biology of the Staphylococci. VCH Publ, New York.
4. Ascencio, F., H.-A. Hansson, O. Larm, and T. Wadström. 1995. *Helicobacter pylori* interacts with heparin and heparin-dependent growth factors. FEMS Immunol. Med. Microbiol. 12:265-272.
5. Ascencio, F., O. D. Liang, L.-Å. Fransson, and T. Wadström. 1993. Affinity of the gastric pathogen *Helicobacter pylori* for the N-sulphated glycosaminoglycan heparan sulphate. J.Med.Microbiol. 38:240-244.
6. Bazzoni, G., A. B. Nunez, G. Mascellani, P. Bianchini, E. Dejana, and A. Del Maschio. 1993. Effect of heparin, dermatan sulfate, and related oligo-derivatives on human polymorphonuclear leukocyte functions. J. Lab. Clin. Med. 121:268-275.
7. Berge, A. and U. Sjöbring. 1993. PAM, a novel plasminogen-binding protein from *Streptococcus pyogenes*. J. Biol. Chem. 268:25417-25424.
8. Carlsson Wistedt, A., U. Ringdahl, W. Müller-Esterl, and U. Sjöbring. 1995. Identification of a plasminogen-binding motif in PPAM, a bacterial surface protein. Molec. Microbiol. 18:569-578.
9. Evans, D. G., D. J. Evans,Jr, J. J. Mouls, and D. Y. Graham. 1988. N-acetylneuraminyllactose-binding fibrillar hemagglutinin of Campylobacter pylori: a putative colonization factor antigen. Infect. Immun. 56:2896-2906.
10. Fabrizius-Homan, D. J. and S. L. Cooper. 1991. Competitive adsorption of vitronectin with albumin, fibrinogen and fibronectin on polymeric biomaterials. J. Biomed. Mat. Res. 25:953-971.
11. Flock, J.-I., S. A. Hienz, A. Heimdahl, and T. Schennings. 1996. Reconsideration of the role for fibronectin binding in endocarditis caused by *Staphylococcus aureus*. Infect. Immun. in press.
12. Gaffney, P. R., J. J. O'Leary, C. T. Doyle, A. Gaffney, J. Hogan, F. Smew, and P. Annis. 1991. Response to heparin in patients with ulcerative colitis. Lancet 337:238-239.
13. Gatermann, S. and H.-G. W. Meyer. 1994. *Staphylococcus saprophyticus* hemagglutinin binds fibronectin. Infect. Immun. 62:4556-4563.
14. Gristina, A. G. 1987. Biomaterial-centered infection: Microbial adhesion versus tissue integration. Science 237:1588-1595.
15. Herrmann, M., S. J. Suchard, L. A. Boxer, F. A. Waldvogel, and P. D. Lew. 1991. Thrombospondin binds to *Staphylococcus aureus* and promotes staphylococcal adherence to surfaces. Infect. Immun. 59:279-288.
16. Hirmo, S., S. Kelm, R. Schauer, B. Nilsson, and T. Wadström. 1996. Adhesion of *Helicobacter pylori* strains to α-2,3-linked sialic acids. Glycoconjugate J, in press.

17. Hirmo, S., M. Utt, M. Ringnér, and T. Wadström. 1995. Inhibition of heparan sulphate and other glycosaminoglycans binding to *Helicbacter pylori* by various polysulphated carbohydrates. FEMS Immunol. Med. Microbiol. 10:301-306.
18. Höök, M., L. M. Switalski, T. Wadström and M. Lindberg. 1989. Interactions of pathogenic microorganisms with fibronectin. p. 295-308. In D. F. Mosher. (ed.), Fibronectin. Academic Press, San Diego.
19. Jönsson, K., C. Signäs, H.-P. Müller, and M. Lindberg. 1991. Two different genes encode fibronectin-binding proteins in *Staphylococcus aureus*. The complete nucleotide sequence and characterization of a second gene. Eur. J. Biochem. 292:1041-1048.
20. Korhonen, T. K., R. Virkola, K. Lähteenmäki, Y. Björkman, M. Kukkonen, T. Raunio, A.-M. Tarkkanen, and B. Westerlund. 1992. Penetration of fimbriate enteric bacteria through basement membranes: A hypothesis. FEMS Microbiol. Immunol. 100:3072-3125.
21. Kreis, T. and R. Vale. 1993. Guidebook to the Extracellular Matrix and adhesion proteins. Oxford University Press, Oxford, New York and Tokyo.
22. Kukkonen, M., T. Raunio, R. Virkola, K. Lähteenmäki, P. H. Mäkelä, B. Westerlund, and T. K. Korhonen. 1993. Basement membrane carbohydrate as a target for bacterial adhesion: binding of type I fimbriae of *Salmonella enterica* and *Escherichia coli* to laminin. Molec. Microbiol. 7:229-237.
23. Kuusela, P. 1978. Fibronectin binds to *Staphylococcus aureus*. Nature 276:718-720.
24. Kuusela, P., S.-F. Kaukoranta-Tolvanen, M. Ullberg, G. Kronvall and O. Saksela. 1994. Surface associated activation of plasminogen on staphylococci and streptococci, and novel mechanisms for bacteriemia to use the activator system of the host. p. 97-102. In T. Wadström, I. A. Holder, and G. Kronvall. (ed.), Molecular pathogenesis of surgical infections. Gustav Fischer Verlag, Stuttgart.
25. Kuypers, J. M. and R. A. Proctor. 1989. Reduced adherence to traumatized rat heart valves by a low fibronectin-binding mutant of *Staphylococcus aureus*. Infect. Immun. 57:2306-2312.
26. Lelwala-Guruge, J., F. Ascencio, A. S. Kreger, Å. Ljungh, and T. Wadström. 1993. Isolation of a sialic acid-specific surface hæmagglutinin of *Helicobacter pylori* strain NCTC 11637. Zbl Bakt 280:93-106.
27. Liang, O. D., F. Ascencio, L.-Å. Fransson, and T. Wadström. 1992. Binding of heparan sulfate to *Staphylococcus aureus*. Infect. Immun. 60:899-906.
28. Liang, O. D., F. Ascencio, R. Vazquez-Juarez, and T. Wadström. 1993. Binding of collagen, fibronectin, lactoferrin, laminin, vitronectin and heparan sulfate to *Staphylococcus aureus* strain V8 at various growth phases and under nutrient stress conditions. Zbl Bakt 279:180-190.
29. Ljungh, Å., T. Kronevi and T. Wadström. 1990. Fibronectin-substituted gels for treatment of experimental wound infections in a pig model. p. 163-168. In T. Wadström, I. Eliasson, I. A. Holder, and Å. Ljungh. (ed.), Pathogenesis of wound and biomaterial-associated infections. Springer Verlag, London.
30. Ljungh, Å. and T. Wadström. 1995. Binding of extracellular matrix proteins by microbes. p. 501-514. In R. J. Doyle and I. Ofek. (ed.), Methods of Enzymology. Microbial adhesion. Academic Press, New York.
31. Lottenberg, R., D. Minning-Wenz, and M. D. P. Boyle. 1994. Capturing host plasmin(ogen): a common mechanism for invasive pathogens? Trends Microbiol 2:20-24.
32. Lowrance, J. H., L. M. Baddour, and W. A. Simpson. 1990. The role of fibronectin binding in the rat model of experimental endocarditis caused by *Streptococcus sanguis*. J. Clin. Invest. 86:7-13.
33. Mamo, W., P. Jonsson, J.-I. Flock, M. Lindberg, H.-P. Muller, T. Wadström, and L. Nelson. 1994. Vaccination against *Staphylococcus aureus* mastitis: immunological response of mice vaccinated with fibronectin-binding protein (FnBP-A) to challenge with *S.aureus*. Vaccine 12:988-992.
34. Marshall, K. C. 1992. Biofilms: An overview of bacterial adhesion, activity, and control at surfaces. ASM News 58:202-207.
35. McDevitt, D., P. Francois, P. Vaudaux, and T. J. Foster. 1994. Molecular characterization of the clumping factor (fibrinogen receptor) of *Staphylococcus aureus*. . Molec. Microbiol. 11:237-248.
36. Modun, B., P. Williams, W. J. Pike, A. Cockayne, J. P. Arbuthnott, R. Finch, and S. P. Denyer. 1992. Cell envelope proteins of *Staphylococcus epidermidis* grown in vivo in a peritoneal chamber implant. Infect. Immun. 60:2551-2553.
37. Pascu, C., S. Hirmo, Å. Ljungh, and T. Wadström. 1995. Inhibition of extracellular matrix and serum protein binding to *Staphylococcus aureus* by heparin and various polysulphated agents. Med. Microbiol. Lett. 4:397-405.
38. Pascu, C., Å. Ljungh, and T. Wadström. 1996. Staphylococci bind heparin-binding host growth factors. Curr. Microbiol. 32:000-000.
39. Patti, J. M., B. L. Allen, M. J. McGavin, and M. Höök. 1994. MSCRAMM-mediated adherence of microorganisms to host tissues. Annual Review of Microbiology 48:585-617.
40. Patti, J. M., T. Bremell, D. Krajewska-Pietrasik, A. Abdelnour, A. Tarkowski, C. Rydén, and M. Höök. 1994. The *Staphylococcus aureus* collagen adhesin is a virulence determinant in experimental septic arthritis. Infect. Immun. 62:152-161.

41. Patti, J. M., H. Jonsson, B. Guss, L. M. Switalski, K. Wiberg, M. Lindberg, and M. Höök. 1992. Molecular characterization and expression of a gene encoding a *Staphylococcus aureus* collagen adhesive protein. J. Biol. Chem. 267:4766-4772.
42. Paulsson, M., I. Gouda, O. Larm, and Å. Ljungh. 1994. Adherence of coagulase-negative staphylococci to heparin and other glycosaminoglycans immobilized on polymer surfaces. J. Biomed. Mat. Res. 28:311-317.
43. Paulsson, M., M. Kober, C. Freij-Larsson, M. Stollenwerk, B. Wesslén, and Å. Ljungh. 1993. Adhesion of staphylococci to chemically modified and native polymers and the influence of preadsorbed fibronectin, vitronectin and fibrinogen. Biomaterials 14:845-853.
44. Paulsson, M., Å. Ljungh, and T. Wadström. 1992. Rapid identification of fibronectin, vitronectin, laminin and collagen cell surface binding proteins on coagulase-negative staphylococci by particle agglutination assays. J. Clin. Microbiol. 30:2006-2012.
45. Pekna, M., R. Larsson, B. Formgren, U. R. Nilsson, and B. Nilsson. 1993. Complement activation by polymethylmethacrylate minimized by end-point heparin attachment. Biomaterials 14:189-192.
46. Peters, G., R. Locci, and G. Pulverer. 1982. Adherence and growth of coagulase-negative staphylococci on surfaces of intravascular catheters. J. Infect. Dis. 146:479-482.
47. Ringnér, M., K. H. Valkonen, and T. Wadström. 1994. Binding of vitronectin and plasminogen to *Helicobacter pylori*. FEMS Microbiol. Immunol. 9:29-34.
48. Rozalska, B. and T. Wadström. 1993. Protective opsonic activity of antibodies against fibronectin-binding proteins (FnBPs) of *Staphylococcus aureus*. Scand. J. Immunol. 37:575-580.
49. Sakata, N., B. Rozalska, and T. Wadström. 1994. Immunological recognition of fibronectin-binding proteins of *Staphylococcus aureus* and *Staphylococcus capitis, strain LK499*. Microbiol. Immunol. 38:359-366.
50. Schierholz, J., B. Jansen, L. Jaenicke, and G. Pulverer. 1994. *In-vitro* efficacy of an antibiotic releasing silicone ventricle catheter to prevent shunt infection. Biomaterials 15:996-1000.
51. Schulze-Koops, H., H. Burkhardt, J. Heesemann, T. Kirsch, B. Swoboda, C. Bull, S. Goodman, and F. Emmrich. 1993. Outer membrane protein YadA of enteropathogenic Yersiniæ mediates specific binding to cellular but not plasma fibronectin. Infect. Immun. 61:2513-2519.
52. Shen, W., H. Steinrück, and Å. Ljungh. 1995. Expression of binding of plasminogen, thrombospondin, vitronectin, and fibrinogen, and adhesive properties by *Escherichia coli* strains isolated from patients with colonic diseases. Gut 36:401-406.
53. Sjöbring, U., G. Pohl, and A. Olsén. 1994. Plasminogen, adsorbed by Escherichia coli expressing curli or by *Salmonella enteritidis* expressing thin aggregative fimbriae, can be activated by simultaneously captured tissue-type plasminogen activator (t-PA). Molec. Microbiol. 14:443-452.
54. Slomiany, B. L., V. L. N. Murty, J. Piotrowski and A. Slomiany. 1994. Effects of antiulcer agents on the physicochemical properties of gastric mucus. p. 179-192. In E. Chantler and N. A. Ratcliffe. (ed.), Mucus and related topics. Soc Experim Biol, Cambridge.
55. Slomiany, B. L., J. Piotrowski, S. Sengupta, and A. Slomiany. 1991. Inhibition of gastric mucosal laminin receptor by *Helicobacter pylori* lipopolysaccharide. Biochem. Biophys. Res. Commun. 175:963-970.
56. Sodeinde, O., Y. Subrahmanyam, K. Stark, T. Quan, Y. Bao, and J. Goguen. 1992. A surface protease and the invasive character of plague. Science 258:1004-1007.
57. Switalski, L. M., J. M. Patti, W. Butcher, A. G. Gristina, P. Speziale, and M. Höök. 1993. A collagen receptor on *Staphylococcus aureus* strains isolated from patients with septic arthritis mediates adhesion to cartilage. Molec. Microbiol. 7:99-107.
58. Toba, T., R. Virkola, B. Westerlund, Y. Björkman, J. Sillanpää, T. Vartio, N. Kalkkinen, and T. K. Korhonen. 1995. A collagen-binding S-layer protein in Lactobacillus crispatus. Appl. Environm. Microbiol. 61:2467-2471.
59. Trust, T. J., P. Doig, L. Emödy, Z. Kienle, T. Wadström, and P. O'Toole. 1991. High-affinity binding of the basement membrane proteins collagen type IV and laminin to the gastric pathogen *Helicobacter pylori*. Infect. Immun. 59:4398-4404.
60. Tyrrell, D. J., S. Kilfeather, and C. P. Page. 1996. Therapeutic uses of heparin beyond its traditional role as an anticoagulant. Trends in Polysaccharide Science 16:198-204.
61. Valkonen, K. H., M. Ringnér, Å. Ljungh, and T. Wadström. 1993. High-affinity binding of laminin by *Helicobacter pylori*: Evidence for a lectin-like interaction. FEMS Immunol. Med. Microbiol. 7:29-38.
62. Valkonen, K. H., T. Wadström, and A. P. Moran. 1994. Interaction of lipopolysaccharide of *Helicobacter pylori* with basement membrane protein laminin. Infect. Immun. 62:3640-3648.
63. Vaudaux, P., D. Pittet, A. Haeberli, P. G. Lerch, J.-J. Morgenthaler, R. A. Proctor, F. A. Waldvogel, and D. P. Lew. 1993. Fibronectin is more active than fibrin or fibrinogen in promoting *Staphylococcus aureus* adherence to inserted intravascular catheters. J. Infect. Dis. 167:633-641.

64. Virkola, R., J. Parkkinen, J. Hacker, and T. K. Korhonen. 1993. Sialyloligosaccharide chains of laminin as an extracellular matrix target for S fimbriæ of *Escherichia coli*. Infect. Immun. 61:4480-4484.
65. Wadström, T. 1995. An update on *Helicobacter pylori*. Current Science 11:69-75.
66. Wadström, T., S. Hirmo, and T. Borén. 1996. Biochemical aspects of *Helicobacter pylori* colonization of the human gastric mucosa. Alimentary Pharmacology and Therapy 10:in press
67. Wadström, T., M. Paulsson and Å. Ljungh. 1994. Molecular pathogenesis of staphylococcal infections: Microbial adhesion to extracellular matrix and colonization of wounded tissues and biomaterial surfaces. p. 343-352. In R. Möllby, J.-I. Flock, C. E. Nord, and B. Christensson. (ed.), Staphylococcal Infections. Gustav Fischer Verlag, Stuttgart.
68. Wadström, T., J. Rydberg, B. Rozalska, and J. Lelwala-Guruge. 1994. Intravenous *Helicobacter pylori* induces low levels of TNF-α and IL-1α in a murine model. APMIS 102:49-52.
69. Westerlund, B., I. van Die, C. Kramer, P. Kuusela, H. Holthöfer, A.-M. Tarkkanen, R. Virkola, N. Riegman, H. Bergmans, W. Hoekstra, and T. K. Korhonen. 1991. Multifunctional nature of P fimbriæ of uropathogenic *Escherichia coli*: mutations in *fsoE* and *fsoF* influence fimbrial binding to renal tubuli and immobilized fibronectin. Molec. Microbiol. 5:2965-2975.
70. Yu, J., R. Andersson, L.-Q. Wang, S. Bengmark, and Å. Ljungh. 1995. Fibronectin on the surface of biliary drain materials - a role in bacterial adherence. J. Surg. Res. 59:000-000.
71. Yu, J., M. Nordman-Montelius, M. Paulsson, I. Gouda, O. Larm, L. Montelius, and Å. Ljungh. 1994. Adhesion of coagulase-negative Staphylococci and adsorption of plasma proteins to heparinized polymer surfaces. Biomaterials 15:805-814.
72. Yu, J. L., R. Andersson, and Å. Ljungh. 1995. Protein adsorption and bacterial adhesion to biliary stent materials. J Surg Res, in press.

16

PROTEINS F1 AND F2 OF *STREPTOCOCCUS PYOGENES*

Properties of Fibronectin Binding

Emanuel Hanski, Joseph Jaffe, and Vered Ozeri

Department of Clinical Microbiology
The Hebrew University-Hadassah Medical School
Jerusalem 91010, Israel

INTRODUCTION

Microbial adhesion to host tissues is the initial event in the pathogenesis of most infections and, as a such, is an attractive target for the development of new antimicrobial therapeutics. Fibronectin (Fn) is a multifunctional glycoprotein present in soluble form in plasma and other body fluids and in insoluble form in extracellular matrices (ECM) and basement membranes (Mosher, 1989; Hynes, 1990). Since it binds specifically to a variety of receptors and substrates molecules, many bacterial pathogens including staphylococci, *Escherichia coli*, *Treponema pallidum*, mycobacteria and streptococci have exploited the binding properties of Fn to acquire a mechanism for adherence, colonization and subsequent invasion of host tissues (Patti *et al.*, 1994). Thus, an understanding of the molecular details of bacteria-Fn interactions will not only provide considerable insight to the pathogenesis of many bacterial diseases, but will also serve as a powerful model system for structural and functional analyses of Fn itself.

In Gram-positive bacteria, Fn binding is mediated through surface proteins. These proteins, as many other surface proteins, are linked to the bacterial cell wall via their C-terminus (Schneewind *et al.*, 1992). This includes: a proline-rich domain associated with the cell wall; a conserved LPXTGX sequence motif which serves as a signal for cell wall sorting, a hydrophobic membrane-spanning domain, and a short charged tail at the very C-terminus end which is exposed to the cytoplasm (Schneewind *et al.*, 1992; Navarre and Schneewind, 1994). The Fn binding domain consists of a tandem repeat element that is adjacent to the conserved proline-rich domain, and contains of up to 6 highly homologous repeats of 32-44 amino acids that mediate the binding to Fn (McGavin *et al.*, 1991, 1993).

The interaction between the *Streptococcus pyogenes* and the host epithelium involves a highly regulated and complex interplay of multiple bacterial adhesins and their cognate host cell receptors (Hasty *et al.*, 1992). We have recently cloned two genes *prt*F1 and *prt*F2 encoding two Fn binding surface proteins: protein F1 and protein F2 (Hanski and Caparon,

Toward Anti-Adhesion Therapy for Microbial Diseases, edited by Kahane and Ofek
Plenum Press, New York, 1996

1992; Sela *et al.*, 1993; Jaffe *et al.*, 1996). A gene encoding a Fn binding protein analogous to protein F1 has been cloned and characterized (Talay *et al.*, 1991, 1992, 1994). Insertional inactivation of *prt*F1 generated a mutant which lost Fn binding activity and the ability to adhere to respiratory epithelial cells (Hanski and Caparon, 1992). Moreover, expression of *prt*F1 in *Enterococcus faecalis*, conferred upon these organisms the ability to bind Fn and to adhere to respiratory epithelial cells (Hanski *et al.*, 1992). These studies have provided strong evidence for the role of protein F1 in promoting Fn binding and adherence.

Study of the distribution of *prt*F1-related genes among GAS of different M-types revealed that among 65 strains which bind Fn at high levels, 56 strains of 42 different M-types possessed *prt*F1-related genes; the remaining nine strains representing four different M-types expressed *prt*F2-related genes encoding a distinct family of Fn binding proteins, termed F2 (Natanson *et al.*, 1995; Jaffe *et al.*, 1996).

Here we describe the structures of protein F1 and F2 with respect to their Fn binding properties.

RESULTS

Two Fn binding domains of protein *prt*F1 were characterized; the presence of both was shown to be required for maximal Fn binding (Sela *et al.*, 1993). These two domains are encoded by two types of DNA sequences. One type consists of a repetitive sequence named RD2 (repeat domain type 2), encoding five almost identical repeats of 37 amino acid each, except that the C-terminal one contains only 32 amino acids. The second type is a non-repetitive DNA sequence named UFBD (upstream Fn binding domain), which resides immediately upstream to RD2 and encodes a protein of 43 amino acids (Fig. 1).

Precise Localization and Properties of the Fn Binding Domains of Protein F1

To characterize further the properties of the two Fn binding domains of *prt*F1, we constructed a series of proteins that were tagged with 6 consecutive histidine residue and purified on metal-chelating column. This included UFBD or RD2 as separate molecules, or

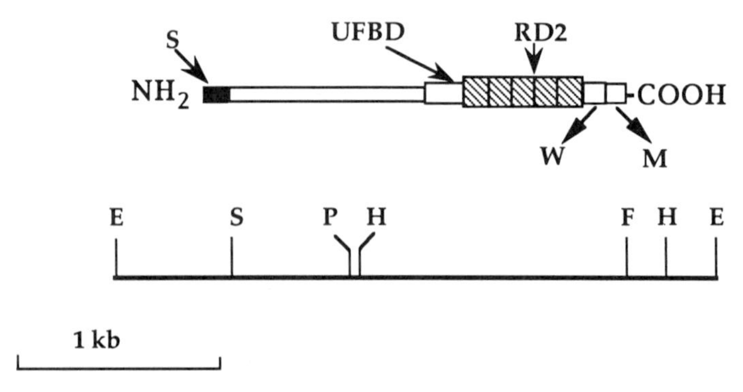

Figure 1. Structure of *prt*F1. Above the restriction map is a summary of the structure of *prt*F1 as deduced from the DNA sequence analysis conducted by Sela *et al.* (1993). The symbols: S, UFBD, RD2, W, and M represent: signal sequence, upstream Fn binding domain, repeat domain type 2, the cell wall binding domain, and the hydrophobic membrane-spanning domain, respectively. The letters on the restriction map represent the following restriction sites: E-*Eco*RV; S-*Sph*I; P-*Pst*I; H-*Hin*dIII; F-*Fsp*I.

a single molecule that contained both domains. It was discovered that a single protein composed of UFBD + RD2 (5 repeats) is much more potent in inhibiting Fn binding to the parental JRS145 strain than either domain alone, or when the two domains on separate molecules are admixed (Ozeri *et al.*, 1996). Furthermore, a protein which is composed of UFBD and a single repeat of RD2 is as efficient at inhibiting binding as a protein that consisted of UFBD and all 5 RD2 repeats. Since a single RD2 repeat by itself has no Fn binding activity (Ozeri *et al.*, 1996), this results suggest that a high affinity Fn binding site resides on this stretch of amino acids, that contains UFBD and a single RD2 repeat. To further localize this Fn binding domain, which we termed UR, a series of proteins were constructed that were truncated at the C-terminal end of the RD2 repeat, they were purified and tested for inhibition of Fn binding to the native bacterium. The smallest protein which retained full activity is composed of 49 amino acids of which 43 are located towards the N-terminus of the first RD2 repeat (previously termed UFBD) and the rest are the first six amino acids of this repeat (Figs. 2, 3).

Since among *prt*F1-related genes the number of RD2 repeats changes between one to six (Natanson *et al.*, 1995), and since one repeat (composed of 37 amino acids) fails to express binding activity (Ozeri *et al.*, 1996), we identified the precise composition of a functional RD2 unit. The smallest active RD2 functional unit is composed of 44 amino acids,

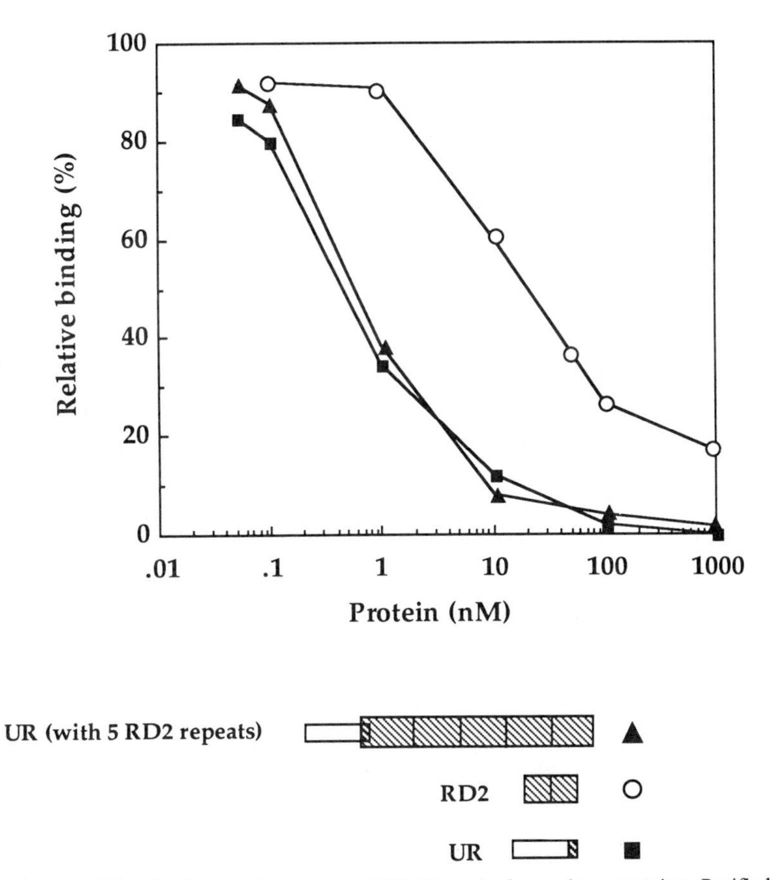

Figure 2. Inhibition of Fn binding to *S. pyogenes* JRS145 strain by various proteins. Purified proteins representing UR, UR expressed together with 5 RD2 repeats, and a functional unit of RD2 were tested for their ability to inhibit Fn binding to *S. pyogenes* JRS145.

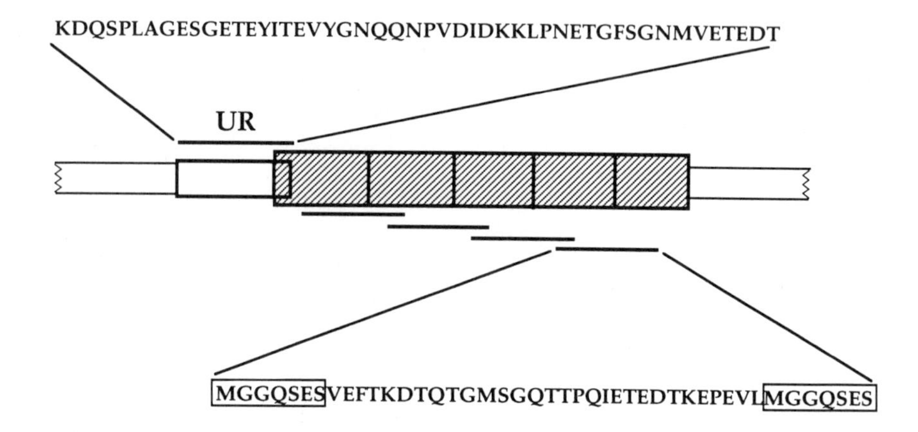

Figure 3. Localization of functional domains of UR and RD2 on *prt*F1. The localization of a functional UR domain and possible localization of functional RD2 domains are shown together with their corresponding amino acid sequences. The motif MGGQSES flanking a functional unit of RD2 and essential for its activity is indicated.

and overlaps the junction between two contiguous sequence repeats. The presence of the motif 'MGGQSES' at both ends of the functional binding unit is essential for Fn binding activity. A schematic presentation of the locations of functional UR and RD2 units on the Fn binding moiety of protein F1 is shown in Fig. 3.

UR and RD2 differ not only by their amino acid composition, but also by their relative affinities for Fn. The apparent affinity of a functional UR unit for Fn is 0.5 nM, whereas the apparent affinity of a functional unit of RD2 is 50 nM. Thus, UR dictates the binding of protein F1 to soluble Fn (Fig. 2).

Localization of RD2 and UR Binding Sites on Fn

To determine the localization of UR and RD2 binding sites on Fn, we have expressed these domains on the streptococcal cell surface by constructing fusion proteins with the secretion and anchoring domains of M protein. The chimerical proteins were expressed in a *S. pyogenes* strain (SAM2) in which the genes encoding both protein F and M protein have been inactivated (Norgren *et al.*, 1989; Hanski and Caparon, 1992). The procedure for the construction and expression of the chimerical proteins is detailed elsewhere (Hanski *et al.*, 1995). By this procedure we expressed UR (SAM27), and RD2 composed of five repeats (SAM17). A schematic presentation of *prtF1* domains expressed within the surface exposed region of M protein is described in Fig. 4.

In a previous study we have demonstrated that RD2 effectively blocked the binding of the radiolabeled N-terminal 29-kDa fibrin binding domain (Fib.1) of Fn to *S. pyogenes* JRS145 strain, whereas UFBD had no effect on Fib.1 binding (Sela *et al.*, 1993). These results led us to propose that RD2 binds to Fib.1 and UFBD binds to a different region of Fn. The experiment depicted in Fig. 5A shows that while Fib.1 effectively blocked the binding of Fn to SAM17 (expressing RD2 only), it had no effect on the binding of Fn to

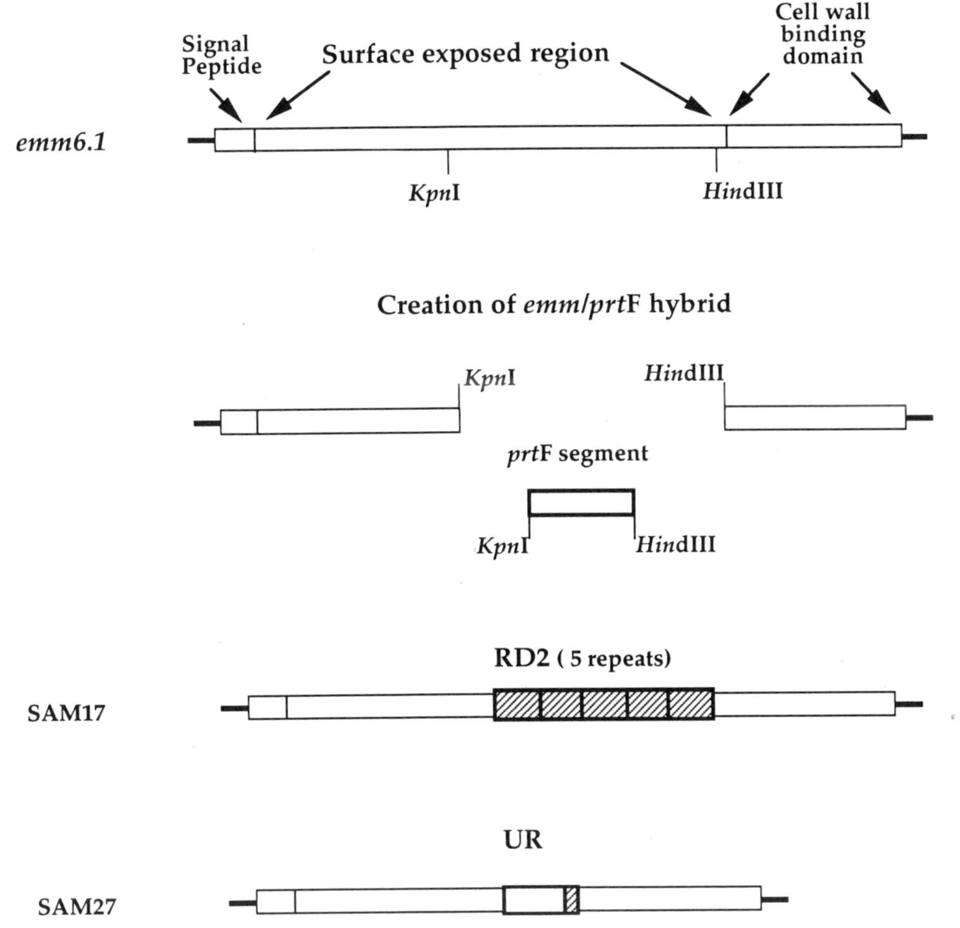

Figure 4. Expression of *prt*F1 domains in hybrid proteins exposed on the streptococcal cell surface. A schematic representation that illustrates the construction of the *emm/prt*F1 hybrids: SAM17 and SAM27.

SAM27 (expressing UR domain), and reduced by only 20% or less the binding of Fn to JRS145 (expressing the intact protein F1). The latter observation underscores the conclusion that the binding of soluble Fn to protein F1 is mediated primarily by the UR domain.

Our initial attempts to block Fn binding to SAM27 by admixing six distinct fragments of Fn produced by the procedure of Borsi *et al.* (1986) failed, even though the mixture included: Fib. 1, collagen binding domain (Col.), cell binding domain, heparin binding domains (two fragments of 38 and 29-kDa) and the C-terminal fibrin binding domain. In spite of the fact that Col. had no effect by itself on the binding of Fn to JRS145, a mixture of Fib. 1 and Col. fragments completely blocked the binding (Fig. 5B), yet with an apparent affinity of 10 nM, which is lower by at least 10 fold than the affinity of JRS145 for Fn (Fig. 5C). In contrast, as has already been mentioned, Fib. 1 + Col., did not affect the binding of Fn to SAM27 (Fig. 5C). These results led us to hypothesize that the 70-kDa N-terminal fragment of Fn, composed of Fib. 1 and the collagen domains, might be responsible for binding of Fn to UR. The binding of Fib. 1 to RD2, which is adjacent to UR on protein F1, might facilitate the binding of the Fib. 1 and Col. domains of Fn to UR, even when added

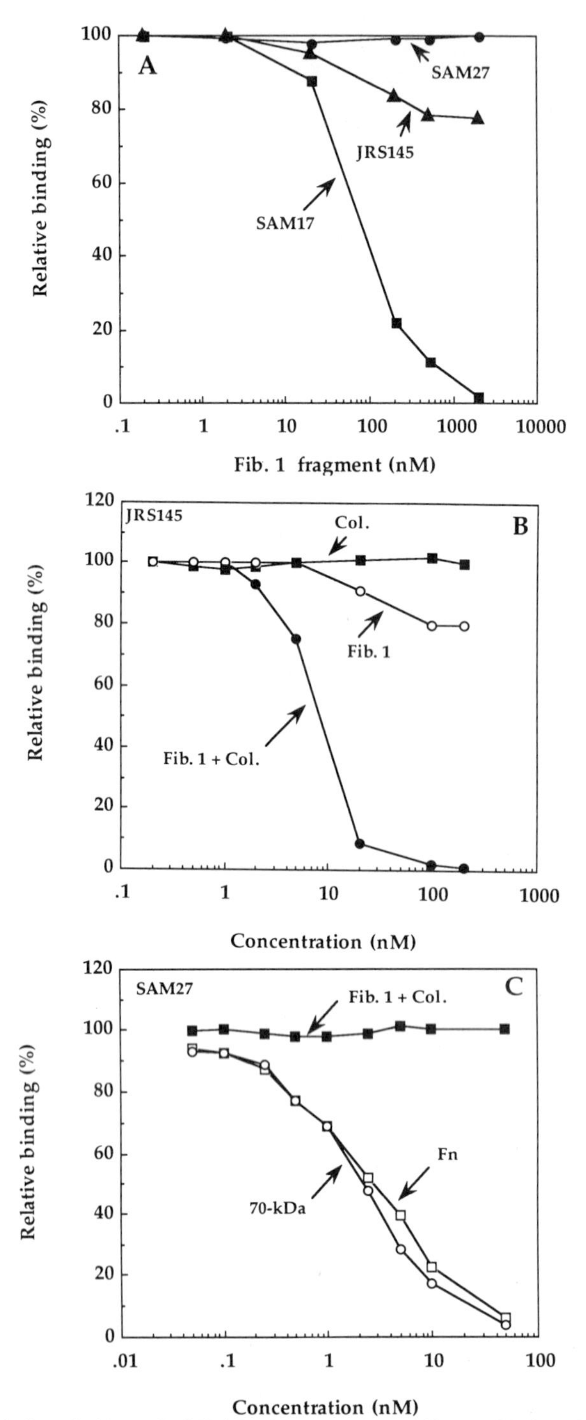

Figure 5. UR binds to the N-terminal 70-kDa fragment of Fn while RD2 binds to the 29-kDa N-terminal fibrin binding domain. (A) Displacement of Fn by the 29-kDa N-terminal fragment (Fib. 1) from the indicated bacteria. The specific binding values of SAM17, SAM27 and JRS145 in the absence of Fib. 1 were: 10,000, 134,000 and 110, 000 cpm, respectively. These values were denoted as 100% binding. (B) Displacement of Fn from JRS145 by Fib. 1, 40-kDa collagen binding fragment (Col.) and a mixture of the two fragments. The specific binding of JRS145 that was denoted as 100% binding was of 115,000 cpm. (C) Displacement of Fn from SAM27 by a mixture of Fib. 1 + Col. ; Fn and the 70 kDa N-terminal fragments of Fn containing the Fib. 1 and the Col. fragments. The specific binding of SAM27 which was denoted as 100% binding was of 138,000 cpm.

separately. These two fragments fail to block Fn binding to SAM27, since the latter expresses a functional UR only.

To test this hypothesis further, we produced the 70-kDa N-terminal fragment of Fn (McKeown-Longo and Mosher, 1985) and examined its ability to block Fn binding to SAM27. As shown in Fig. 5C, the 70-kDa fragment blocked the binding of iodinated Fn with a potency similar to that of a non-iodinated Fn. These results establish that the fragment of Fn that binds to UR is composed of Fib. 1 and Col. domains linked together in a single fragment. The binding of UR to the N-terminal 70-kDa fragment of Fn must be dependent on the conformation of the latter, since UR does not bind to Fib. 1 and Col. domains when added separately and comprise the entire N-terminal 70-kDa fragment.

Expression of Protein F2 Is Essential for High Affinity Fn Binding in *S. Pyogenes* 100076 Strain

S. pyogenes 100076 strain of M type 49 binds high level of Fn with high affinity but lacks *prt*F1. This suggested that it expresses a Fn binding protein different than protein F1 (Natanson *et al.*, 1995; Jaffe *et al.*, 1996). We have cloned the gene (*prt*F2) encoding this Fn binding protein, termed protein F2 (Jaffe *et al.*, 1996). To unequivocally demonstrate that protein F2 is responsible for Fn binding in 100076 strain, *prt*F2 was insertionally inactivated to produce the isogenic mutant SAM101. As shown in Fig. 6, inactivation of *prt*F2 completely diminished Fn binding, indicating that protein F2 is responsible for the high affinity Fn binding activity in 100076 strain.

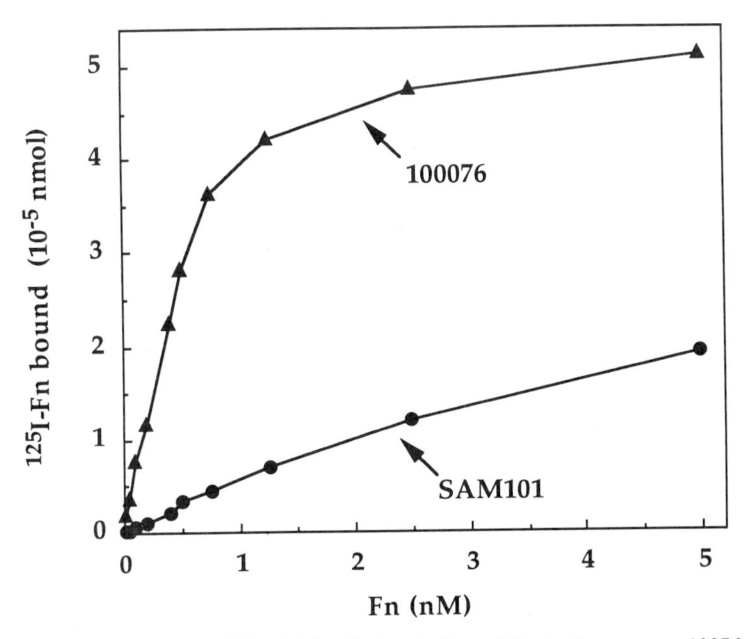

Figure 6. Inactivation of *prt*F2 abolishes high-affinity binding of Fn to *S. pyogenes* 10076 strain. Dose response curves for binding of iodinated Fn to 10076 and to its isogenic inactivated-*prtF2* mutant, SAM101.

Localization of *prt*F2 Binding Domains

Since *prt*F2 possesses a highly homologous repeat region to that of *Streptococcus dysgalactia* (Lingren *et al.*, 1993) and that of *Streptococcus equisimilis* (Lingren *et al.*, 1994), we assumed that its Fn binding domain is confined to that region, which we termed FBRD, Fn binding repeat domain (Fig. 7). However, to our surprise, a DNA fragment containing the 5′ portion of *prt*F2 but lacking completely FBRD, expressed Fn binding activity. This led us to the conclusion that *prt*F2 contains an additional binding domain encoded by an upstream sequence, termed UFBD, upstream Fn binding domain (Fig. 7). Progressive deletions of 5′ portion of *prt*F2 indicated that UFBD resides on a stretch of approximately 100 amino acids, and is separated by 100 amino acids from FBRD (Fig. 7). More precise localization and characterization of Fn binding by UFBD is in progress.

CONCLUDING REMARKS

Several surface components of *S. pyogenes* have been implicated in Fn binding (Hasty *et al.,* 1992). Perhaps the best evidence for the role of a specific surface protein in promoting the ability of *S. pyogenes* to bind Fn has come from studies of a family of highly homologous proteins which include sfb (Talay *et al.*, 1991, 1994) and protein F1 (Hanski and Caparon, 1992). These proteins bind Fn very strongly with apparent Kd of 1 nM (Hanski and Caparon, 1992). They inhibit the binding of Fn to intact streptococcal cells (Talay *et al.,* 1992; Hanski and Caparon, 1992) and can inhibit the adherence of *S. pyogenes* to several epithelial cells (Talay *et al.,* 1992). Protein F2 represents a distinct family of Fn binding proteins that although displaying some degree of amino acid sequence homology with protein F1 family, has similar properties with respect binding of Fn at high affinity and acting as adhesins of *S. pyogenes* (Hanski and Caparon, 1992; Jaffe *et al.*, 1996). These two type of Fn binding proteins are also similar in the organization of their Fn binding domains. Both proteins have a repeat domain and an additional non-repeated domain, each capable of binding Fn independently. This strategy which involves two independent separable binding domain located on the same adhesin molecule, that bind to two different sites on the target molecule (so far demonstrated for protein F1), may account for the adherence of *S. pyogenes* to such a variety of cells and tissues.

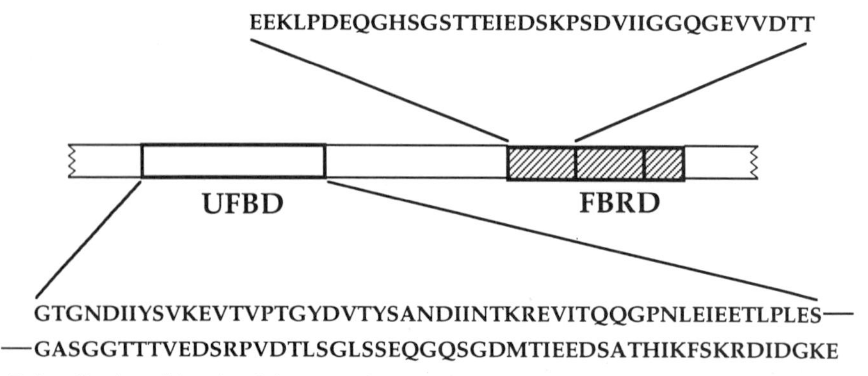

Figure 7. Localization of functional domains of UFBD and FBRD on *prt*F2. This representation shows the locations of functional domains of UFBD and FBRD on the Fn binding moiety of protein F2, together with their amino acid sequences.

While the repeated domains of proteins F1 and F2 and other Fn binding adhesins of Gram positive bacteria, such as *Staphylococcus aureus*, *S. dysgalactiae* and *S. equisimilis*, possess conserved motifs, the second non-repeated domains of protein F1 and protein F2 seem to be more divergent. These non-repeated domains also differ in localization with respect to their corresponding repeated domains. In protein F1 the two domains are adjacent and the UR contains residues of the first repeat, whereas in protein F2 the two domains are separated by 100 amino acids. The significance of these differences with respect to Fn binding and adherence to cells and extracellular matrix remains to be elucidated.

ACKNOWLEDGMENTS

This work was supported in part by United States-Israel Binational Science Foundation, Ministry of Science and Arts and Ministry of Health Foundations (to E. H.)

REFERENCES

Borsi, L., Castellani, P., Balza, E., Siri, A., Pellecchia, C., De Scalzi, F., and Zardi, L. (1986). Large-scale procedure for the purification of fibronectin domains. *Anal. Biochem.*, 155: 335-345.

Hanski, E. Horwitz, P. A., and Caparon M. G. (1992). Expression of protein F, the fibronectin-binding protein of Streptococcus pyogenes JRS4, in heterologous streptococcal and enterococcal strains promotes their adherence to respiratory epithelial cells. Infect. Immun. 60: 5119-5125.

Hanski, E., and Caparon, M. G. (1992) Protein F, a fibronectin-binding protein, is an adhesin of the group A streptococcus—*Streptococcus pyogenes*. *Proc. Natl. Acad. Sci. USA*. 89: 6172-6176.

Hanski, E., Fogg, G., Tovi, A., Okada, N., Burstein, I., and Caparon M. G. (1995). Molecular analysis of *Streptococcus pyogenes* adhesion. *Methods Enzymol*. 253: 269-310.

Hasty, D. L., Ofek, I., Courtney, H. S., and Doyle R. J. (1992). Multiple adhesins of streptococci. *Infect. Immun.* 60: 2147-2152.

Hynes, R. O., (1990). Fibronectins. Springer-Verlag, Berlin.

Jaffe, J., Natanson-Yaron, S., Caparon, M. G., and Hanski, E. (1996) Protein F2, a novel fibronectin binding protein from *Streptococcus pyogenes,* possesses two binding domains. *Mol. Microbiol.* (in press).

Lindgren, P.-E., McGavin, M., Signäs, C., Guss, B., Gurusiddappa, S., Höök, M., and Lindberg, M. (1993) Two different genes coding for fibronectin-binding proteins from *Streptococcus dysgalactiae*: the complete nucleotide sequence and characterization of binding domains. *Eur. J. Biochem.* 214: 819-827.

Lindgren, P.-E., Signäs, C., Rantamäki, L., and Lindberg, M. (1994) A fibronectin-binding protein from *Streptococcus equisimilis*: characterization of the gene and identification of the binding domain. *Vet. Microbiol.* 41: 235-247.

McGavin, M. J., Gurusiddappa, S., Lindgren, P.-E., Lindberg, M., Raucci, G., and Höök, M. (1993). Fibronectin receptors from *Streptococcus dysgalactiae* and *Staphylococcus aureus*. *J. Biol. Chem.* 268: 23946-23953.

McGavin, M. J., Raucci, G., Gurusiddappa, S., and Höök, M. (1991). Fibronectin binding determinants of the *Staphylococcus aureus* fibronectin receptor. *J. Biol. Chem.* 266: 8343-8347.

McKeown-Longo, P. J., and Mosher, D. F. (1985). Interaction of the 70,000-mol-wt amino-terminal fragment of fibronectin with the matrix-assembly receptor of fibroblasts. *J. Cell Biol.* 100: 364-374.

Mosher, D. F., (1989). Fibronectin. Academic Press. New York.

Natanson, S., Sela, S., Moses, A. E., Musser, J. M., Caparon, M. G., and Hanski, E. (1995) Distribution of fibronectin-binding proteins among Group A streptococci of different M-Types. *J. Infect. Dis.* : 171: 871-878.

Navarre, W. W., and Schneewind, O. (1994) Proteolytic cleavage and cell wall anchoring at the LPXTG motif of surface proteins in Gram-positive bacteria. *Mol. Microbiol.* 14: 115-121

Norgren, M., Caparon, M. G. , Scott, J. R. (1989). A method for allelic replacement that uses the conjugative transposon Tn916: deletion of the *emm6.1* allele in *Streptococcus pyogenes* JRS4. *Infect. Immun.* 57: 3846-3850.

Ozeri, V., Tovi, A., Burstein, I., Natanson-Yaron, S., Caparon, M. G., Yamada, K. M., Akiyama, S. K., Vlodavsky, I., and Hanski, E. (1996) A Two-domain mechanism for Group A Streptococcal adherence through protein F to extracellular matrix. *EMBO J.*, 15: (in press).

Patti, J. M., Allen, B., L., McGavin, M., J., and Höök, M. (1994) MSCRAMM-mediated adherence of microrganisms to host tissues. *Annu. Rev. Microbiol.* 48: 586-617.

Schneewind, O. Model, P., and Fischetti, V. A. (1992) Sorting of protein A to the Staphylococcal cell wall. *Cell* 70: 267-281.

Sela, S., Aviv, A., Tovi, A., Burstein, I., Caparon, M. G., and Hanski, E. (1993). Protein F: an adhesin of *Streptococcus pyogenes* binds fibronectin via two distinct domains. *Mol. Microbiol.* 10: 1049-1055.

Talay, S. R., Ehrenfeld, E., Chhatwal, G. S., and Timmis K. N. (1991) Expression of the fibronectin-binding components of *Streptococcus pyogenes* in *Escherichia coli* demonstrates that they are proteins. *Mol. Microbiol.* 5: 1727-1734.

Talay, S. R., Valentin-Weigand, P., Jerlström , P. G., Timmis, K. N., Chhatwal, G. S. (1992). Fibronectin-binding protein of *Streptococcus pyogenes:* sequence of the binding domain involved in adherence of streptococci to epithelial cells. *Infect. Immun.* 60: 3837-3844.

Talay, S. R., Valentin-Weigand, P., Timmis, K. N., Chhatwal, G. S. (1994) Domain structure and conserved epitopes of Sfb protein, the fibronectin-binding adhesin of *Streptococcus pyogenes*. *Mol. Microbiol.* 13: 531-539.

17

ADHESION MOLECULES IN LEUKOCYTE ENDOTHELIAL INTERACTION

Amos Etzioni

Department of Pediatrics and Clinical Immunology
Rambam Medical Center
Faculty of Medicine
Technion, Haifa, Israel

INTRODUCTION

Neutrophils are the front line defense against most microbial pathogens. They provide a rapid, relatively nonspecific defense mechanism, after which a more long lasting, antigen-specific response is established by T and B lymphocytes. To fulfill this role successfully, the neutrophil must be able to migrate from the blood into the area of inflammation, a process which involves activation of endothelial cells and leukocyte by inflammatory stimuli, adherence to the endothelial surface at the site of inflammation, and migration through endothelium to extravascular tissue. The orchestration of these steps must by precisely regulated to ensure a rapid response to isolate and destroy the invading pathogen yet cause minimal damage to healthy tissues (Etzioni and Douglas 1993). The process of leukocyte accumulation at sites of inflammation is a dynamic one, involves multiple steps and is mediated by several families of adhesion molecules (Springer 1995). These include the Integrins, the Selectins and members of the Immunoglobulin (Ig) superfamily. Each is involved in a different phase of leukocyte emigration through the endothelium and the synchronization of their expression and function is crucial for the normal recruitment of leukocyte from the blood stream to the tissue (Etzioni 1996).

In this chapter we will discuss the various adhesion molecules involved in leukocyte emigration through the endothelial layer in the blood vessels, the multiple steps in the adhesion cascade and then we will focus on the consequences of defects in these molecules and the therapeutical use of blocking adhesion molecules in several clinical pathological conditions.

LEUKOCYTE INTEGRINS

Integrins are transmembrane cell surface proteins that bind to cytoskeletal proteins and communicate extracellular signals. Each integrin consists of a noncovalently linked,

Toward Anti-Adhesion Therapy for Microbial Diseases, edited by Kahane and Ofek
Plenum Press, New York, 1996

heterodimeric α and β chains. Integrins have been arranged in subfamilies according to the β subunits, and each β subunit may have from one to eight different α subunits associated with it (Hynes 1992). The specificity of binding to various cell adhesion factors appears to depend primarily on the extracellular portion of the α subunit for all the integrins, Within the integrin family of adhesion receptors only the β2 leukocyte integrins are involved in leukocyte adhesion to the endothelial surface.This subfamily comprises three membrane glycoproteins with a common β subunit, designated β_2(CD18). The α subunits of each of the three heterodimer members - lymphocyte associated antigen-1 (LFA-1), macrophage antigen-1 (Mac-1), and p150,95- are designated CD11a, CD11b, and CD11c, respectively. Both the α and β subunits have a relative small cytoplasmic domain which contains regions capable of binding to cytoskeletal elements. β_2 integrin expression is restricted to leukocytes, and they participate in many leukocyte adhesion-related functions in addition to migration through the endothelial cells, such as phagocytosis, killing of bacteria and antibody-dependent cell mediated cytotoxicity(Kishimoto and Rothlein 1993).

An important characteristic of the leukocyte integrins is that under baseline conditions they exist in a relatively inactive conformation, rendering the leukocyte non-adhesive. One of the key events at the adhesion cascade is the activation and deactivation of these integrins at the proper times and places (Springer 1995). Leukocytes mediate adhesion through binding to their ligands, members of the Ig superfamily integrins, which are expressed on all cells and may affect leukocyte adherence and interactions with all cells in the body, particularly during inflammatory conditions.

Ig SUPERFAMILY

The intercellular adhesion molecules (ICAM) were originally defined functionally as LFA-1 ligands. This family includes three molecules, ICAM-1, 2, and 3. The gene for ICAM-1 is located on chromosome 19 and is a ligand for both LFA-1 and Mac-1. The binding sites for the integrins are distinct; LFA-1 binds to domains 1 and 2 while Mac-1 binds to domain 3. Human ICAM-2 is a single copy gene located on chromosome 17 and will bind only LFA-1. Other integrin molecules do not bind ICAM-2. The distribution and regulation of the ICAMs are quite distinct. ICAM-1 is expressed only at low levels on some vascular endothelial cells and on lymphocytes under normal conditions and is dramatically upregulated by endotoxin, Interleukin-1 (IL-1), and Tumor Necrosis Factor (TNF). ICAM-2 expression, in contrast, is constitutive and is not regulated by the various cytokines. ICAM-3 is constitutively and strongly expressed on resting lymphocytes but not on endothelial cells. Once the cell is activated, ICAM-3 will accumulate in the uropod of the cell thus facilitating interaction and aggregation with more leukocytes, and recruiting cells to the inflammatory area.

SELECTINS

This family of adhesion molecules was discovered in 1989, when the cDNA sequences of three distinct cell surface glycoproteins found on endothelium (E-selectin, CD62E), platelets (P-selectin, CD62P), and leukocytes (L-selectin, CD62L) were reported (Bevilacqua 1993). The genes for the selectins family are closely linked on chromosome 1. All three members have common structural features, most prominently an N-terminal lectin like domain, which is central to the carbohydrate binding properties of all three selectins (Bevilacqua 1993). The term selectin was proposed to highlight the amino-terminal lectin domain and to indicate the selective function and expression of these molecules.

All three selectins are involved in the first step of the adhesion cascade, the rolling phase, but there are fundamental differences in their distribution, activation, and mode of expression (Tedder 1995).

E-selectin is restricted to endothelial cells, and its expression is induced when the cells are activated by IL-1 or TNF. Although E-selectin expression *in vitro* is typically transient, it is chronically expressed in certain inflammatory conditions and is also detected in the serum.

P-selectin is expressed on platelets as well as on endothelial cells. Its expression does not necessarily require *de novo* synthesis because it is stored in the α granules in platelets and in secretory granules (Weibel-Palade bodies) in endothelial cells. Thus, within minutes of activation of either cell type by thrombin or histamine, P-selectin is rapidly redistributed to the surface of the cells. Its expression *in vitro* is typically very short-lived, up to 15 minutes. However, studies *in vivo* suggest that endothelial P-selectin may also be regulated at the level of protein synthesis, providing a mechanism for more prolonged expression.

In contrast to the E- and P-selectins, L-selectin is constitutively expressed on leukocytes but not on endothelial cells. Although originally described as a lymphocyte homing receptor, it was subsequently shown to be expressed on most other leukocytes. After a transient increase of this selectin during activation, it is shed rapidly.

Selectin Ligands

Studies of the molecular basis of selectin adhesion have fucosed mainly on carbohydrate recognition by the lectin domains. Identification of the physiological ligands for the selectins has been challenging because like many lectins, the selectins can bind a variety of carbohydrate structures in virto (Varki 1994). These ligands are carbohydrate groups which are typically found as a terminal structures of one or more glycoproteins and /or glycolipids. The lectin and the epidermal growth factor domains in the selectins play a crucial role in mediating this binding. One major selectin ligand is a member of a class of sialylated and fucosylated tetrasaccharides related to the sialylated Lewis X blood group (SLeX, CD15).

While the selectin bind weakly with small sialylated, fucosylated oligosaccharides, they appear to bind with higher avidity to carbohydrate determinants on a limited number of glycoproteins or proteglycants. Recently specific glycoproteins for E and P-selectins were also identified. E-selecin ligand (ESL-1) is a 150,000 Mr glycoprotein expressed mainly on neutrophils. P-selectin has also been reported to selectively bind a 160,000Mr glycoprotein, PSGL-1 which is expressed on all leukocytes. A mAb specific for PSGL-1 completely inhibited P-selecin mediated rolling of leukocytes under a range of physiologic shear stress.

THE ADHESION CASCADE

The migration of leukocytes from the blood stream to the tissue occurs in several steps. First, loose adhesion to the vessel wall, primary in post-capillary venules, under conditions of flow, causes the leukocytes to roll on the endothelium. This transient and reversible step is a prerequisite for the next stage, the activation of leukocytes. This is followed by firm adhesion after which migration occurs. Each of these steps involves different adhesion molecules, and can be differentially regulated.

Step 1: Rolling, selectin-dependent. Leukocytes in the circulation must resist tremendous shear forces in order to stop along the vascular endothelium. Under normal conditions, leukocytes move rapidly and do not adhere to the endothelium. The phenomenon of leukocyte rolling has been known for more than a century, but its molecular basis was delineated only recently. Monoclonal antibodies (MoAbs) to selectins, markedly reduced the rolling process *in vivo* and *in vitro*. Moreover, rolling is dramatically diminished in P-selectin deficient mice. The selectins expressed on the endothelial cells, will bind to the leukocytes through their ligands, mainly SLeX. MoAbs to SLeX will also significantly reduce rolling and patients who lack SLeX molecules will show defective *in vivo* rolling.

Step 2: Activation, Integrins-dependent. The initial rolling interaction of leukocyte is reversible unless the leukocyte are activated and become firmly adherent. Binding to the selectins bring the leukocytes into the local microenvironment. Chemoattractants that are released from the surrounding tissue can trigger the leukocyte to activation-dependent sticking and arrest.

Increased adhesive capacity is mediated through both qualitative (conformational) and quantitative (up regulation of surface expression) changes in the integrin receptors.

Step 3: Firm adhesion and Transendothelial Migration, Integrins and Ig-like ligands. Activation of integrins results in increased affinity for their Ig-like ligands on the endothelial cells. This ensures that binding is firm enough to withstand the continuous shear forces in the blood vessels. MoAbs to both integrins and their Ig-like ligands block the firm adhesion of leukocyte to endothelial cells. The final event, the transendothelial migration of the leukocytes to the site of inflammation, is in part dependent on integrins/Ig-like interactions and can be blocked by MoAbs to these molecules. Another member of the endothelial Ig-superfamily, PECAM-1 (CD31), was found to be important mainly in transmigration. CD31 is constitutively expressed on endothelial cells. Heterotypic interaction between leukocyte and endothelial CD31 was found to be crucial for neutrophil and monocyte diapedesis between endothelial cells. Anti-PECAM-1 MoAbs blocked leukocyte migration across cytokine-activated endothelial monolayers.

ADHESION MOLECULE DEFECTS

Studies of genetic deficiency syndromes have provided important insights into the molecular basis and biology of leukocyte emigration. Both animal and human deficiency syndromes have been described.

Animals Models

A spontaneous mutation in the gene coding for the β_2 integrin has been described both in dogs and cattle. These animals suffered from recurrent infections and persistent mature neutrophilia associated with poor growth performance. The carrier frequency of the defective gene among Holstein cattle is approximately 15% in bulls and 6% in cows.

Using methods of homologous recombination in embryonic stem cells, it is possible to generate strains of mice deficient in specific adhesion molecules. A "knockout" mouse with partial CD18 deficiency was found to be viable and fertile, with mild granulocytosis. The mutant mice showed an impaired inflammatory response to chemical peritonitis and delayed rejection of cardiac transplants.

In another model, ICAM-1 deficient mice were generated. The animals developed normally and had mild granulocytosis. Deficient mice exhibited abnormalities of inflammatory response, especially in impaired neutrophil emigration.

Mice lacking P-selectin develop normally. However, they exhibit striking leukocytosis, diminished rolling of leukocytes in mesentric vessels and delayed recruitment of neutrophils to the peritoneal cavity after experimentally induced peritonitis. Rolling can be visualized at time points beyond an hour, but the frequency of rolling leukocytes remains below those of wild type mice. (Ley 1995). In contrast, mice deficient of expression of L-selectin initially have normal levels of leukocyte rolling, but show a significant decline in rolling within the first hour (Ley 1995). Therefore, spontaneous leukocyte rolling is initiallly P-selectin dependent, but shows a prominent L-selectin dependent component at later time points. E-selectin deficient mice have a milder defect in neutrophil recruitment and do not have leukocytosis. Still, administration of anti P-selectin MoAbs to the E-selectin knockout mice completely blocks neutrophil recruitment, suggesting that in this model some redundancy of function may exist between the two endothelial cell selectins.

Human Leukocyte Adhesion Deficiency (LAD) Syndromes

The best way to appreciate the *in vivo* importance of the various adhesion molecules is by looking at those rare "experiments of nature" in which a specific defect in adhesion molecule exists. Currently, two such syndromes are described (Etzioni 1994). In the first, LAD I, the β_2 integrin family is deficient, while in the second, LAD II, the fucosylated ligands for selectins are absent.

LAD I

Before 1980, several reports documented a group of patients with recurrent bacterial and fungal infections, defective leukocyte motility and phagocytosis, impaired wound healing and delayed separation of the umbilical cord. Over the years multiple reports (around 100 patients) with a similar picture have been described. Several years later it was found that the syndrome is a result of a defect in the expression of the β2 integrin molecule on leukocyte surface (Anderson 1995). The disorder is inherited in an autosomal recessive manner with heterozygotes exhibiting no significant clinical manifestations. The gene encoded for the CD18 is located on chromosome 21. Heterogeneous mutation in the common β2 chain - splicing, frame shift, missense, and initiation codon have been demonstrated to be the molecular basis of the syndrome (Anderson 1995). As a consequence neutrophils are unable to emigrate from the blood vessels to the tissue as the adhesion and transmigration through the endothelium is severely impaired (Fig. 1). Although neutrophil motility is the most pronounced defect, the opsonophagocytic activity as well as lymphocyte function was also found to be defective.

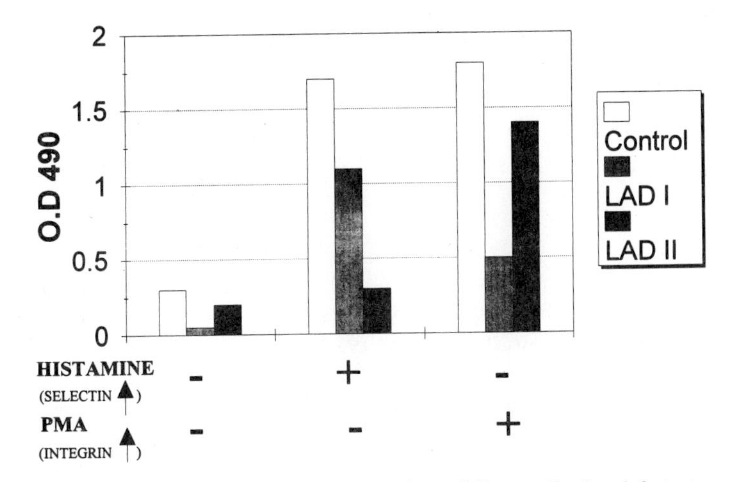

Figure 1. LAD I and LAD II patients have different adhesion defects.

Clinically, LAD I has been devided into two groups, a severe phenotype and a moderate one (Harlan 1993). In the severe form no expression of CD18 on leukocyte can be detected and the patients will suffer from life threatening infection and will die early in life if definite therapy (bone marrow transplantation) is not instituted. In the moderate form some surface expression (2-10%) exists and these patients have fewer serious infections and survive into adulthood. Since LAD I is a monogenic disorder involving hematopoietic cells, it is obviously an attractive candidate for curative treatment with gene therapy.

LAD II

This new adhesion molecule deficiency syndrome was described in 1992 (Etzioni 1992). It is also characterized by recurrent infections, failure to form pus, gingivitis, and pronounced neutrophilia. The severity of the infecious complications resembeles that of the moderate type of LAD I. In contrast to LAD I, this syndrome is also characterized by mental retardation, growth retardation and exhibits the rare Bombay blood group phenotype. LAD II, which is transmitted in an autosomal recessive way, is the result of a general defect in endogenous fucose metabolism. This results in the inability to synthesize fucosylated carbohydrate structures, such as the H antigen on the erythrocytes or the Sialyl Lewis X, the ligand for the selectins, on the neutrophils.

The expression of CD18 is compeletly normal and only fucose containing surface antigens are defective. Neutrophils from these patients do not bind to E or P-selectin expressed on cytokine activated endothelial cells, and exhibit a marked decrease in migration to skin window *in vivo*.

Compelling evidence in support of the current paradigma of leukocyte-endothelial interaction was obtained by using intravital microscopy to study neutrophil behaviour from LAD I and LAD II patients (von-Adrian 1993). Fluorescein labeled cells were observed during interactions with venules in rabbit peritonium treated with interleukin-1 to induce E-selectin. Neutrophils from the LAD I showed normal rolling, but were unable to stick and emigrate upon chemotactic stimulation. Neutrophils from LAD II patients rolled poorly, and failed to stick and emigrate under the shear forces proved by the flow. However, when flow was reduced the cells adhered and emigrated in response to a chemoattractant.

Although the clinical picture in both adhesion molecule deficiencies is somewhat similar, the distinct function of the CD18 integrin and the fucosylated ligands for the selectins has been clearly demonstrated in these rare syndromes (Fig. 1).

ADHESION MOLECULES—THERAPEUTIC APPLICATIONS (TABLE 1)

As well as being involved in host defense, leukocyte-endothelial interactions can generate pathological inflammation in a variety of conditions. Diseases characterized by acute inflammation with infiltration of neutrophils are often associated with increased expression of E- and P-selectins, whereas in chronic conditions ICAM-1 expression predominates.This has suggested several avenues for inhibiting the adhesion process. MoAbs against the various adhesion molecules have been developed and tested in many animal models with beneficial effects. More recently the used of saccarides that block selectin-mediated rolling has been investigated (Kubes 1995).

Table 1. Anti-adhesion molecules therapy

Diabetes	α–VLA-4
Asthma	α–ICAM-1
Rheumatoid arthritis	α–Integrin
Graft rejection	α–ICAM-1
Meningitis	α–CD18 and fucoidin
Reperfusion injury	α–CD18 and saccaride

In the ischemic/reperfusion injury model in cats it was shown that administration of Sialyl Lewis X containing oligosaccaride attenuted the myocardial injury and preserved coronary endothelium function.

Using the experimental meningitis model in the rabbit, it was found that administration of the polysaccaride fucoidin can prevent pleocytosis and may be an effective therapeutic approach to attenuate leukocyte-dependent central nervous system damage in bacterial meningitis.

REFERENCES

Anderson DC, Kishimoto TK, Smith CW. 1995 Leukocyte adhesion deficiency and other disorders of leukocyte adherence and motility. In: Scriver CR, Beaudet AL, Sly WS, Valle D. Eds: The metabolic and molecular bases of inherited diseaes. 7th Ed. McGraw-Hill, New-York pp3955-95.

Bevilaqua MP.1993 Endothelial leukocyte adhesion molecules. Ann Rev Immunol 11:767-804.

Etzioni A, Frydman M, Pollack S, Avidor I, Phillips ML, Pualsom JC, Gershoni-Baruch R 1992 Severe recurrent infections due to a novel adhesion molecule defect. New Engl J Med 327:1789-92.

Etzioni A, Douglas DS. 1993 Microbial phagocytosis and killing in host defense. In Spirer Z, Roifman CM, Branski D.Eds: Pediatric Immunology. Karger, Basel pp17-27.

Etzioni A 1994 Adhesion molecule deficiencies and their clinical significance. Cell Adhes Comm 2:257-60.

Etzioni A. 1996 Adhesion molecules - their role in health and disease. Ped Res 39:191-8.

Harlan JM 1993 Leukocyte adhesion deficiency syndrome: Insights into the molecular basis of leukocyte emigration. Clin Immunol Immunopathol 67:s16-s24.

Hynes RO 1992 Integrins: versatility, modulation, and signaling in cell adhesion. Cell 69:11-25.

Kishimoto TK, Rothlein R 1993 Adhesion molecules which guide neutrophil endothelial cell interaction at site of inflammation. In: Gupta S, Griscelli C, eds. New concepts in immunodeficiency diseases. New-York, John Wiley & Sons.131-52.

Kubes P, Jutila M, Payne D. 1995 Therapeutic potential of inhibiting leukocyte rolling in ischemia/reperfusion. J Clin Invest 95:2510-9.

Ley K, Bullard DC, Arbones ML, Bosse R, Vestweber D, Tedder TF, Beaudet Al. 1995 Sequential contribution of L- and P-selectin to leukocyte rolling in vivo. J Exp Med 181:669-75.

Springer TA. 1995 Traffic signals on endothelium for lymphocyte recirculation and leukocyte emigration. Ann Rev Physiol. 57:827-72.

Tedder TF, Steeber DA, Chen A, Engel P. The selectins: vascular adhesion molecules. FASEB J 9:866-73.

Varki A, 1994 Selectin ligands. Proc Natl Acad Sci USA 91:7390-97.

von-Andrian UH, Berger EM, Ramezani L, Chambers JD, Ochs HD, Harlan JM, Paulson JC, Etzioni A, Arfors KE 1993 In vivo behavior of neutrophils from two patients with distinct inherited leukocyte adhesion deficiency syndromes. J Clin Invest 91:2893-7.

18

MULTIPLE STAGES OF VIRUS-RECEPTOR INTERACTIONS AS SHOWN BY SIMIAN VIRUS 40

Leonard C. Norkin[1] and Howard A. Anderson[2]

[1] Department of Microbiology
University of Massachusetts
Amherst, Massachusetts 01003
[2] Experimental Immunology Branch
National Cancer Institute
Bethesda, Maryland 20892

1. INTRODUCTION

None of the many antibiotics that are clinically effective in the fight against bacteria are active against viruses. Furthermore, the search for clinically effective antiviral agents has thus far produced relatively few useful drugs. Nevertheless, adhesion based therapies may be particularly promising in the fight against viruses. First, whereas adhesion may enhance the infectivity of certain bacteria, the interaction of a virus with its receptor is absolutely critical for infection. Furthermore, since virus-receptor interactions are highly specific, a virus can not easily mutate its receptor binding site and remain viable. Thus, the anti-adhesion approach to antivirals is not likely to lead to drug resistant variants. Also, whereas viruses were once thought to interact with a single cell membrane component, an increasing number of viruses are appearing to interact sequentially with multiple distinct cell surface components (Haywood, 1994; Norkin, 1995). Initial binding may be followed by secondary interactions that strengthen adhesion, promote entry, or both. This complexity may be a cause for optimism since it produces additional opportunities for the rational development of drugs that might target initial binding, adhesion strengthening, or entry.

Studies of simian virus 40 (SV40) illustrate how viral binding and entry can be facilitated by distinct cell surface components. Binding was dependent on expression of surface major histocompatibility complex (MHC) class I molecules, whereas entry was dependent on cell surface factors that were sensitive to the cholesterol - binding drug, nystatin, and the phorbol ester, PMA. Entry may also depend on the transmission of a signal activated by SV40 at the cell surface.

Note that the primary receptors for a variety of viruses have been identified during the past few years(Geleziunas, Bour, and Wainberg, 1994).

Toward Anti-Adhesion Therapy for Microbial Diseases, edited by Kahane and Ofek
Plenum Press, New York, 1996

2. RESULTS

2.1. Class I MHC Molecules Are a Component of the SV40 Receptor

Several lines of evidence show that class I molecules encoded by the MHC (human lymphocyte antigens [HLA] in humans) are a necessary component of the SV40 receptor (Atwood and Norkin, 1989; Breau, Atwood, and Norkin, 1992). First, preadsorption of SV40 to CV-1 monkey kidney cells specifically inhibited the binding of monoclonal antibody (mAb) against class I molecules. Second, preadsorption of anti-HLA mAbs, but not mAbs against other cell surface proteins, inhibited binding and infection by SV40. Third, SV40 selectively binds to class I molecules in cell surface extracts. Fourth, preincubation of SV40 with purified class I molecules inhibits infectivity. Finally, SV40 does not bind to two different human lymphoblastoid cell lines which do not express surface class I MHC molecules. One line has a defect in the β2-microglobulin gene. The other line does not express the three major human classical transplantation antigens, HLA-A, HLA-B, and HLA-C, because of gamma-ray-induced mutations in the MHC. Transfection of these cell lines with the cloned genes for β2 microglobulin and HLA-B8, respectively, restored expression of surface class I MHC molecules and resulted in SV40 binding. SV40 binding

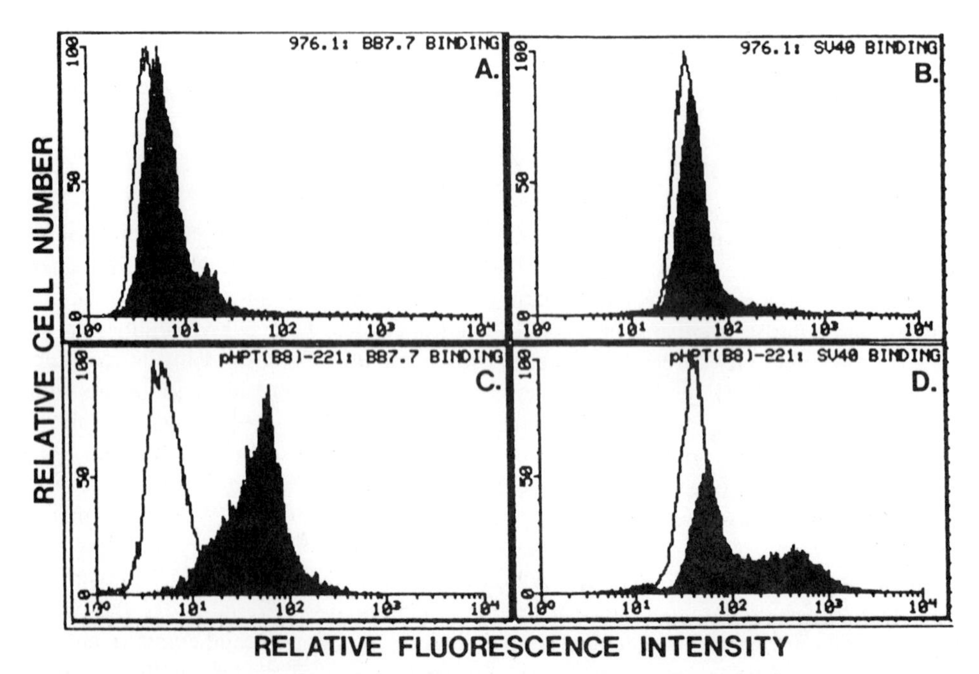

Figure 1. Expression of class I HLA and the SV40 receptor on the HLA-A, HLA-B, HLA-C null mutant 976.1 cells and on the HLA-B8 transfectant line pHPT(B8)-.221. Expression of surface HLA and SV40 binding were assessed by flow cytometry. HLA was detected with anti-HLA mAb BB7.7 by using an FITC-labeled secondary antibody. SV40 was detected by using rabbit anti-SV40 sera and FITC-labeled goat anti-rabbit IgG. (A and B) BB7.7 binding (A) and SV40 binding (B) to 976.1 cells; (C and D) BB7.7 binding (C) and SV40 binding (D) to pHPT(B8)-.221 cells. Open peaks, control fluorescence of cells reacted with FITC-labeled secondary antibody only (A and C) and with rabbit anti-SV40 sera plus FITC-labeled goat anti-rabbit IgG (B and D); shaded peaks, binding of BB7.7 and SV40 as indicated above. Reprinted with permission from Breau, Atwood, and Norkin (1992).

to the HLA-A, HLA-B, HLA-C null mutant cells and to the HLA-B8 transfectant line is shown in figure 1.

2.2. Class I MHC Molecules Do Not Enter with SV40

The role of a virus receptor in entry per se depends on the particular case (Marsh and Helenius, 1989). The possibility that class I molecules might have a direct role in SV40 entry was interesting for the following reasons. First,viruses enter cells either by endocytosis or, in the cases of some enveloped viruses, by membrane fusion (Marsh and Helenius, 1989). Endocytic virus entry is generally mediated by clathrin-coated vesicles. Similar vesicles mediate the internalization of a variety of ligands (e.g. growth factors and peptide hormones) with their receptors. However, SV40 follows a novel entry pathway in which the majority of particles enter by endocytosis through noncoated vesicles. Furthermore, these vesicles deliver the virus to the endoplasmic reticulum (Kartenbeck, Stuckenbrok, and Helenius, 1989). This is particularly interesting since the ER is not considered to be a target for endocytic traffic. Targeting of SV40 to the ER was originally demonstrated by ultrastructure (Kartenbeck, Stukenbrok, and Helenius, 1989). Our results from double-label confocal microscopy (Fig. 2), which gives a less detailed but broader overview of the cell, are in agreement with the earlier findings. Second, class I MHC molecules acquire in the ER the antigenic peptides that they present at the cell surface (reviewed in Yewdell and Bennink, 1992). Since it was not yet certain whether class I molecules at the cell surface recycle to the ER, it was possible that SV40 might target the ER through its association with class I molecules. Finally, although receptor-mediated endocytosis through clathrin-coated vesicles has been extensively studied, little is known about endocytosis through noncoated vesicles.

To ask whether class I molecules internalize either spontaneously or after exposing cells to SV40, cell surface proteins were labeled with 125I using the lactoperoxidase

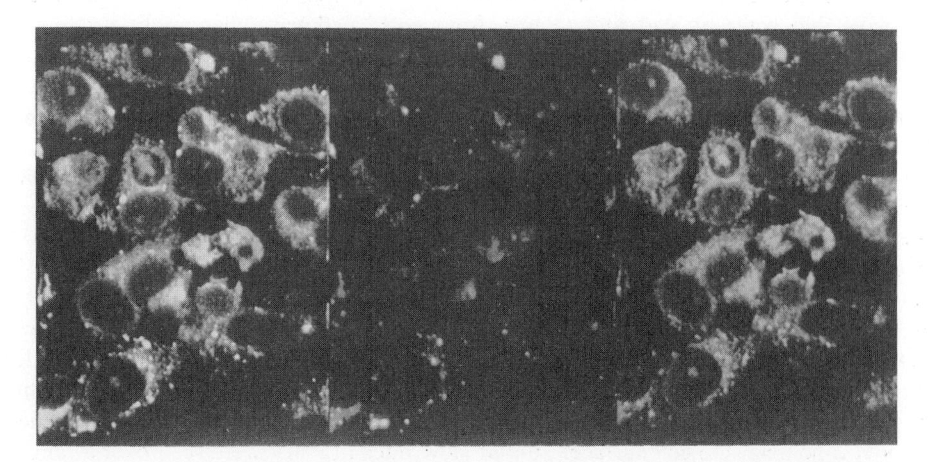

Figure 2. Input SV40 accumulates in the endoplasmic reticulum. CV-1 cells were infected with SV40 (500 PFU/cell) in the presence of cyclohexamide (to prevent synthesis of progeny virions) for 20 hours, fixed, and stained by indirect immunofluorescence with a monoclonal antibody specific for the ER protein, protein disulfide isomerase, and a Texas Red-labeled secondary antibody (left panel) and rabbit antiserum specific for SV40, and an FITC - labeled secondary antibody (middle panel). These panels represent the same field of cells. The two images are shown superimposed in the right panel. The points of overlap of SV40 and PDI appear yellow in the original color micrograph. Since all SV40 is found in regions of the cytoplasm that are labeled by the ER-specific mAB, the figure shows that SV40 accumulates in the ER. Similar results were obtained in the absence of cyclohexamide, as expected since progeny virions do not appear until later.

technique. Surface-labeled class I proteins that might internalize would be resistant to release by papain, which cleaves class I molecules at a specific site just proximal to the plasma membrane. Thus, after various periods of incubation, cells were solubilized and extracts were immunmoprecipitated with antisera specific for the cytoplasmic tail of class I molecules (anti-CT). As late as three hours, neither control cells nor cells exposed to high SV40 inputs (400 PFU/cell) contained papain-resistant 125I-surface-labeled class I molecules (H.A. Anderson and L.C. Norkin, submitted). Thus, class I molecules on CV-1 monkey kidney cells do not appear to internalize either spontaneously or after exposure to SV40. Similar results were obtained when this experiment was done in the presence of brefeldin A, which prevents the exit of class I proteins from the ER. This shows that our results are not explained by rapid recycling of class I molecules between the ER and the cell surface.

In other studies we found that native class I molecules on CV-1 cells spontaneously dissociated into their transmembrane heavy chains and free 2 microglobulin, followed by cleavage and release of the heavy chains by a metalloprotease (Anderson, 1995; H.A. Anderson and L.C. Norkin, submitted). Although there is not yet a concensus on the fate of surface class I molecules, an earlier study showed that class I molecules do not internalize on lymphoid cells (Neefjes et al., 1992). Also, class I molecules are shed from activated T cells by a metalloprotease (Demaria et al., 1994).

2.3. SV40 Entry Is Delayed at the Cell Surface

SV40 bulk entry was examined by adsorbing the virus to cells at 4°C, shifting the cultures to 37°C, and then measuring the amount of virus at the cell surface at various times by flow cytometry. SV40 had a half-life at the cell surface of about 2.5 hours (not shown). This is remarkable since other viruses generally enter within minutes (Marsh and Helenius, 1989). Furthermore, only a small percentage of bound virus left the cell surface during the first hour. Most virus appeared to enter between 1.5 and 3 hours.

The above experiment measured bulk virus entry. It did not distinguish entry pathways that actually lead to infection from entry pathways that might be abortive. Furthermore, it had to be done under conditions of high virus inputs, that might have resulted in artifacts (e.g. from overloading the cells). Thus, we designed an assay to specifically measure infectious entry that would also be accurate under conditions of low virus inputs. The assay was based on our finding that SV40 at the cell surface can be neutralized by anti-SV40 antiserum. This enabled us to determine the time required for preadsorbed SV40 to internalize, since infection would then be resistant to neutralization by antiserum. Infected cells were identified by immunofluorescent staining for the SV40 T antigen, an early SV40 gene product. An assay for infectious entry, based on post-adsorption neutralization, was previously developed for adenovirus (Varga, Weibull, and Everitt, 1991). Our assay for infectious entry showed that at virus inputs as high as 100 PFU/cell, SV40 had infectiously entered only 30% of susceptible CV-1 cells as late as two hours (H.A. Anderson and L.C. Norkin, submitted). Slow SV40 entry did not reflect a property unique to CV-1 cells, since similar results were obtained using HeLa cells. Entry into HeLa cells occurred after a delay of about 2 hours at the cell surface. Similar slow rates of entry were seen over a range of viral inputs from 0.1 to 100 PFU/cell.

Transferrin is considered to be a representative ligand that enters cells by receptor-mediated endocytosis through clathrin-coated vesicles. To establish a control for the above experiments we examined the time course of transferrin internalization. 125I-labeled transferrin was preadsorbed to CV-1 cells at 4°C. Cultures were then shifted to 37°C and the amount of internalized transferrin was determined at various times by measuring the amount of radioactivity resistant to removal by pronase. In contrast to the slow rate of SV40 internalization, most transferrin internalized within 4 minutes.

2.4. SV40 Infects Cells through Noncoated Vesicles

Since SV40 infectious entry is notably slow, and since the majority of SV40 enters cells in noncoated vesicles (Kartenbeck, Stukenbrok, and Helenius, 1989), we asked whether this unusual entry pathway actually leads to productive infection. Cytosol acidification, which disassembles clathrin (Sandvig et al., 1987), was used to block entry via coated vesicles. SV40 infectious entry was monitored using our post-adsorption neutralization assay. Cytosol acidification did not impair SV40 infectious entry. Instead, it actually appeared to enhance infection (H.A. Anderson and L.C. Norkin, submitted). As a control for this experiment, we measured the effect of cytosol acidification on transferrin internalization. As expected, cytosol acidification under these conditions led to a greater than 50% inhibition of transferrin internalization.

2.5. SV40 Infection, But Not Binding, Is Blocked by Drugs That Selectively Disrupt Caveolae

We were interested in the possibility that SV40 infectious entry is mediated by caveolae for several reasons. First, SV40 infectious entry does not appear to occur through coated vesicles. Second, although the function of caveolae is not entirely clear (Severs, 1988), these noncoated vesicles have been implicated in clathrin-independent endocytosis (Milice et al., 1987; Tran et al., 1987; Montesano et al., 1982; Simionescu et al., 1982). Finally, an earlier ultrastructure analysis of SV40 entry showed the presence of SV40 in membrane invaginations that bear a remarkable resemblance to caveolae (Maul et al., 1987) (Fig. 3).

As a first step to determine whether SV40 enters cells through caveolae, we carried out our infectious entry assay under conditions that selectively disrupt caveolae. Nystatin is a cholesterol-binding drug that selectively disrupts caveolae at the cell surface, while not affecting clathrin-coated pits (Rothberg et al., 1992). Activators of protein kinase C, such as phorbol esters, also disrupt caveolae and block their invaginations (Smart et al., 1994). Treatment of cells with nystatin and phorbol 12-myristate 13-acetate (PMA) each reduced SV40 infectious entry by more than 50%. Results with nystatin are shown in figure 4. As a control, we measured the effects of these treatments on the internalization of transferrin under the same conditions as above. PMA and nystatin had no effect on the internalization of transferrin.

Nystatin and PMA treatment did not affect SV40 binding, as measured directly by flow cytometry. Furthermore, the effects of nystatin and PMA on infection were reversible

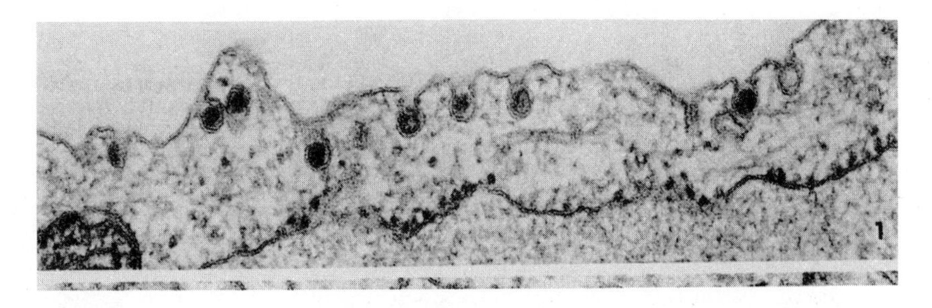

Figure 3. Cross-sectioned cultured mouse embryo cell 24 hrs after infection with SV40. There are no adsorbed virions; all are membrane-enveloped and partially internalized. The RER is intact with ribosomes. X100,000. Reprinted with permission from Maul et al. (1987).

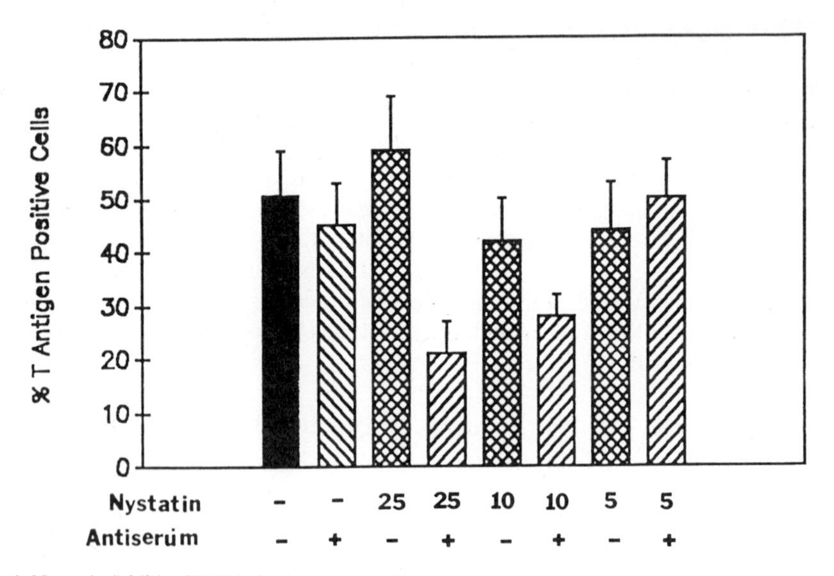

Figure 4. Nystatin inhibits SV40 infectious entry. CV-1 cells were treated with nystatin (25, 10, and 5 μg/ml) or mock-treated as indicated. Cells were then infected with SV40 (10 PFU/cell) for 1 hr on ice and then incubated for 4 hrs at 37°C. Nystatin was continuously present or not present as indicated. Cultures were then treated with anti-SV40 antiserum as indicated. All cultures were then incubated in the absence of nystatin and stained for SV40 T antigen at 48 hrs.

if the cells were not treated with neutralizing antisera prior to removal of the drugs. Thus, caveolae-disrupting drugs block SV40 entry, but not binding.

2.6. SV40 Entry May Be Enhanced by a Transmembrane Signal

Mouse polyomavirus binds to a receptor that is not yet identified, but which is different from the SV40 receptor (Clayson and Compans, 1989). Note that it is not unusual for closely related viruses to use different receptors (Norkin, 1995). Polyomavirus appears to activate cellular early response genes through an intracellular signal transmitted by polyomavirus at the cell surface (Zullo, Stiles, and Garcia, 1987). The functional significance of the polyomavirus signal is not known. SV40 also transmits a signal from the cell surface that leads to the upregulation of early response genes (Breau and Norkin, 1994; W.C. Breau and L.C. Norkin, unpublished results). The SV40 signal can be transmitted by UV-inactivated virus and by noninactivated virus in the presence of cyclohexamide. This shows that SV40 can induce the upregulation of early response genes in the absence of either viral or cellular gene expression. Furthermore, the SV40 signal is sensitive to genistein, a specific inhibitor of protein tyrosine kinases. Activation of a tyrosine kinase is an early step in the signaling pathways induced by a variety of ligands that bind to eukaryotic cells. We do not know whether class I molecules act in transmitting the SV40 signal beyond facilitating binding. However, it is possible since anti-HLA mAbs also activate a genistein-sensitive signal that upregulates early response genes in CV-1 cells (W.C. Breau and L.C. Norkin, unpublished results).

We asked whether the SV40-induced signal might be a factor in SV40 entry by determining if infectious entry is sensitive to genistein. Using our post-adsorption neutralization assay, we found that infectious entry was essentially complete in control cells by 4 hours (Fig. 5). However, when cells were treated with genistein from 0.5 hours before to 4

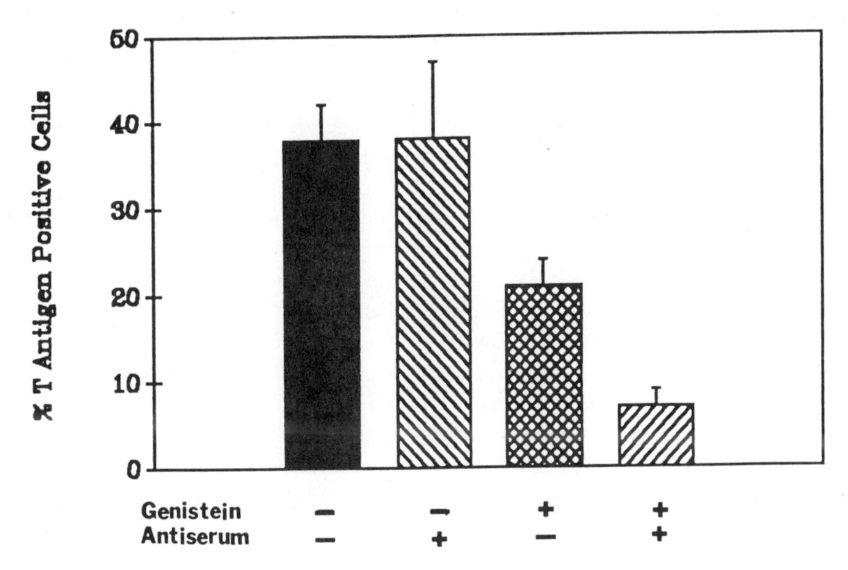

Figure 5. Genistein inhibits SV40 infectious entry. CV-1 cells were treated with genistein (200 μM) or mock-treated as indicated. Cells were then infected with SV40 (1 PFU/cell) for 1 hr on ice, and then incubated for 4 hours at 37°C. Genistein was continuously present or not present as indicated. Cultures were then treated with anti-SV40 antiserum, as indicated. All cultures were then incubated in the absence of genistein and stained for SV40 T antigen at 48 hrs.

hours after infection, and then exposed to anti-SV40 antiserum, there was an 85% reduction in the percentage of infected cells (Fig. 5). The genistein affect on SV40 entry was partly reversible (Fig. 5). Also, genistein did not affect SV40 binding to cells as indicated by flow cytometry, nor did it affect the uptake of transferrin (not shown).

3. CONCLUDING REMARKS

MHC class I molecules are a necessary component of the SV40 receptor, being required for binding that leads to infection (Atwood and Norkin, 1989; Breau, Atwood, and Norkin, 1992). However, our results and those of others show that class I molecules are not sufficient to insure infectious entry. For example, whereas class I molecules are expressed on both the apical and basolateral surfaces of polarized Vero C1008 epithelioid cells, SV40 preferentially infects the apical surfaces of those cells (Clayson and Compans, 1988).

Results described here show that class I molecules do not have a direct role in SV40 entry (although it remains to be determined whether they act in transmitting the SV40 signal). We also show that the integrity of caveolae might be required for SV40 entry, but not binding. Although our pharmacological-based studies implicating caveolae in SV40 entry need to be confirmed by more direct methods, our results show that SV40 uses different cellular factors to mediate binding and entry.

Important examples of other viruses that interact sequentially with multiple cell surface components to bind and enter include HIV and the herpesviruses (Norkin, 1995). Although CD4 is the primary cell surface receptor for HIV, additional target cell membrane components are necessary for entry. A recent study suggests that the interaction of the HIV envelope glycoprotein gp120-gp41 with CD4 induces conformational changes in CD4 that lead to the association of CD4 with accessory transmembrane molecules that can target the virus to clathrin-coated pits (Golding et al., 1995). Herpes simplex virus (HSV) also interacts sequentially with distinct cell surface components to enter cells. Initially, the virus adsorbs to heparan sulfate glycosaminoglycans. This interaction is mediated by HSV glycoprotein C (Herold et al., 1991) and may serve to concentrate the virus at the cell surface or promote a conformational change necessary for entry. HSV glycoprotein D may then interact with a more limited number of cell surface sites to facilitate entry (Johnson and Ligas, 1988). SV40 thus far appears unique in entering cells through noncoated endocytic vesicles, and only after a notable delay at the cell surface. Furthermore, transmission of an extracellular signal may be necessary for SV40 entry, possibly accounting in part for delayed entry.

The virus-receptor interaction is a potential target for the rational design of drugs that might be used to treat virus diseases. Despite the potential of this approach, it has not yet resulted in the development of clinically effective antiviral therapies (Norkin, 1995). We emphasized here that virus-receptor interactions can be complex, involving multiple distinct components on both the virus and the cell. From a practical perspective this complexity may be a cause for optimism since it provides more potential targets for the rational development of antiviral drugs that might act at the points of virus binding, adhesion strengthening, or entry.

4. ACKNOWLEDGMENTS

This work was supported by National Institutes of Health grant CA50532 and National Science Foundation Grant MCB-9219207. We are grateful to Maryanne Wells for expert preparation of the manuscript.

5. REFERENCES

Anderson, H.A., 1995. Ph.D. Thesis, University of Massachusetts, Amherst, MA.

Atwood, W.J., and Norkin, L.C., 1989. Class I major histocompatibility proteins as cell surface receptors for simian virus 40. J. Virol 63:4474-4478.

Breau, W.C., Atwood, W.J., and Norkin, L.C., 1992. Class I major histocompatibility proteins are an essential component of the simian virus 40 receptor. J. Virol. 66:2037-2045.

Breau, W.C., and Norkin, L.C., 1994. Extracellular SV40 activates primary response genes. J. Cell Biochem. Suppl. 18B:254.

Clayson, E.T., and Compans, R.W., 1988. Entry of simian virus 40 is restricted to apical surfaces of polarized epithelial cells. Mol. Cell. Biol. 8:3391-3396.

Clayson, E.T., and Compans, R.W., 1989. Characterization of simian virus 40 receptor moieties on the surface of vero C1008 cells. J. Virol. 63:1095-1100.

Demaria, S., Schwab, R., Gottesman, S.R.S., and Bushkin, Y., 1994. Solubule ß2-microglobulin-free class I heavy chains are released from the surface of activated and leukemia cells by a metalloprotease. J. Biol. Chem. 269:6689-6694.

Geleziunas, R., Bour, S., and Wainberg, M.A., 1994. Human immunodeficiency virus type I-associated CD4 downmodulation. Adv. Virus Res. 44:203-266.

Golding, H., Dimitrov, D.S., Manischewitz, J., Broder, C.C., Robinson, J., Fabian, S., Littman, D.R., and Lapham, C.K., 1995. Phorbol ester-induced down modulation of tailles CD4 receptors requires prior binding of gp120 and suggests a role for accessory molecules. J. Virol. 69:6140-6148.

Haywood, A.M., 1994. Virus receptors: binding, adhesion strengthening, and c lenius, A., 1989. Endocytosis of simian virus 40 into the endoplasmic reticulum. J. Cell Biol. 109:2721-2729.

MacKay, R., and Consigli, R.A., 1976. Early events in polyoma virus infection: attachment, penetration and nuclear entry. J. Virol. 19:620-636.

Marsh, M., and Helenius, A., 1989. Virus entry into animal cells. Adv. Virus Res. 36:107-151.

Maul, G.G., Rovera, G., Vorbrodt, A., and Abramczuk, J., 1978. Membrane fusion as a mechanism of simian virus 40 entry into different cellular compartments. J. Virol. 28:936-944.

Milici, A.J., Waltrous, N.W., Stukenbrok, H., and Palade, G.E., 1987. Transcytosis of albumin in capillary epithelium. J. Cell Biol. 105:2603-2612.

Montesano, R., Roth, J., Robert, A., and Orci, L., 1982. Non-coated membrane invaginations are involved in binding and internalization of tetanus toxins. Nature 296:651-653.

Neefjes, J.J., Smit, L., Gehrmann, M., and Ploegh, H.L., 1992. The fate of the three subunits of major histocompatibility class I molecules. Eur. J. Immunol. 22:1609-1614.

Norkin, L.C., 1995. Virus receptors: implications for pathogenesis and the design of antiviral agents. Clin. Microbiol. Rev. 8:293-315.

Rothberg, K.G., Heuser, J.E., Danzell, W.C., Ying, Y.-S., Glenney, J.R., and Anderson, R.G.W., 1992. Caveolin, a protein component of caveolae membrane coats. Cell 68:673-682.

Sandvig, K., Olsnes, S., Beterson, O.W., and van Deurs, B., 1987. Acidification of the cytosol inhibits endocytosis from coated pits. J. Cell Biol. 105:679-689.

Severs, N.J., 1988. Caveolae, static inpocketings of the plasma membrane, dynamic vesicles, or plain artifact? J. Cell Sci. 90:341-348.

Simionescu, M., Simionescu, N., and Palade, G.E., 1982. Differentiated microdomains on the luminal surface of capillary endothelium: distribution of lectin receptors. J. Cell Biol. 94:406-413.

Smart, E.J., Foster, D.C., Ying, Y., Kamen, B.A., and Anderson, R.G.W., 1994. Protein kinase activators inhibit receptor-mediated potocytosis by preventing internalization of caveolae. J. Cell Biol. 124:307-313.

Tran, D., Carpenter, J.L., Sawano, F., Gorden, G., and Orci, L., 1987. Ligands internalized through coated or non-coated invaginations follow a common intracellulr pathway. Proc. Natl. Acad. Sci. USA 84:7947-7961.

Varga, M.J., Weibull, C., and Everitt, E., 1991. Infectious entry pathway of adenovirus type 2. J. Virol. 65:6061-6070.

Yewdell, J.W., and Bennink, J.R., 1992. Cell biology of antigen processing and presentation to major histocompatibility complex class I molecule-restricted T lymphocytes. Adv. Immunol. 52:1-124.

Zullo, J., Stiles, C.D., and Garcia, R.L., 1987. Regulation of c-myc and c-fos mRNA levels by polyomavirus: distinct roles for the capsid protein VP1 and the viral early proteins. Proc. Natl. Acad. Sci. USA 84:1210-1214.

19

THE EFFECT OF RESPIRATORY VIRUS INFECTION ON EXPRESSION OF CELL SURFACE ANTIGENS ASSOCIATED WITH BINDING OF POTENTIALLY PATHOGENIC BACTERIA

O. R. Elahmer, M. W. Raza, M. M. Ogilvie, C. C. Blackwell, D. M. Weir, and R. A. Elton

Department of Medical Microbiology
University of Edinburgh
Edinburgh, Scotland

1. RESPIRATORY VIRUSES AND INVASIVE BACTERIAL DISEASES

Serious secondary bacterial infection can occur following illness due to respiratory viruses and viral infections have also been suggested to be predisposing factors for bacterial meningitis [Moore *et al.*, 1990; Cartwright *et al.*, 1991]. Respiratory virus infection can compromise host defences against bacterial infection in a number of ways: immune suppression; diminished phagocytosis by polymorphonuclear leucocytes; local tissue injury; loss of mucociliary function and decreased bacterial clearance; formation of exudates that enhance bacterial growth; and increased bacterial binding to virus infected cells. Most investigators have studied associations of influenza virus and respiratory pathogens such as *Streptococcus pneumoniae* [Plotkowski *et al.*, 1986], *Staphylococcus aureus* [Musher and Fainstein, 1981] and *Haemophilus influenzae* [Bakeletz *et al.*, 1988; Fainstein *et al.*, 1980].

Respiratory syncytial virus (RSV) is an important pathogen in infants and young children. It is a major cause of hospitalization in infants during the first 6 months of life and secondary bacterial infections such as acute otitis media and pneumonia are common [Ruuskanen and Ogra, 1993]. It is also a significant cause of serious lower respiratory tract infection in immunocompormised adults and the elderly [Falsey *et al.*, 1995]. In a Finnish study, RSV infections were associated with significant rise of antibodies to *Haemophilus influenzae*, *Branhamella (Moraxella) catarrhalis* and pneumococcal antigens [Korppi *et al.*, 1989]. In the United States, RSV infection was associated with isolation of *Bordetella pertussis* from infants during a winter outbreak [Nelson *et al.*, 1986] and recent studies indicate that RSV infection was significantly associated with invasive pneumococcal disease

Toward Anti-Adhesion Therapy for Microbial Diseases, edited by Kahane and Ofek
Plenum Press, New York, 1996

in both adults and children [Kim *et al.*, 1996]. While influenza virus has been associated with outbreaks of disease due to *Neisseria meningitidis* [Young *et al.*, 1972; Cartwright *et al.*, 1991], RSV is much more prevalent among young children in whom the majority of bacterial meningitis occurs.

Although animal models have been used to study respiratory infections and meningitis, we chose to use a human epithelial cell line which can be infected with RSV. In this model system, we have used flow cytometry methods to demonstrate enhanced binding of the following bacterial species labelled with fluorescein isothiocynate (FITC) to a human epithelial cell line (HEp-2) infected with respiratory syncytial virus (RSV) serogroup A: *Neisseria meningitidis; H. influenzae* serotype b [Raza *et al.*, 1993]; *Staphylococcus aureus* [Saadi *et al.*, 1993]; *B. pertussis* [Saadi *et al.*, 1994; Saadi *et al.*, submitted for publication].

We proposed two hypotheses to explain the increased binding: 1) the virus encoded antigens might be new receptors for the bacteria; 2) viral infection upregulated some "native" cell surface antigens involved in bacterial binding. We found evidence that RSV coded glycoprotein G (attachment) but not glycoprotein F (fusion) acts as an additional receptor for meningococci [Raza *et al.*, 1994]. The G glycoprotein is heavily glycosylated and we predicted that bacterial surface proteins might bind to the carbohydrate moiety of this viral antigen; however, periodate treatment or growth of the virus infected cells in monensin which specifically blocks O-glycosylation [Pressman, 1976], enhanced rather than reduced binding of the bacteria. This suggested that the protein moiety of the glycoprotein might be involved in bacterial binding [Raza, 1992]. The glycoprotein G molecules of RSV subtypes A and B differ in their protein moieties; therefore, we needed to examine both subtypes in the binding studies. If lectin-like interactions contributed to enhanced binding of meningococci, we predicted that carbohydrate containing antigens such as capsules or endotoxin might be involved in these interactions.

2. OBJECTIVES OF THE STUDY

The objectives of this study were: 1) to determine if as observed for serotype A, serotype B of RSV enhanced bacterial binding; 2) to assess binding of other bacterial species associated with invasive disease following RSV infection; 3) to examine the effect of viral infection on expression of host cell antigens proposed to be involved in bacterial binding; 4) to assess bacterial surface antigens that might be involved in interactions with virus infected cells.

3. BINDING OF BACTERIA TO HEP-2 CELLS INFECTED WITH RSV SEROTYPES A OR B

Bacterial binding was assessed by the flow cytometry method described previously. Results were expressed as the binding index calculated as the proportion of cells with fluorescence greater than that of the control multiplied by the mean fluorescence of the population [Raza *et al.*, 1993]. Binding of FITC-labelled and unlabelled bacterial species tested to monolayers of HEp-2 cells was assessed by light microscopy; there were no differences in binding associated with FITC-labelling of the bacteria.

The enhanced binding observed for meningococci and *H. influenzae* type b to cells with RSV subtype A (Edinburgh strain) was also found with RSV subtype B (18573). In addition to meningococci, several species of Gram-negative bacteria exhibited enhanced binding to cells infected with RSV A or RSV B: *B. pertussis*; *Neisseria lactamica*; *Moraxella*

Table 1. Summary of 6 experiments on binding of Gram- negative species to HEp-2 cells and HEp-2 cells infected with RSV-A or RSV-B (200 bacteria : cell, except for *B. pertussis*, 500 : cell)

			% binding to uninfected cells (95%CI)			
Species	Strain	BI uninfected cells	RSV-A	P	RSV-B	P
N. meningitidis	C:2b:P1.2	15971	335 (280-400)	0.001	286 (263-310)	0.001
	P:2b:P1.10	16769	320 (255-400)	0.001	301 (277-327)	0.001
	B:2b:P1.10	13604	352 (225-548)	0.001	292 (225-379)	0.001
N. lactamica	LO1	23675	271 (250-295)	0.001	167 (139-201)	0.001
H. influenzae	type b	20670	267 (199-359)	0.001	197 (157-246)	0.001
B. pertussis	type 1,2	908	234 (169-324)	0.001	171 (144-203)	0.001
B. pertussis	type 1,3	1099	222 (149-330)	0.01	163 (136-197)	0.01

catarrhalis MC1 (Table 1). Binding of *M. catarrhalis* to RSV-infected cells was enhanced for the strain (MC1) which grew in the presence of the selective antibiotics in New York City Medium. Isolate MC2 did not grow on the selective medium, and it was the only exception to the pattern found with the other species tested. The binding to RSV infected cells was significantly reduced compared with the uninfected HEp-2 cells. An additional 6 isolates of this species (MC3-MC8) that did not grow on the selective medium were examined in the assays. The mean binding indices for the antibiotic resistant strain were consistently lower than those for the sensitive strains, and the binding indices of the antibiotic sensitive strains to RSV-infected cells were significantly reduced compared with binding to uninfected cells (Table 2).

Binding indices for strains of *Streptococcus pneumoniae* associated with meningitis were much greater than those for strains associated with respiratory infection; however, binding indices for each strain were significantly greater for cells infected with either RSV A or RSV B (Table 3).

4. Are Host Cell Antigens that Act as Receptors for Meningococci Enhanced by RSV Infection?

While there is evidence that the Lewis[a] antigen is a receptor for *S. aureus* [Saadi *et al.*, 1993; Saadi *et al.*, 1994; Essery *et al.*, 1993] and *Helicobacter pylori* [Alkout *et al.*, 1996], this antigen was not a candidate for investigation in these studies. The monoclonal antibody to Lewis[a] did not bind to HEp-2 cells [Raza, 1992]. It has been suggested that

Table 2. Summary of 6 experiments on binding of *M. catarrhalis* isolates (400 bacteria : cell) to HEp-2 cells and Hep-2 cells infected with RSV-A or RSV-B

			% binding to uninfected cells (95 % CI)			
Bacteria	Isolates	BI uninfected cells	RSV-A	P	RSV-B	P
M. catarrhalis	MC1	597	387 (143-1042)	0.05	298 (139-641)	0.05
	MC2	1761	51 (40-78)	0.01	30 (11-80)	0.05
	MC3	1133	66 (51-86)	0.01	38 (22-65)	0.01
	MC4	3896	64 (50-84)	0.01	31 (12-78)	0.05
	MC5	5101	72 (63-82)	0.01	48 (31-75)	0.01
	MC6	1747	52 (45-60)	0.001	25 (11-55)	0.01
	MC7	4295	74 (64-86)	0.01	50 (31-81)	0.05

Table 3. Summary of 6 experiments on binding of *Strep. pneumoniae* (200 bacteria : cell) to HEp-2 cells and HEp-2 cells infected with RSV-A or RSV-B.

			% binding to uninfected cells (95% CI)			
Species	Strain	BI uninfected cells	RSV-A	P	RSV-B	P
Meningitis strains	7F	12586	164 (128-210)	0.01	153 (122-193)	0.01
	12F	15374	154 (124-192)	0.01	152 (131-177)	0.001
	18C	15330	163 (127-207)	0.01	163 (133-200)	0.01
Respiratory isolates	type 6	265	169 (148-193)	0.001	188 (143-247)	0.01
	type 10	422	176 (129-242)	0.01	164 (120-222)	0.01
	type 23	223	191 (156-236)	0.001	188 (159-223)	0.001
	type 33	219	199 (142-278)	0.01	188 (140-254)	0.01
	type 42	234	182 (150-221)	0.001	191 (164-222)	0.001

integrins involved in cell adhesion and recognition might be used by microbial pathogens as a means of entry into the host cell [Tuomanen, 1993].

By flow cytometry assays we demonstrated binding of monoclonal antibodies to HEp-2 cells and to HEp-2 cells infected with RSV serotype A by the method described previously [Saadi *et al.*, 1993]. The data for 7 experiments is summarized in Table 4. Monoclonal antibodies to CD11a, CD14, CD18 and CD29 but not CD11b or CD11c bound to uninfected HEp-2 cells. Infection of the cells with RSV A resulted in non-significant levels of increased binding of anti-CD11a and anti-CD29, but significantly enhanced binding of anti-CD18 ($P < 0.01$) and anti-CD14 ($P <0.001$). These experiments were repeated with monoclonal antibodies from a second source and similar results obtained.

The results of 7 experiments in which bacterial binding was assessed following pre-treatment of uninfected and RSV-infected HEp-2 cells with monoclonal antibodies is summarized in Figure 1. Treatment of RSV-infected cells with anti-CD29 did not significantly affect binding of *N. meningitidis* strain C:2b:P1.2, but anti-CD18 reduced binding of meningococci to uninfected ($P < 0.05$, 95%CI 56-100) and RSV-infected cells ($P < 0.01$, 95%CI 57-81). There was a slight decrease noted for bacterial binding following pre-treatment of uninfected cells with anti-CD14, but anti-CD14 treatment significantly reduced binding of meningococci to RSV-infected cells ($P < 0.001$, 95%CI 48-73)..

Table 4. Summary of 6 experiments on binding of *N. meningitidis* LOS immunotype strains (400 bacteria : cell) to HEp-2 cells and HEp-2 cells infected with RSV-A or RSV-B

N. meningitidis		% binding to uninfected cells (95% CI)			
immunotype strains	BI uninfected cells	RSV-A	P	RSV-B	P
L1	348	213 (184 -246)	0.001	314 (244 - 406)	0.001
L2	335	188 (153 - 232)	0.001	226 (177 - 290)	0.001
L3	296	211 (183 - 244)	0.001	242 (205 - 286)	0.001
L4	284	209 (184 - 239)	0.001	249 (197 - 316)	0.001
L5	266	212 (179 - 251)	0.001	236 (187 - 300)	0.001
L6	350	183 (161 - 207)	0.001	222 (186 - 264)	0.001
L7	340	193 (161 - 229)	0.001	239 (195 - 294)	0.001
L8	567	260 (180 - 374)	0.01	302 (211 - 433)	0.001
L9	952	188 (132 - 268)	0.01	243 (219 - 269)	0.001
L10	334	167 (156 - 179)	0.001	225 (190 - 266)	0.001
L11	469	351 (219 - 564)	0.01	476 (297 - 762)	0.001
L12	339	174 (130 - 234)	0.01	266 (202 - 350)	0.001

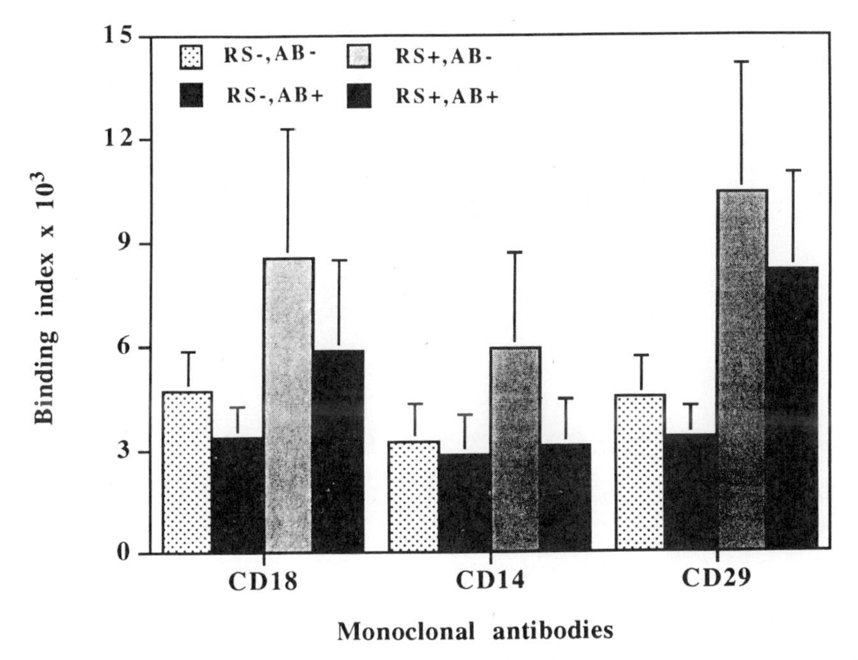

Figure 1. Binding of meningococci to HEp-2 cells and RSV-infected cells teated with monoclonall antibodies to host cell antigens.

5. THE ROLE OF ENDOTOXIN IN BINDING OF MENINGOCOCCI TO RSV INFECTED CELLS

CD14 and CD18 can act as receptors for endotoxin. Binding of sheep erythrocytes and erythrocytes coated with meningococcal LOS [Wright and Jong, 1986 to uninfected and RSV-infected cells in monolayers was assessed microscopically for the number of HEp-2 cells with attached erythrocytes in a total of 100 cells and the numbers of erythrocytes attached to the individual cells counted. A binding index was calculated by multiplying the percentage of cells with erythrocytes attached by the mean number of erythrocytes per 100 cells for each coverslip.

Sheep erythrocytes did not bind to HEp-2 cells in monolayers; however, erythrocytes coated with purified salmonella endotoxin or lipooligosaccharide (LOS) from three strains of meningococci (C:2a:NT, B:15:p1.16; B:4:NT) bound to the monolayers. RSV-infected cells bound more of the endotoxin-coated erythrocytes. Inhibition of binding of the erythrocytes coated with salmonella endotoxin was observed following treatment of the HEp-2 cells with anti-CD14, anti-CD18 or anti-CD29. With RSV-infected cells, binding was reduced by both antibodies. Figure 2 summarizes results from an experiment with LOS from strain C:2a:NT in which binding was reduced by anti-CD14 and anti-CD29.

Previous studies indicated that enhanced binding of *N. meningitidis* strains was not associated with particular serogroup (capsular), serotype (outer membrane protein 2/3) or subtype (outer membrane protein 1) antigens on the bacteria. Strains with different LOS immunotype antigens were assessed for binding to determine if the carbohydrate composition of the endotoxin affected binding to virus infected cells. Each of the 12 immunotype strains obtained from Dr. W. Zollinger, Walter Reed Army Medical Institute exhibited enhanced binding to cells infected with RSV A or RSV B (Table 4).

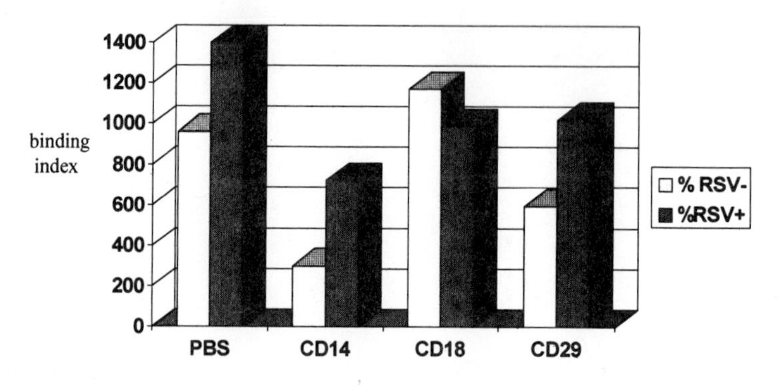

Figure 2. Binding of LOS coated sheep erythrocytes to HEp-2 cells and HEp-2 cells infected with RSV treated with monoclonal antibodies to CD14, CD18 or CD29.

6. ARE THERE DIFFERENCES IN CELL ENVELOPES OF ANTIBIOTIC SENSITIVE STRAINS OF *M. CATARRHALIS* COMPARED WITH THE RESISTANT STRAIN?

Differences in outer membrane proteins are associated with antibiotic resistance of some Gram-negative species [Gutmann *et al.*, 1985]. Outer membranes of *M. catarrhalis* strains were extracted and compared by polyacrylamide gel electrophoresis [Hancock and Poxton, 1988]. Preliminary experiments demonstrated the 7 antibiotic-sensitive strains of *M. catarrhalis* had similar outer membrane profiles which differed from the antibiotic-resistant strain at several bands. The sensitive strains had prominent protein bands at 81 and 66 kDa which were not present in the resistant strain which had a band at 19 kDa that was absent in the sensitive strains.

7. CONCLUSIONS

The first objective of the study was to determine if subtype B of RSV also enhanced binding of bacterial species previously studied in experiments with subtype A. Enhanced binding of meningococci, type B *H. influenzae* and *B. pertussis* was demonstrated in experiments in which both strains of the virus were tested at the same time. Pneumococci cause a significant proportion of bacterial meningitis cases as well as invasive respiratory tract infections. The higher binding indices observed with the pneumococcal strains associated with meningitis were not anticipated and require further investigation to determine which cell surface components are involved in the enhanced binding. The antibiotic sensitive strains of *M. catarrhalis* were the only ones that did not exhibit enhanced binding to the RSV infected cells.

Our results indicate that RSV infection enhances bacterial binding in two ways, production of new surface antigens coded for by the virus (glycoprotein G) and upregulation of two endogenous cell surface antigens. RSV-infected cells had significantly higher binding indices for monoclonal antibodies to CD14 and CD18. These changes might be due to the direct effect of the virus on the cells or mediated through cytokines secreted by infected cells: interleukin 4 decreases CD14 expression on blood monocytes while TNF and interleukin 6 induce moderate increases [reviewed Ziegler-Heitbrock and Ulevitch, 1993].

It has been suggested that cellular antigens normally involved in cell to cell recognition might be "hijacked" by bacteria [Tuomanen, 1993]. There are reports that the complement receptors CR3 (CD11b/CD18) and CR4 (CD11c/CD18) are receptors on myeloid cells for *Escherichia coli* [Wright and Jong, 1986] and a number of intracellular microorganisms [Schlesinger and Horwitz, 1991]. Binding of erythrocytes coated with pertussis toxin to macrophages was inhibited by capping with anti-Lewis^{a} or anti-Lewis^{x} (CD15) antibodies (van t'Wout *et al.*, 1992]; and pre-treatment of epithelial cells with these antibodies inhibited binding of *S. aureus* and *B. pertussis* [Saadi et al., 1993; 1994; submitted for publication]. CD14 acts as a receptor for bacterial endotoxins [Ziegler-Heitbrock and Ulevitch].

Binding of meningococci to HEp-2 cells and RSV-infected cells could be inhibited by pre-treatment of the cells with antibodies to CD14 and CD18. Endotoxins of meningococci appear to be involved in these interactions as demonstrated in the binding of erythrocytes coated with salmonella or meningococcal LOS to HEp-2 cells. Each of the 12 meningococcal immunotype strains exhibited enhanced binding to RSV infected cells as did the other Gram-negative species tested (Table 4).

The decrease in binding of antibiotic-sensitive *M. catarrhalis* strain MC2 to RSV infected cells was the first exception to the pattern of increased binding observed with all other species tested. In a study in Denmark, significantly more children in the 1-48 month age range with upper or lower respiratory tract infections were colonized with *M. catarrhalis* (68%) compared with children without such infections (36%) ($P < 0.001$). After recovery, the isolation rate in the infected group fell to that of the uninfected group [Ejlertsen *et al.*, 1994]. RSV infection is associated with *M. catarrhalis* infections in children; four of 8 patients with seroconversion to *M. catarrhalis* exhibited a concomitant RSV infection [Korppi *et al.*, 1992]. The outer membrane protein profiles of the antibiotic resistant strain and the antibiotic sensitive strains differed at three bands. The 81 kDa protein (CopB) is associated with serum sensitivity and clearance of this organism from the lungs of mice. A mutant strain that did not express this protein acquired wild type levels of serum resistance and the ability to resist pulmonary clearance *in vivo* [Helminen *et al.*, 1993]. A monoclonal antibody to the protein bound to the majority (70%) of *M. catarrhalis* strains tested [Helminen *et al.*, 1993b]. Our strains have not yet been tested for serum sensitivity; however, if the 81 kDa protein is associated with both serum sensitivity and reduced binding to RSV-infected epithelial cells, these strains might be less likely to cause serious disease following RSV infection. Further investigation on contributions of the various outer membrane proteins, particularly the 19 kDa protein present in the MC1 strain that exhibited enhanced binding to the RSV infected cells are needed.

ACKNOWLEDGMENTS

This work was supported in part by the Meningitis Association of Scotland and Chest Heart and Stroke, Scotland. ORA is the recipient of a postgraduate studentship from the Ministry of Higher Education through the Libyan Interests Section of the Saudi Arabian Embassy.

REFERENCES

Alkout AH, Blackwell CC, Weir DM, Luman W, Palmer K. Adhesin of *Helicobacter pylori* that binds to H type 2 and Lewis blood groups: an explanation of increased susceptibility of blood group O and non-secretors to peptic ulcers. In Toward Anti-Adhesin Therapy of Microbial Diseases Ed. I. Kahane and I. Ofek Plenum

Cartwright KAV, Jones DM, Smith AJ, Stuart JM, Kaczmarski EB, Palmer SR. Influenza A and meningococcal disease. Lancet 1991; 338: 554-557.

Ejlertsen T, Thisted E, Ebbesen F, Olesen B, Renneberg J. *Branhamella catarrhalis* in children and adults. A study of prevalence, time of colonisation, and association with upper and lower respiratory tract infections. J Infect 1994; 29: 23-31.

Falsey AR, Cunningham CK, Barker WH, Kovides RW. Respiratory syncytial virus and influenza A infection in hospitalized elderly. J Infect Dis 1995; 172: 389-394.

Gutmann L, Williamson R, Moreau N, Kitzis M-D, Collatz E, Acar JF, Goldstein FW.. Cross resistance to naladixic acid, trimethoprim and chloramphenicol associated with alterations in outer-membrane proteins of *Klebsiella, Enterobacter* and *Serratia.* J Infect Dis 1985; 151: 501-507.

Hancock IC, Poxton IR. Bacterial cell surface techniques. John Wiley and Sons Ltd. Chichester: 1988.

Helminen ME, McIver I, Latimer JL, Cope LD, McCracken GH, Hansen EJ. A major outer membrane protein of *Moraxella catarrhalis* is a target for antibodies that enhance pulmonary clearance of the pathogen in an animal model. Infect Immun 1993; 61: 2003-2100.

Helminen ME, McIver I, Paris J, Latimer JL, Lumbley SL, Cope LD, McCracken GH, Hansen EJ. A mutation affecting expression of a major outer membrane protein of *Moraxella catarrhalis* alters serum resistance and survival in vivo. J Infect Dis 1993; 168: 1194-1201.

Kim PE, Musher DM, Glezen WP, Rodriguez-Barradas MC, Nahm WK, Wright CE. Association of invasive pneumococcal disease with season, atmospheric conditions, air pollution and the isolation of respiratory viruses. Clin Infect Dis 1996; 22: 100-106.

Korppi M, Matila ML, Jaaskelainen J, Leinonen M. Role of *Moraxella (Branhamella) catarrhalis* as a respiratory pathogen in children. Acta Pediatrica 1992; 81: 993-996.

Korppi M, Leinonen M, Koskala M, Makela H, Launiala K. Bacterial coinfection in children hospitalised with respiratory syncytial virus. Pediatr Infect Dis 1989; 8: 687-692.

Moore PS, Hierholzer J, De Witt W, Gount K, Lippveld T, Plikaytis B, Broome CV. Respiratory viruses and mycoplasma as cofactors for epidemic group A meningococcus meningitis J Am Med Assoc 1990; 264: 1271-1275.

Nelson WJ, Hopkins RS, Roe MH, Glode MP. Simultaneous infection with *Bordetella pertussis* and respiratory syncytial virus in hospitalized children. Pediatr Infect Dis 1986; 5: 540-544.

Pressman BC. Biological application of ionophores. Ann Rev Biochem 1976; 45: 501-530.

Raza MW. Viral infections as predisposing factors for bacterial meningitis. PhD Thesis University of Edinburgh 1992.

Raza MW, Ogilvie MM, Blackwell CC, Stewart J, Elton RA, Weir DM. Effect of respiratory syncytial virus infection on binding of *Neisseria meningitidis* and type b *Haemophilus influenzae* to human epithelial cell line (HEp-2). Epidemiol Infect 1993; 110: 339-347.

Raza MW, Blackwell C., Ogilvie MM, Saadi AT, Stewart J, Elton RA, Weir DM. Evidence for the role of glycoprotein G of respiratory syncytial virus in binding to *Neisseria meningitidis* to HEp-2 cells. FEMS Immunol Med Microbiol. 1994; 10: 25-30.

Ruuskanen O, Ogra PL. Respiratory syncytial virus Curr Prob Pediat 1993; 23: 50-79.

Saadi AT, Blackwell CC, Mackenzie DAC, Busuttil A, Raza MW, Essery SD, Weir DM, Elton RA, Brooke H, Gibson AAM. Immunisation against *Bordetella pertussis* and the decline in SIDS in southeast Scotland. Third SIDS International Congress. 1994; p. 118.

Saadi AT, Blackwell CC, Raza MW, James VS, Stewart J, Elton RA, Weir DM. Factors enhancing adherence of toxigenic *Staphylococcus aureus* to epithelial cells and their possible role in sudden infant death syndrome. Epidemiol Infect 1993; 110: 507-517.

Saadi AT, Weir DM, Poxton IR, Stewart J, Essery SD, Raza MW, Blackwell CC, Busuttil A . Isolation of an adhesin from *Staphylococcus aureus* that binds Lewis^a blood group antigen and its relevance to sudden infant death syndrome. FEMS Immunol Med Microbiol 1994; 8: 315-320.

Saadi AT, Blackwell CC, Essery SD, Raza MW, Weir DM, Elton RA, Busuttil, A, Keeling JW. Developmental and environmental factors that enhance binding of *Bordetella pertussis* to human epithelial cells in relation to sudden infant death syndrome. (submitted for publication)

Schlesinger LS, Horwitz MA. Phagocytosis of *Mycobacterium leprae* by human monocyte derived macrophages is mediated by complement receptors CR1 (CD35), CR3 (CD11b/CD18) and CR4(CD11c/CD18) and INFγ activation inhibits complement receptor function and phagocytosis of this bacterium. J Immunol 1991; 147: 1983-1994.

van t'Wout J, Burnette WN, Mar VL, Rozdzinski E, Wright SD, Tuomanen E. Role of carbohydrate recognition domains of pertussis toxin in adherence of *Bordetella pertussis* to human macrophages. Infect Immun 1992; 60: 3303-3308.

Wright SD, Jong MTC. Adhesion-promoting receptors on human macrophages recognize *Escherichia coli* by binding to lipopolysaccharide. J Exp Med 1986; 164: 1876-1888.

Ziegler-Heitbrock WL, Ulevitch RJ. CD14 cell surface receptor and differentiation marker. Immunol Today 1993; 14: 121-125.

20

ANTI-*ESCHERICHIA COLI* ADHESIN ACTIVITY OF CRANBERRY AND BLUEBERRY JUICES

I. Ofek,[1] J. Goldhar,[1] and N. Sharon[2]

[1] Department of Human Microbiology
Sackler School of Medicine
Tel-Aviv University
Tel Aviv 69978
Israel
[2] Department Membrane and Biophysics
The Weizmann Institute of Science
Rehovot 76100, Israel

INTRODUCTION

For many decades, cranberry juice has been recommended by physicians in North America for the treatment or prevention of urinary tract infections (UTI). However, until recently, no experimental evidence has been presented for the purported beneficiary effect of the juice, nor for its mechanism of action. For a while, it was believed that the antibacterial effect of the juice is due to its ability to acidify urine and thus to inhibit bacterial growth but no support for this was found (Kunin, 1987; Bodel et al., 1959).

A new approach to protect against bacterial infections has been proposed based on agents which inhibit the adhesion of bacteria to epithelial cells lining the mucosal surfaces of the body (Aronson et al., 1979; also reviewed in Ofek and Doyle, 1994). This has prompted the search for anti-adhesion agents in cranberry juice.

Sobota (1984) and Schmidt and Sobota (1988) were the first to report that cranberry juice cocktail inhibited the adherence of urinary tract isolates of *Escherichia coli* to human uroepithelial cells. The strains used in these studies, however, were not defined with respect to the type and receptor specificity of fimbrial adhesins they might have expressed. *E. coli* can make at least two type of adhesins (Duguid and Old, 1980 ; Ofek and Doyle , 1994). The organisms are capable to switch off and on the expression of each of the type of adhesins to gain a better foothold when moving into new environments, such as when they move from the intestinal system to the urinary tract. One type of adhesin is mannose specific (MS) associated with type 1 fimbriae and shared by most *E.coli* isolates. The other class of adhesins, collectively termed mannose resistant (mannose resistant) adhesins, may be in fimbrial and non-fimbrial form and differ in their specificity among the various isolates.

Toward Anti-Adhesion Therapy for Microbial Diseases, edited by Kahane and Ofek
Plenum Press, New York, 1996

Virtually all strains of *E.coli* are capable of expressing a mannose specific fimbrial lectin which mediates the adherence of the bacteria to uroepithelial cells. In addition, most pyelonephritogenic isolates of *E.coli* express a Galα1-4Gal specific lectin associated with P fimbriae, a mannose resistant adhesin which also mediates the adherence of the bacteria to uroepithelial cells.

We employed the yeast aggregation assay and hemagglutination of human group A erythrocytes to monitor the activity of the mannose sensitive type 1 fimbrial adhesin and of mannose resistant adhesins, respectively (Zafriri et al, 1989; Ofek et al, 1991; 1993).

ANTI-TYPE 1 FIMBRIAL ADHESIN OF CRANBERRY JUICE

All juices tested (guava, pineapple, mango, grapefruit, blueberry juice (BJ), as well as cranberry juice cocktail (CJC), inhibited activity of the MS type 1 fimbrial adhesin (Zafriri et al, 1989). This inhibition was abolished by dialysis and was shown to be due to the fructose content of the juices, suggesting that fructose is the sole inhibitor of this adhesin in the juices. Fructose was previously reported to inhibit the MS type 1 fimbrial adhesins, albeit with one tenth the potency of D-mannose (Old, 1972; Salit and Gottsclich, 1977). All the juices tested, however, contained at least 10 times or more (e.g.>1% fructose) the concentration needed to cause 50% inhibition in the yeast aggregation assay.

HIGH MOLECULAR WEIGHT INHIBITOR OF CRANBERRY JUICE

Cranberry and blueberry juices (from berries of plants belonging to the genus of *Ericaceae*), were found to contain an adhesion-inhibitor not shared by the other juices tested and which acted specifically on the P fimbriae expressed by uropathogenic *E. coli* (Zafriri et al., 1989; Ofek et al., 1991). These juices at 1:4 dilution inhibited activity of mannose resistant adhesins expressed by all 30 urinary isolates and by one blood isolate (probably originating from a patient with urosepsis). Only four of the 22 non-urinary isolates (20 fecal and two meningitis) tested were inhibited by the juices. As the source of all urinary isolates is the gut, we cannot exclude the possibility that some or all four fecal isolates whose mannose resistant adhesin was inhibited by the juices were urinary pathogens colonizing the gut of patients suffering from diarrhea caused by other agents.

The inhibitory activity in both cranberry and blueberry juice was non-dialyzable (M.W. cut off >15000), acid resistant (1N HCl, 30 min. room temp.) and sensitive to alkali (pH >9). The high molecular weight inhibitor obtained from cranberry after extensive dialysis of the juice or cocktail was resistant to tyrosine and heat (100°C , 30 min.) and nitrogen free, showing that it is not a protein. Other tests have led to the conclusion that it is not of lipid nor of polysaccharide nature.

The anti-adhesin activity of the high M.W. inhibitor required preincubation of the bacteria with the inhibitor and was readily absorbed by strains expressing the mannose resistant adhesin, but not by strains that do not express it, suggesting that the bacterial adhesin is probably the target for the inhibitory activity of the juices. This notion was confirmed by showing that the heamagglutinating activity of the soluble adhesin obtained from NFA-1-carrying *E. coli* was decreased more than four-fold by its preincubation with cranberry juice cocktail (Table 1).

Table 1. Effect of cranberry juice cocktail (CJC) on hemagglutination caused by soluble non-fimbrial adhesin NFA I from *E. coli* 827[a]

	Titer of inhibition of CJC (4 mg/ml)	
Preincubation time of CJC (4 mg/ml) with NFA (2 mg/ml)	Unabsorbed LE, 392+ [b]	Absorbed with LE392-[b]
0	1:2	<1:1
2	1:4	<1:1
30	1:8	<1:1

[a]NFA, non fimbrial hemagglutinin was extracted as described elsewhere (Goldhar et al, 1984).
[b] *E. coli* (strain LE, 392) bearing plasmid possessing or lacking genes coding for NFA adhesin (Hale et al, 1988).

Table 2. Preparation of PF-1 fraction from cranberry juice cocktail

Step	Procedure
1	Cranberry juice cocktail, 1.5 lit. (or juice concentrate, 25 ml)
2	Dialysis (~15000 M.W. cut-off)
3	Nondialysable material (1g)
4	Fractionation on polyacrylamide resin (biogel P-2) in phosphate buffer pH 7.5 solution and elution with distilled water
5	Water eluted fraction, lyophilize (PF-1 100 mg)

PURIFICATION OF THE HIGH M.W. INHIBITOR

The dialyzed inhibitor derived from cranberry juice cocktail was adsorbed to Bio-Gel P2 equilibrated with phosphate buffered saline. A partially purified inhibitor was obtained by elution with distilled water from Bio-Gel P2 column. The product , PF-1, inhibited the mannose resistant adhesins of all 20 urinary isolates tested in concentrations of 12 to 25 µg/ml. A typical procedure for obtaining PF-1 fraction is shown in Table 2 where 100mg of PF-1 could be obtained from 1.5 lit. of cranberry juice cocktail. This correspond roughly to an inhbitory titer of 1:16-1:18 of the cocktail (Table 3), suggesting that most of the inhibitory activity of the juice against P fimbrial adhesin is contained in PF-1

Table 3. Chemical analysis of PF-1 fraction from cranberry juice cocktail

Type of analysis	Results
Carbon content	56.6% (w/w)
Hydrogen content	4.14 % w/w)
Nitrogen, Sulfur, Chlorine	Undetectable
Special analysis[a]	Decolorize an aqueous solution of mthylene blue giving water-insoluble precipitate
	Do not migrate on TLC developed with 7% acetic acid and sprayed with ferric chloride and give greenish gray color
	Turned to red upon heating in a mixture of n-butanol and concentrated hydrochloric acid

[a] Performed by Okuda (1995); All characteristics are typical to those observed for condensed tannins (proanthocyanidins, possibly a mixture of flavan derivatives (catechin or analogs) condensed with each other to various extents).

Table 4. Inhibition of *E. coli* adhesins by cranberry juice fractions

Strains and adhesins	Minimum concentration (μg/ml) or dilution needed to cause 50% inhibition of			
	Cranberry juice fraction			
	Nondialysable material	PF-1	Tannic acid (Sigma)	Cranberry juice cocktail
IHE, P fimbriae	23.5±7.8	14.5±4.5	>100[a]	1:8-1:1:6
H-10407, CFA/I	>500[a]	280±93	>500[a]	<1:3
GOR75, CFA II	>500[a]	470±23	>500[a]	<1:3
346, Type 1 fimbria	>500[a]	>500	>500[a]	1:10-1:16[b]

a, Higher concentrations caused agglutination of the bacteria
b, Due to fructose concentration in the cocktail

Analysis of PF-1 by Okuda (1995) led to the conclusion that it is comprised mainly of condensed tannins (proanthocyanidins) (Table 4). However, commercial tannin (Sigma) lacked inhibitory activity at concentrations as high as 100μg/ml (Table 3).

CONCLUSIONS

The results suggest that both cranberry and blueberry juice contain a high molecular weight constituent which selectively inhibits mannose resistant adhesins produced by urinary isolates of *E. coli* by binding to the bacterial surface, possibly to the adhesin itself.

It thus appears that the claimed beneficiary effect of cranberry juice in prevention of recurrent urinary tract infections may be due to the presence of both mannose sensitive and mannose resistant anti-adhesive agents in the juice. The first well controlled large scale clinical study as described in this meeting confirmed this notion (Avorn et al., 1994). Because the source of most uropathogenic *E. coli* is the gut, the possibility that cranberry juice (and blueberry juice) may act in the gut or in the bladder or both should be considered.

ACKNOWLEDGMENTS

We thank Prof. Okuda from Okayama University, Japan for the analysis of PF-1.

REFERENCES

Aronson, M., Medalia, O., Schori, L., Mirelman, D., Sharon, N., and Ofek, I. (1979). Prevention of *E. coli* colonization of the urinary tract by blocking bacterial adherence with α-methyl-D-mannopyranoside. J. Infect. Dis. 139: 329-332.

Avorn, J., Monane, M., Gruwitz, J. H., Glynn, R. J., Choodnovskiy, I. and Lipsitz, A. (1994). Reduction of bacteriuria and pyuria after ingestion of cranberry juice. J. Amer. Med. Assoc. 271: 751-754

Bodel, P. T., Cotran, R., and Kass, E. H. (1959). Cranberry juice and the antimicrobial action of hippuric acid. J. Lab. Clin. Med. 54: 881-888.

Duguid, J. P., and Old, D. C. (1980). Adhesive properties of Enterobacteriaceae.186-217. In E.H. Beachey (ed), Bacterial Adherence: Receptor and Recognition., Series B, Vol. 6.London: Chapman & Hall.

Goldhar, J., Peri, R., and Ofek, I. (1984). Extraction and properties of non-fimbrial mannose resistant hemagglutinin from urinary isolate of Escherichia coli. Infect. Immun.2, 49-54

Hales, B.A., Beverley-Clarke, H., High, N.J. Jann, K., Perry, R. Goldhar, J. and Boulnois. G.Y. (1988). Molecular cloning and charactherization of the genes for a non-fimbrial adhesin from *Escherichia coli*. Microbial Pathogenesis 5:9-15.

Kunin, C. M. (1987). Detection, prevention and management of urinary tract infections. Lea & Febiger, Philadelphia, pp 299-323.

Ofek I., Goldhar J., Zafriri D., Lis H., Adar R., and Sharon N. (1991). Anti-*Escherichia coli* adhesin activity of cranberry and blueberry juices. New Eng. J. Med. 324: 1599.

Ofek, I., Zafriri, D., Goldhar, J. , Heiber, R., Lis, H. and Sharon, N. (1993). Effect of various juices on activity of adhesins expressed by urinary and nonurinary isolates of *Escherichia coli*. In America's Foods Health Messages and Claims: Scientific, Regulatory, and Legal Issues (J. Tilloston, ed) CRC press, Inc. pp 193-201.

Ofek, I., and Doyle, R. J. (1994). Adhesion of Bacteria to Cells and Tissues, Chapter 11. Chapman and Hall, New York.

Okuda, T. (1994). Okayama Univ., Japan, personal communication

Old, D. C. 1972. Inhibition of the interaction between fimbrial hemagglutinins and erythrocytes by D-mannose and other carbohydrates. J. Gen. Microbiol. 71:149-157

Salit, I. E., and Gottsclich, E. C. (1977). Haemaglutination by purified type 1 *Escherichia coli* pili. J. Exp. Med. 146:1169-1181

Schmidt, D. R., and Sobota, A. E. (1988). An examination of the anti-adherence activity of cranberry juice on urinary and nonurinary bacterial isolates. Microbios 55: 173-181.

Sobota, A. E. (1984). Inhibition of bacterial adherence by cranberry juice: Potential use for treatment of urinary tract infection. J. Urol. 131: 1031-1016.

Zafriri, D., Ofek, I., Adar, R., Pocino, M., and Sharon, N. (1989). Inhibitory activity of cranberry juice on adherence of type 1 and type P fimbriated *Escherichia coli* to eukaryotic cells. Ant. Microbial Agt. Chem. 33: 92-98.

21

THE EFFECT OF CRANBERRY JUICE ON THE PRESENCE OF BACTERIA AND WHITE BLOOD CELLS IN THE URINE OF ELDERLY WOMEN

What Is the Role of Bacterial Adhesion?

Jerry Avorn

Program for the Analysis of Clinical Strategies
Department of Medicine
Brigham and Women's Hospital
Harvard Medical School
Boston, Massachusetts

For generations, a belief has persisted that the ingestion of cranberry juice somehow protects the urinary tract against infection, but until recently no adequately designed large-scale clinical trials had been performed to test this hypothesis. We conducted a randomized, double-blind, placebo-controlled clinical trial to measure the effect of regular ingestion of cranberry juice on the bacterial flora of older women. The study sample consisted of 153 women (mean age 78.5 years) who were randomly allocated to an experimental group, which drank 300 ml per day of a standard commercially available cranberry beverage containing 26% juice; the controls were instead asked to drink the same amount of a synthetic placebo drink prepared for this study, which had an identical taste and color, but contained no cranberry product. Both drinks contained the same amount of vitamin C. Urines were collected monthly for six months and tested for the presence of bacteria and white blood cells. Bacteriuria was defined as the presence of $\geq 10^5$ organisms per ml.

We found that women randomly assigned to the cranberry beverage group had significantly less bacteriuria with pyuria than controls; their likelihood of having a urine positive by these criteria was only 42% than that seen in the controls (P=.004). Analysis of probabilities of transition indicated that women in the experimental group with a positive urine in a given month had only 27% the likelihood of controls to have their urine remain positive in the following month (P=.006). The pH measurements of urines in both groups were nearly identical, indicating that urinary acidification did not explain these findings. This clinical trial provides an interesting *in vivo* counterpoint to laboratory-based studies of the effect of cranberry juice on bacterial adhesion. However, additional studies are necessary

Toward Anti-Adhesion Therapy for Microbial Diseases, edited by Kahane and Ofek
Plenum Press, New York, 1996

to determine whether this is indeed the mechanism responsible for the observed findings. A follow-up study is currently being completed examining the effect of cranberry juice on the clinical frequency of symptomatic urinary tract infections in younger women.

22

ROLE OF HUMAN MILK CONSTITUENTS IN BLOCKING THE ADHERENCE OF ENTERIC PATHOGENS

Shai Ashkenazi

Schneider Children's Medical Center of Israel
Felsenstein Medical Research Center
Beilinson Campus, Petah Tiqva
Sackler Faculty of Medicine
Tel Aviv University
Tel Aviv, Israel

1. INTRODUCTION

1.1. Human Milk and Enteric Infections

Breast feeding has been associated with a reduced incidence of infantile infections, particularly gastroenteritis (Larsen and Homer, 1978; Howie et al., 1990). The protection might be related to the high content of immunoglobulins, especially secretory IgA, in human milk. Immunoglobulin-mediated activity, however, usually depends on previous exposure of the mother to the causative microorganism and may therefore be of limited importance in developed countries. Other protective mechanisms may also play a role.

1.2. Adherence to the Gastrointestinal Tract

The adherence of microorganisms to the intestinal mucosa is a crucial initial step in the infectious process (Ofek and Doyle, 1994). It enables the organisms to resist expulsion by the peristaltic clearing mechanism with colonization and then expression of other virulent traits (Boedeker, 1982). Adherence is usually mediated by the binding of bacterial surface proteins, usually called adhesins, to intestinal receptors, which are typically sugar residues on cell surface glycoproteins or glycolipids. Certain carbohydrates can therefore competitively inhibit the binding of some bacteria (Ofek and Doyle, 1994). Likewise, enterotoxins bind to intestinal receptors before initiating their functional activity.

Toward Anti-Adhesion Therapy for Microbial Diseases, edited by Kahane and Ofek
Plenum Press, New York, 1996

1.3. Human Milk Carbohydrates

Human milk is unique with regard to its content of complex carbohydrates, containing free oligosaccharides, glycoproteins and glycolipids. The total content of free oligosaccharides in human milk is 3 to 6 gm/l, as compared to only traces in cow's milk. With advanced methods of isolation and characterization, more than 130 oligosaccharides, containing up to about 20 monosaccharides per molecule, have been identified (Kunz and Rudloff, 1993). In addition to free lactose, most oligosaccharides detected so far carry lactose at their reducing end. Elongation is usually performed by an enzymatic attachment of N-acetylglucosamine residues linked in β1-3, β1-4 or β1-6 linkage to galactose residues. Further variations are achieved by adding fucose or sialic acid (Ginsburg and Robbins, 1984, Kunz and Rudloff, 1993).

Many human milk oligosaccharides are present in low quantities or are absent in cow's milk. Moreover, oligosaccharides in the urine of breast-fed and formula-fed infants differ considerably; some compounds, mainly focusylated oligosaccharides, are virtually absent in the urine of formula-fed infants (Kunz and Rudloff, 1993). This might suggest intestinal absorption and urine excretion of intact oligosaccharides from breast milk. Likewise, comparison of the ganglioside content of human, bovine and formula milk showed significant qualitative and quantitative differences (Lagreid et al., 1986).

The glycoconjugates that are found in human milk may act as receptor analogues, thereby interfering with the binding of enteric pathogens or their toxins to the intestine. Several studies have examined this hypothesis.

2. METHODS

2.1. Microorganisms and Toxins

Several classes of diarrheogenic *Escherichia coli* have been studied. These include enterotoxigenic and enterohemorrhagic *E. coli*, which adhere to the intestinal mucosa and then produce toxins; enteropathogenic *E. coli*, which adhere to and efface the microvillous surface of intestinal cells; and *E. coli* containing S fimbriae, which can cause infections outside the intestinal tract. A variety of other organisms, among them *Candida albicans* and Rota virus, have been studied (Table 1). In addition, the effect of human milk on the binding of several enteric toxins, including enterotoxins such as Cholera toxin and heat-labile and heat-stable enterotoxins, and cytotoxins such as Shiga toxin and Shiga-like toxins produced by E. coli, has been evaluated.

2.2. Experimental Systems

In vitro assays, animal models and human specimens have been utilized as experimental systems to study receptor analogues in human milk (Table 2). Binding to erythrocytes and hemagglutination were used as an indicator of intestinal binding. *In vitro* binding of *E. coli* to HeLa or HEp-2 monolayers was used for enteropathogenic *E. coli*, while toxin binding to human milk glycolipids was examined after separating the glycolipids using high-performance thin-layer chromatography (HPTLC). Rabbit or guinea pig enterocytes, rabbit ileal brush border membrane vesicles and infact mice were used in the animal experiments. Human specimens included human ileostomy glycoproteins and buccal epithelial cells.

Table 1. Inhibition of the attachment of enteric microorganisms or toxins by human milk fractions

Microorganism/toxin	Experimental system	Human milk fraction(s)
Enteropathogenic *E. coli*	HeLa/HEp-2 cells	Fucosylated oligosaccharides
Enterotoxigenic *E. coli*	Hemagglutination	Glycoconjugates
CFA/I	Guinea pig enterocytes	Glycoconjugates
CFA/II	Hemagglutination	Mannosylated glycoproteins
Type 1 fimbriae	Rabbit enterocytes	Mannosylated glycoproteins
Enterohemorrhagic *E. coli*	Buccal cells/ Ileostomy	Sialylated oligosaccharides
E. coli-S fimbriae	glycoproteins	Sialylated glycoconjugates
C. pylori	Hemagglutination	Neutral oligosaccharides
V. cholerae	Hemagglutination	Glycoconjugates
Y. enterocolitica	Rabbit brush border	Fucα1-2Gal residues
C. albicans	Buccal cells	Mucin
Rota virus	Infant mice	GM_1 ganglioside
Cholera toxin	Hemagglutination	GM_1 ganglioside
Heat-labile enterotoxin	Hemagglutination	Fucosylated oligosaccharides
Heat-stable enterotoxin	Infant mice	Galα1-4Galβ1-4Glc-ceramide
Shiga toxin	HPTLC-separated binding	Galα1-4Galβ1-4Glc-cermide
Shiga-like toxins	PPTLC-separated binding	

3. RESULTS

The inhibitory effects of human milk on the attachment of enteric pathogens or toxins are summarized in Table 1. The binding of a variety of enteropathogens is inhibited as is the binding of enterotoxins and cytotoxins. The inhibition is sometimes as high as 90%, while in other situations it is of a lower magnitude. For some microorganisms the molecule(s) that interfere with the binding have been well characterized and isolated from human milk or colostrum; for others, there is only indirect evidence of the involvement of glycoconjugates in the inhibitory activity, for instance by periodate or mannosidase treatments. Results with selected pathogens and enterotoxins are described below.

Table 2. Experimental systems utilized to study the effect of human milk on the adherence of enteropathogens and enteric toxins

1. *In vitro* assays
   a. Hemagglutination with human or animal erythrocytes.
   b. Adherence to monolayers of HeLa or HEp-2 cell lines.
   c. Toxin binding to human milk fractions separated by high performance thin layer chromatography (HPTLC)
2. Animal studies
   a. Rabbit or guinea pig enterocytes.
   b. Intact intestinal surfaces of guinea pigs.
   c. Rabbit ileal brush border.
   d. Rabbit ilial loops.
   e. Infant mice.
3. Human specimens
   a. Ileostomy-obtained glycoproteins
   b. Ileostomy-obtained intestinal cells
   c. Buccal epithelial cells.

3.1. Diarrheogenic Escherichia Coli

Enterohemorrhagic *E. coli* (EHEC) has emerged in recent years as an important pathogen and a cause of major outbreaks. This organism can cause nonbloody afebrile diarrhea, hemorrhagic colitis with severe abdominal pain and bloody stools and late toxin-related complications: hemolytic-uremic syndrome and thrombotic thrombocytopenic purpura.

While examining the adherence of EHEC to rabbit intestinal cells, obtained from the small and large intestine, we found significant inhibition by D-mannose, α-methyl mannoside (about 50% inhibition) and L-fucose (about 20% inhibition). Skimmed human milk and colostrum inhibited (about 40%) the attachment of EHEC to the small and large intestine (Ashkenazi, 1994; Ashkenazi et al., 1991). The inhibition was in the protein fraction of human milk and was not affected by boiling or trypsin treatment, but was nearly abolished by treatment with periodate or α-mannosidase. The protective protein fraction bound to concanavalin column and was eluted by methyl mannoside. It was concluded that mannosilated glycoproteins in human milk inhibit the adherence of EHEC.

Among the various classes of diarrheogenic *E. coli*, enterotoxigenic *E. coli* (ETEC) were the first examined for the effect of human milk on their adherence. Holmgren and his colleagues (1981) used hemagglutination as a measure of bacterial adherence, and examined milk and colostrum fractions, whose non-immunoglobulin fractions were prepared by immunosorbent columns. They found significant inhibition by human milk and colostrum, which was more pronounced by the non-immunoglobulin fractions, suggesting receptor analogues. Later studies (Holmgren et al., 1983) suggested that glycoproteins and free oligosaccharides mediated the inhibitory effects of human milk.

Using animal studies, we have examined the effects of human milk on the adherence of ETEC to guinea pig intestinal cells and intact intestinal sections (Ashkenazi and Mirelman, 1987). Non-immunoglobulin fractions of human milk and colostrum caused about 50% inhibition of the adherence of only *E. coli* H10407, which is mediated by colonization factor antigen (CFA) I, and *E. coli* E1392, which is mediated by CFA/II. The inhibitory activity resisted proteolytic digestion with trypsin and boiling, but was nearly completely abolished by periodate treatment, indicating that carbohydrate residues were involved (Ashkenazi, 1994).

Enteropathogenic *E.coli* (EPEC) causes watery diarrhea, mostly in neonates and young infants. The mechanism is through close adherence and damage to the microvillous surface of the small intestine with accumulation of actin and development of other cytoskeletal changes under the adhered bacteria.

The effect of human milk on the adherence of EPEC was studied using the *in vitro* system of adherence to monolayers of HEp-2 cells (Cravioto et al., 1991), or HeLa cells (Silva and Giampaglia, 1992). Morphologically, human milk interfered with the localized adherence to HEp-2 cells. Quantitatively, more than 90% inhibition was achieved by human milk. The inhibition was in the purified oligosaccharide fraction of human milk; further fractionation showed that the pentasaccharide and the difucosullactose subfractions retained the highest inhibitory activity (Cravioto et al., 1991).

3.2. Enteric Toxins

Some confusion exists in the literature regarding the terminology of enteric toxins. Enterotoxins act on the mucosal epithelium of the small intestine, causing fluid and electrolyte secretion by an enzymatic effect, without damage to the intestinal mucosa. The prototype is cholera toxin and others are heat-labile and heat-stable enterotoxins produced by *E. coli* and other enteropathogens. Cytotoxins are defined by their ability to kill mammalian cells, usually by inhibiting protein synthesis. They damage intestinal epithelium, and fluid loss is related to impaired absorption rather than active fluid excretion. The prototype is Shiga toxin and others are Shiga-like toxins produced by *E. coli*.

We examined possible effects of human milk on the binding of Shiga toxin and Shiga-like toxins (Newburg et al., 1992). The receptor of these toxins is a glycolipid, globotriaosyl ceramide (Gb_3). By analyzing the neutral glycolipids of human milk with high-pressure liquid chromatography (HPLC), Gb_3 could be detected. Its content was about 150 ng of Gb_3 per ml human milk. Functionally, labeled Shiga toxin bound to Gb_3 which was separated from human milk.

Cholera toxin and the heat-labile enterotoxin bind to their receptor, the ganglioside GM_1. The receptor was detected in human milk by high-performance thin layer chromatography (Laegreid et al., 1986). Functionally, non-immunoglobulin fraction of human milk inhibited the binding of enterotoxin to GM_1 *in vitro* by an ELISA assay (Holmgren et al., 1981). In rabbit ileal loops, human milk significantly inhibited the fluid accumulation caused by Cholera toxin (Laegreid et al., 1986).

The effect of the heat-stable enterotoxin are usually determined in infant mice, in which rapid intestinal fluid loss is developed, leading to increased gut-to-weight ratio and often to death. Human milk was capable of reducing both the fluid loss and mortality in mice caused by heat-label enterotoxin (Newburg et al., 1990). The inhibitory activity was in the fucosylated oligosaccharide fraction of human milk.

3.3. Rota Virus

Rota virus is a common cause of acute gastroenteritis mainly in infants. Yolken et al. (1992) examined the effect of human milk on the adherence and experimental infections caused by Rota virus. They found that human milk mucin, which was prepared by affinity chromatography, could bind to Rota virus. Reduced mucin lost the activity; further purification identified a 46 kD component of the mucin with highest specific activity. *In vivo* experiments showed that human milk, especially its macromolecular and acidic glycoprotein fractions, protected mice from Rota virus-induced gastroenteritis. Bovine milk or formula had no protection.

4. CONCLUSIONS AND FUTURE TRENDS

Currently there is striking evidence that glycoconjugates in human milk are potent inhibitors of the attachment of enteropathogens or enterotoxins to the intestine, which is crucial for the infectious process. These glycoconjugates can be viewed as soluble receptor analogues of intestinal cell-surface carbohydrates. This has been documented for multiple microorganisms and in several experimental systems. Most of the inhibitory effect is in the free oligosaccharide fractions, but human milk glycoproteins and glycolipids are important as well.

In the future, several additional studies are needed. First, for several inhibitory effects, characterization of the precise human milk fraction which is involved is necessary, and further evidence with human specimens is mandatory. We need more insight into the metabolism of human milk oligosaccharides in the gastrointestinal tract, namely how far they are degraded in the gut and in what location. It has been shown that oligosaccharides typical of human milk are present in the urine of breast-fed infants. This might suggest that intact oligosaccharides are absorbed in the intestine.

Because human milk glycoconjugates, and especially free oligosaccharides, can protect against infections, it is tempting to add them to infant formulas. Nonetheless, considering the large variety of these compounds, additional data are needed before implementing this approach.

REFERENCES

Andersson B, Porras O, Hanson LA, Lagergard T, Svanborg-Eden C. Inhibition of attachment of *Streptococcus pneumoniae* and *Haemophilus influenzae* by human milk. *J Infect Dis* 1986; 153:232-7.

Ashkenazi S. A review of the effect of human milk fractions on the adherence of diarrheogenic *Escherichia coli* to the gut in an animal model. Isr J Med Sci 1994; 30:335-8.

Ashkenazi S, Newburg DS, Cleary TG. The effect of human milk on the adherence of enterohemorrhagic *E. coli* to rabbit intestinal cells. Adv Exp Med Biol 1991; 310:173-7.

Ashkenazi A, Mirelman D. Nonimmunoglobulin fractions of human milk inhibits the adherence of certain enterotoxigenic *Escherichia coli* strains to guinea pig intestinal cells. *Pediatr Res* 1987; 22:130-4.

Boedeker EC. Enterocyte adherence of *Escherichiae coli*: its relation to diarrheal disease. *Gastroenterology* 1982; 83:489-92.

Cravioto A, Tello A, Villafan H, Ruiz J, Vedoro S, Neeser J. Inhibition of localized adhesion of enteropathogenic *Escherichia coli* to HEp-2 cells by immunoglobulin and oligosaccharide fractions of human colostrum and breast milk. J Infect Dis 1991; 163:1247-55.

Ginsburg V, Robbins PW. *Biology of carbohydrates*. Wiley, New York 1984.

Holmgren J, Svennerholm AM, Ahren C. Nonimmunoglobulin fraction of human milk inhibits bacterial adhesion and enterotoxin binding of *Escherichia coli* and *Vibrio cholerae*. *Infect Immun* 1981; 33:136-41.

Holmgren J, Svenner holm AM, Lindband M. Receptor-like glucocompounds in human milk that inhibit classical and El Tor *Vibrio Cholerae* cell adherence (hemagglutination). *Infect Immun* 1983; 39:147-54.

Howie PW, Forsyth JS, Ogston SA, Clark A, Florey CD. Protective effect of breast feeding against infection. *Br Med J* 1990; 300:11-6.

Kunz C, Rudloff S. Biological functions of oligosaccharides in human milk. Acta Paediatr 1993; 82:903-12.

Laegreid A, Otnaess AK, Fuglesang J. Human and bovine milk: comparison of ganglioside composition and enterotoxin inhibitory activity. *Pediatr Res* 1986; 20:416-21.

Larsen SA, Homer DR. Relation of breast versus bottle feeding to hospitalization for gastroenteritis in a middle class US population. *J Pediatr* 1978; 92:417-418.

Newburg DS, Ashkenazi S, Cleary TG. Human milk contains the Shiga toxin and Shiga-like toxin receptor glycolipid Gb_3. J Infect Dis 1992; 166:832-6.

Newburg DS, Pickering LK, McCluer RH, Cleary RG. Fucosylated oligosaccharides of human milk protect suckling mice from heat-stable enterotoxin of *Escherichia coli*. *J Infect Dis* 1990; 162:1075-80.

Ofek I, Doyle RJ. *Bacterial adhesion to cells and tissues*. Chapman and Hall, 1994.

Yolken RH, Peterson JA, Vonderfecht SL, et al. Human milk mucin inhibits rota virus replication and prevents experimental gastroenteritis. J Clin Invest 1992; 90:1984-91.

Silva MLM, Giampaglia CMS. Colostrum and human milk inhibit localized adherence of enteropathogenic *Escherichia coli* to HeLa cells. Acta Paediatr 1992; 81:266-7.

23

THE FIMH PROTEIN OF TYPE 1 FIMBRIAE

An Adaptable Adhesin

Per Klemm,[1] Mark Schembri,[1] and David L. Hasty[2]

[1] Department of Microbiology, Bld. 301
Technical University of Denmark
DK-2800 Lyngby
[2] Memphis VA Medical Center
Memphis, Tennessee

The most common of the enterobacterial adhesive surface organelles, and one of the best charachterized, is type1 fimbriae. Although the exact biological role of these organelles has been somewhat controversial, recent evidence strongly indicates that type1 fimbriae, among other things, are implicated in enhancing the virulence of certain uropathogenic strains. A detailed dissection of the molecular biology of type 1 fimbriae and their role in bacterial adhesin would undoubtedly help to reach a clearer understanding of important aspects of bacterial pathogenesis.

A single type 1 fimbria is a thin 7nm wide and approximately 1 μm long, surface polymer. It consists of about 1000 subunits of a major building element, i.e., the FimA protein, stacked in a helical cylinder. Additionally, a few percent minor components, viz. the FimF, FimG, and FimH proteins, are also present as integral parts of the fimbriae. The 30-kDa FimH protein has been shown direct and indirect tests to be the actual receptor-binding molecule (Klemm & Krogfelt, 1994). Like the other components of type 1 fimbriae the FimH adhesin is encoded in the *fim* gene cluster. This encompasses 9 genes contained within a 9.5 kb DNA segment which encode the structural elements of the fimbriae, the biosynthesis machinery as well as regulatory elements. The structural components of type 1 fimbriae are produced as precursors having an N-terminal signal sequence. Thus, the FimH protein is produced as a precursor of 300 amino acids, which is processed into a mature form of 279 amino acids.

Until recently the FimH adhesin was thought uniquely to target to D-mannosides. It was therefore quite surprising when it was found that highly conserved variants of the FimH protein, differeng in only a few amino acids, were shown additionally to bind to protein targets (Sokurenko *et al.*, 1992; Sokurenko *et al.*, 1994). All amino acid differences, which lead to different target preferences, were located between residues 32 and 120 in the mature FimH protein. Further information on the structure-function architecture of the FimH adhesin was provided by linker insertion analysis of the *fimH* gene. A systematic change of codons in the wild-type sequence in a number of positions, spanning the entire sequence,

Toward Anti-Adhesion Therapy for Microbial Diseases, edited by Kahane and Ofek
Plenum Press, New York, 1996

resulted in a series of FimH variants with a wide range of adhesive phenotypes. Notably, changes in the N-terminal part of the protein resulted in abolishment, or severly reduced binding potential (Schembri *et al.*, 1996). Taken together these data indicate that the fimH protein seems to consist of two major domains, each constituting roughly one half of the molecule; the N-terminal domain seems to contain the receptor binding site whereas the C-terminal domain seems to contain recognition sequences for the transport machinery.

The information accumulated in the structure-function relationship of the FimH adhesin has also been used to study the feasibility of constructing chimeric vaccines based on FimH. An attractive way of assaying heterologous peptide segments for biological activity is to fuse them into a naturally occurring bacterial surface protein which has the ability to accept grafts of exogenous protein segments. If the peptide segment is displayed in an immunologically active form, this may be assayed directly on surface of the bacteria. The biogenesis machinery of type 1 fimbriae has been shown to be quite tolerant in integrating similar but heterologous structural components into fimbrial organelles. thus, minor components from F1C fimbriae exhibiting as little as 34% identity with their equivalents in type 1 fimbriae are nevertheless readily integrated into type 1 fimbriae, resulting in hybrid organelles. This suggested that engineering of the FimH adhesin to contain heterologous sequences representing foreign epitopes would be possible without affecting compatibility with the transport system or adhesive ability.

In a subsequent study we used two different permissive positions in the FimH adhesin for insertion and display of heterologous sequences representing the preS2 sector of the hepatitis B surface antigen and an epitope from cholera toxin. We showed them to be exposed in immunologically active forms on the surface of the chimeric FimH adhesins, which in turn are present on the surface of bacterial hosts. The system seemed to be quite flexible, since chimeric versions of the FimH adhesin containing as many as 56 foreign amino acids were successfully transported to the bacterial surface as components of the fimbrial organelles (Pallesen *et al.*, 1995). The FimH protein only constitutes a small percentage of the total fimbrial protein. However, over-production of the FimH protein has been achieved and this can probably also be realized for chimeric versions of the adhesin.

The ability of the FimH protein to display different heterologous protein segments on the surface of *E. coli* hosts is indicative of the tolerance of the system and points to several future applications. One obvious possibility is to use chimeric FimH proteins for presentations of immunologically relevant epitopes for vaccine purposes. The data presented on the ability of the system to display large foreign segments in immunologically authentic forms seems promising for such an approach. Furthermore, although the reporter-sequences we have used in the present study represent immunologically relevant protein domains, there should also be several other potential applications. A possibility is to use FimH to display segments that represent the receptor binding domains of naturally occuring adhesins, enzymes, or other target-recognition proteins. This could form the basis for an efficient and rational way of studying protein-protein interaction and receptor-recognition in general. Also, one could perhaps envisage the system to be used for the display of randomly created peptide libraries. In connection with modern panning techniques, such a strategy would allow for the selection of interesting, novel target-recognition chimeras.

In conclusion, we believe that the FimH presentation system has a large intrinsic potential. Firstly, because it can confer surface display of immunologically active heterologous protein segments of substantial size. Secondly, since fimbriae, and thereby FimH, are easy to purify it should be possible to study chimeric version not only in the context of bacteria but also as isolated proteins.

REFERENCES

Klemm,P. and K.A. Krogfelt, 1994. Type 1 fimbriae of *Escherichia coli*. In *Fimbriae, adhesion, genetics, biogenesis and vaccines*.pp.9-26. Edited by P.Klemm, CRC press, Boca Raton, Fla.

Pallesen, L., L.K. Poulsen, G. Christiansen, and P.Klemm. 1995. Chimeric FimH adhesin of type 1 fimbriae: a bacterial surface display system for heterologous peptides. Microbiol 141: 2839-2848

Schembri, M., L. Pallesen, H. Connell, D.L. Hasty, and P. Klemm. 1996. Linker insertion analysis of the FimH adhesin of type 1 fimbriae in an *Escherichia coli fimH*-null background. *FEMS Microbiol Lett*, in press.

Sokurenko, E. V., H.S Courtney, S.N. Abraham, P. Klemm, and D.L. Hasty. 1992. Functional heterogeneity of type 1 fimbriae of *Escherichia coli. Infect. Immun* 60:4709-4719.

Sokurenko, E. V., H.S. Courtney, D.E. Ohman, P. Klemm, and D.L. Hasty. 1994. FimH family of type 1 fimbrial adhesins: Functional heterogeneity due to minor sequence variations among fimH genes. *J.Bacteriol* 176: 748-755.

24

INHIBITORS OF *CANDIDA ALBICANS* ADHESION TO PREVENT CANDIDIASIS

Esther Segal

Department of Human Microbiology
Sackler School of Medicine
Tel Aviv University
Tel Aviv, Israel

INTRODUCTION

A significant increase in fungal infections, particularly the systemic mycoses, which pose a serious problem in immunocompromised and otherwise debilitated individuals, has been associated with contemporary medicine. *Candida albicans* and other *Candida* species constitute major fungal pathogens, causing a wide spectrum of infections, ranging from entities involving mucosal and cutaneous tissues to deep-seated forms involving various organs, and possibly evolving into a disseminated disease (Edwards 1995).

Treatment, and particularly prevention, of the different clinical forms of candidiasis still pose a problem, despite the availability of antifungal drugs. A principal problem in the management of the muco - cutaneous entities is the high rate of recurrences, although treatment with specific, topical or systemic antifungal preparations is, generally, effective in clearance of infection. Recurences are seen in a significant proportion of vaginitis patients (Sobel 1993), who are succsesfully treated, but who come back with new episodes; frequent mucosal and dermal candida infections are associated with diabetes; and AIDS patients are particularly prone to repeated oral or dermal candidiasis. Furthermore, therapy of deep-seated candidiasis, affecting primarily, compromised patients, is as well, not devoid of problems. Amphotericin B, the currently available, gold standard therapeutic agent for treatment of the severe systemic forms is associated with significant adverse side effects (Bennett 1995), and not always effective.

Therefore, prevention of the infection, particularly in susceptible individuals, would seem a reasonable approach in management of both superficial and deep-seated candidiasis. However, effective preventive measures are currently unavailable: 1. no clinically effective vaccine, the tradinional prophylactic means for prevention of infectious diseases, is so far in existance; 2. amphotericin B is not adequate for prophylactic use due to its toxicity; 3. various trials to use prophylactically oral antifungals, such as nystatin tablets or the newer azoles, e.g. fluconazole, have not revealed significant efficacy in preventing systemic candidiasis in risk populations. On the other hand, prolonged prophylactic antifungal

Toward Anti-Adhesion Therapy for Microbial Diseases, edited by Kahane and Ofek
Plenum Press, New York, 1996

treatment in AIDS patients for prevention of mucosal candidiasis resulted in appearance of resistant *C. albicans* strains and emergence of previously unrecognized species, which are less sensitive to the antifungal drugs. Thus, a different approach for prevention of candida infections seems to be warranted.

INHIBITION OF ADHESION

Adhesion of candida to host tissues, which, as in other microbial systems, represents the initial interaction between the microbe and the host, is believed to be involved in initiating the infectious process (Calderone & Braun 1991; Kennedy et al. 1992;). In mucosal and dermal candidiasis the role of adhesion in pathogenesis was shown by epidemiological investigations (Segal et al. 1984; Srebrnik & Segal 1990), demonstrating a correlative relation between the ability of epithelial cells to bind in vitro the fungus and the tendency for infection. In addition, experimental studies revealed that mutants exhibiting reduced in vitro adhesion were less virulent in experimental models (Lehrer et al. 1986), and that the noted hirarchy in adhesion of *Candida* species is compatible with the pathogenic potential of these species. An interaction between candida and mucosal or endothelial surfaces is involved also in development of systemic candidiasis, which can originate from either an endogenous source, the GI tract, where the fungus thrives as a commensal, or from an exogenous source following introduction into the blood stream. Thus, in view of the role of adhesion in pathogenesis of candidiasis, interception of the host-microbe interaction by inhibiting adhesion could be considered a rational approach to prevent the infection.

It is generally believed that microbial attachment to host cells involves, primarily, an interaction between microbial cell-wall components, the adhesins, and their compatible host-cell surface binding components, and that adhesion can also be mediated by unspecific mechanisms, such as hydrophobicity, expressed particularly in binding to innert surfaces. Based on this, widely accepted, concept the logical strategies to intercept adhesion would include use of analogs of the putative adhesins or of the putative cell receptors to competitively bind to the microbial or host-cell sites, respectively, and thereby hinder the interaction. An additional approach to interfere in the process could be based on modulating or affecting the adhesins or the host-cell receptors, which could be achieved through various treatments, e.g. subinhibitory concetrations of antibiotics or substances affecting the normal biosynthesis of the surface components, or by molecular mimmicry.

In analogy to bacterial species, it is assumed that *C. albicans*, apparently, possesses more than a single adhesin system. It was shown (Kennedy 1988) that environmental factors, such as growth conditions, may affect significantly the candida cell surface, which could probably lead also to expression or presentation of different adhesins in dependance with the environmental factors. In addition, evidence from *in vitro* experiments suggests that *C. albicans* adhesin systems may differ among various strains (Douglas 1987). Such findings could possibly explain the differences obtained in studies from various research groups. Moreover, recent research (Hostetter 1994) revealed also differences in adhesive mechanisms between *candida* species.

The current approach regarding the adhesive interactions between candida and the mammalian host, recognizes according to Calderone and collegues (Calderone et al 1994) five systems. Four represent systems associated primarily, with the pathogenesis of candidiasis involving interactions between the fungus and epithelial or endothelial cells. The fifth represents a system associated principally, with the host's defense mechanisms, involving an interaction between candida and macrophages.

Of the candida cell wall components considered as putative adhesins, most studies pointed to the mannoprotein, both the proteinous and the carbohydrate moieties, and/or to chitin, although other components have also been suggested (Kennedy et al 1992; Segal & Sandovsky-Losica 1996). According to Douglas and collegues (Douglas 1987) the protein moiety of the mannoprotein seems to be the adhesive component binding to a glycoside receptor on epithelial cells, involving apparently, a lectin type interaction. These investigators isolated a mannoprotein from *C. albicans* which mediates fungal adhesion, and whose activity was abolished following treatment with proteolytic enzymes. Miyakawa and collegues (Miyakawa et al 1992) on the other hand, attributed to mannan the ability of candida to attach to epithelial cells, as did Cutler's group (Li & Cutler 1993) in respect to interactions with splenic macrophages. The role of chitin in candidal adhesion to epithelial cells will be elaborated in detail later in this article.

As to the nature of the host-cell surface binding components the available information is more scarce. Evidence indicates that these are apparently in dependance to the cell type involved. Thus, in interactions with epithelial-mucosal cells the receptor could be a glycoside, such as fucosyl (Critchley and Douglas 1987); there are also indications that fibronectin might be involved in binding of *candida* to epithelial cells (Kalo et al 1988). In the case of endothelial cells, RGD containing molecules seem to be the receptors (Hostetter 1994).

Most of the efforts to inhibit adhesion concentrated on the use of analogs of the putative candida adhesin(s), or modulation of their biosynthesis and/or activity, while only relatively, limited attempts focused on inhibition by host-cell receptor ligands (Klotz et al 1992). Inhibition was attempted through several methodologies, including : 1. competitive binding with analogs; 2. binding to specific lectins; 3. use of anti-adhesin antibodies; 4. modulations induced by drugs, enzymes or other substances; 5. use of inhibitors of enzymes involved in the biosynthesis of the candida cell wall components, such as tunicamycin, polyoxin D or nikkomycins. The attempts to block adhesion, whether by inhibitors of adhesins or of host-cell surface ligands, have been explored, generally, in *in vitro* systems (Table 1), and corroborated *in vivo*, only in a limited number of investigations.

INHIBITION OF ADHESION BY CHITIN DERIVATIVES

The remainder of this chapter will concentrate on inhibition of *candida* adhesion and prevention of candidiasis by chitin and its derivatives.

Based on the previously indicated strategies to attempt inhibition of adhesion by analogs of putative adhesins, we supplemented in our initial studies (Lehrer et al 1983) the *in vitro* adhesion reaction mixtures of human exfoliated vaginal epithelial cells (VEC) and

Table 1. Examples of inhibitors of candida adhesion acting as analogs of the microbial or host-cell surface

Inhibitor	System		
	in vitro	*ex vivo*	*in vivo*
Mannoprotein	+		
Mannan	+	+	
Mannose	+		
Chitin	+		+
Chitin derivative	+		+
NAG	+		+
Fucose	+		
RGD	+		+
Fibronectin	+		

C. albicans suspensions with analogs of the candida cell wall components. The polysaccharides - mannan, glucan and chitin, and their monomers - mannose, glucose and N-acetylglucoseamine (NAG), respectively, were added to the adhesion mixtures, and adhesion values were compared to those of the controls devoid those substances. It was noticed that a reduction in the adhesion values was obtained when mixtures were supplemented with chitin or NAG. It was subsequently found that derivatives of chitin, such as a hydrolysate or a water extract - chitin soluble extract (CSE), both when prepared from chitin isolated from candida or from chitin purchased commercially (source - Crustaceae shells), exerted an inhibitory effect on *in vitro* adhesion of *C. albicans* to VEC. Further experiments, in which VEC or yeasts, were pretreated with chitin, its derivatives and NAG, prior to the adhesion reaction, and then mixed with untreated counterparts revealed that inhibition of adhesion occured only when the VEC were pretreated. This pointed to the possibity that these substances act, apparently, as competitive analogs of the yeast cell wall and cover the binding sites on the cells, thereby inhibiting the adhesion.

Following these observations, additional *in vitro* adhesion studies to various target cells of human or animal origin (Table 2) have been carried out. These included exfoliated human buccal epithelial cells (BEC), corneocytes, HeLa or HaCat cell lines, murine gastrointestinal (GI) tissue tisks. Adhesion to GI tissue disks was evaluated as a model of significance for studying the initial interaction between candida and the host, which could lead to systemic infection in compromised hosts, since the GI tract serves as a reservoir for the fungus from where it can spread. Adhesion was also studied to innert surfaces, which might have relevance to clinical situations associated with candida infections, such as the resins of dentures and contact eye lenses. The fungal adhesion to the different target systems was evaluated by various methodologies (Table 3), such as microscopy, enumeration of colony forming units (CFU), use of isotope or fluorescence labelled yeasts, direct and indirect ELISA techniques.

Thus, using different evaluation systems we assesed the effect of chitin and chitin derivatives, particularly CSE, on the *in vitro* adhesion of candida to various targets, as summarized in a concise form in Table 4. Significant inhibition of *in vitro* adhesion to exfoliated human VEC, BEC and corneocytes, as well as to tissue line and to murine GI tissue disks was obtained. It should be emphasized, that inspite the increase in the level of adhesion to GI tissue disks from compromized animals (irradiated, treated with methotroxate

Table 2. *In vitro* target systems

I. Exfoliated human epithelial cells
Vaginal epithelial cells (VEC)
Buccal epithelial cells (BEC)
Dermal epithelial cells (Corneocytes)
II. Tissue cell lines
HeLa
HaCat
III. Animal tissue disks
Gastrointestinal tissue disks (various sites) from:
a. Naive (untreated) mice (inbred & outbred strains)
b. Irradiated mice (outbred strain)
C. Mice (outbred strain) treated with methotroxate
D. Mice (outbred strain) treated with 5-Fluorouracil
IV. Innert surfaces
Methyl acrylate (Dentures resin)
Contact lenses

Table 3. Evaluation systems of candida *in vitro* adhesion

I. Microscopic evaluation
  Studied with exfoliated epithelial cells; tissue lines (HeLa); innert surfaces (denture resin):
  a. Enumeration (%) of epithelial cells with adherent candida
  b. Enumeration of adherent candida:
  1. Per given number/individual epithelial cell(s)
  2. Per surface area (mm^2)
II. Radioisotop labelling of candida
  Studied with exfoliated epithelial cells; animal tissue disks
III. Determination of colony forming units (CFU) in culture
  Studied with animal tissue disks; innert surfaces (contact lenses)
IV. Indirect and direct ELISA technique
  Studied with exfoliated epithelial cells; cell lines (HaCat)
V. Fluoroscent labelling of candida
  Studied with exfoliated epithelial cells; tissue lines

[MTX] or 5Fluorouracil [5-FU]) (Sandovsky-Losica and Segal 1990), exposure to CSE had a marked inhibitory effect. In addition, the inhibitory effect was exhibited towards adhesion of various *C. albicans* strains, and to some extent also towards other candida species. The inhibitory effect of CSE on the binding of *C. albicans* to the resins of dentures and eye contact lenses involves, probably, a different mechanism(s) than in the case of attachment to mammalian tissues, possibly hydrophobic forces.

As chitin is a polymer of the aminosugar NAG and the enzyme chitin synthetase, which is a membrane bound enzyme, mediates the polymerization step, it was of interest to

Table 4. Inhibitory effect of chitin derivatives on *in vitro* adhesion of candida

A. Mammalian cells/tissues

candida	Target	Inhibition (%) Chitin	NAG	+ CSE
C. albicans(CBS562)	BEC	70	62	63[a]
C. albicans various isolates	BEC			43-88[b]
C. tropicalis	BEC			38
C. parapsilosis	BEC			28
C. krusei	BEC			30
C. albicans (CBS-562)	VEC	65	59	68
C. albicans (4918)	VEC			72
C. albicans (CBS-562)	Corneocytes			56
C. albicans (CBS-562)	Keratinocytes tisssue line			62
C. albicans (CBS-562)	Murine GI tissue			73
	Naive Compromised			56-79[c]

B. Innert Surfaces

candida	Target	Inhibition % (by CSE)	+
C. albicans(CBS-562)	Denture resin	82	
C. albicans (CBS-562)	Eye lenses resin	47	

assess the effect of exposure of *C. albicans* during the growth period to inhibitors of the enzyme, on the subsequent fungal adhesion. Yeast cultures were exposed to polyoxin D and nikkomycins, inhibitors of chitin synthetase, and *in vitro* adhesion to BEC was assessed, following FACS analysis to verify the reduction in chitin in the treated organisms (Gottlieb et al 1991). The adhesion assays, evaluated both by microscopy and by a radioactive assay, showed that yeasts exposed to polyoxin or nikkomycin exhibited reduced adhesion ability, which was concentration dependent (Fig.1). It was also shown that exposure of yeasts to sub MIC concentrations of the azole drug, bifonazole, which affects the fungal cytoplasmic

A

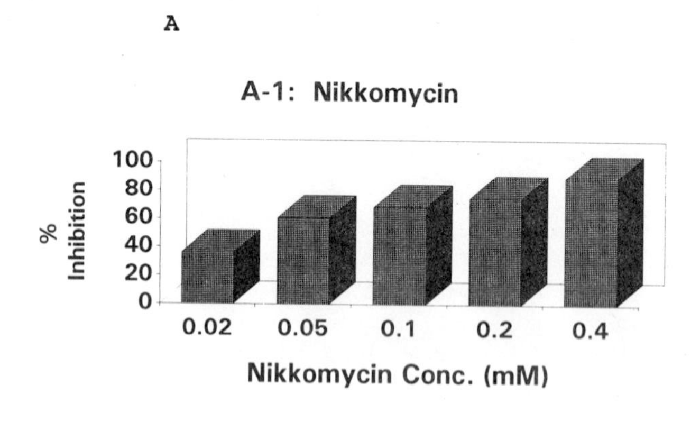

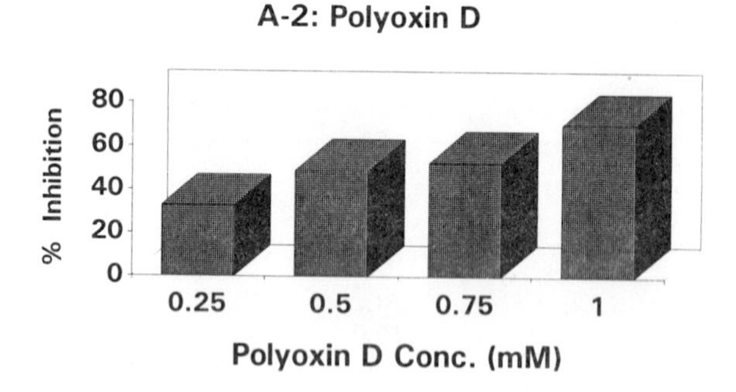

B

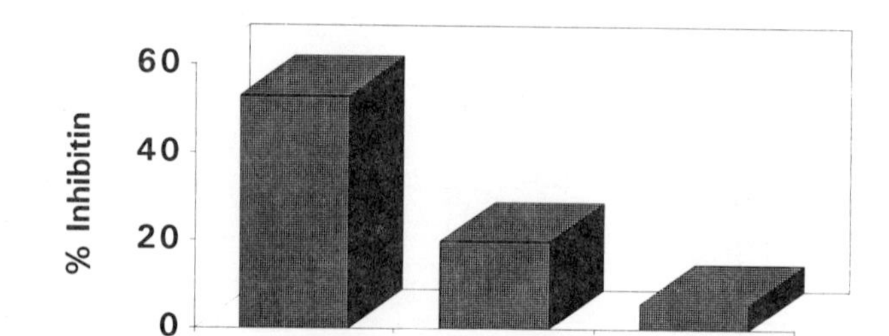

Figure 1. Effect of chitin synthetase inhibitors and bifonazol on *Candida albicans* adhesion. A. Chitin synthetase inhibitors. B. Bifonazol (subinhibitory concentration-Sub MIC).

membrane, and thereby the membrane bound enzymes, results in a reduction of adhesion, while exposure to another antifungal, with a different mode of activity, such as 5-Fluorocytosine, did not affect candidal adhesion. These effects were confirmed by scanning electron microscopical observations.

So we believe that the cummulative information gathered from the *in vitro* studies, suggests that chitin has a role in adhesion and it or its derivatives can inhibit the process.

Following the *in vitro* studies we focused on corroboration of the inhibitory effect for prevention of candidiasis in *in vivo* models. The *in vivo* target systems explored were animal models and also limited human trials (Table 5). The animal models were chosen to represent the major clinical entities of the two groups of *candida* infections: muco - cutaneous vs. systemic.

Vaginitis was induced following intravaginal inoculation with *C. albicans*, in naive mice, in mice with constant oestrus due to hormonal pretreatment (Segal et al 1988), and in mice rendered diabetic by streptozotocin (Segal and Josef-Lev 1995). Murine vaginitis in the diabetic animals or in those with constant oestrus could be induced by a lower fungal inoculum than in naive animals, and affected a higher number of mice, presenting a severe, long lasting, persistent infection. These experimental data are compatible with clinical situations in humans. Experimental oral candidiasis was induced in mice pretreated with antibiotics and experimental dermal candidiasis in guinea pigs.

GI colonization and the eventually following systemic infection was assessed in naive and compromised mice (pretreated by MTX, 5-FU or irradiated) inoculated by the oral route. GI colonization could be demonstrated through histopathology revealing adhesion to mucosa and tissue penetration and quantitative *candida* CFU enumeration in culture. Systemic infection could be demonstrated by involvement of internal organs (kidneys, liver, spleen), as assessed by macro- and histopathology, and CFU enumeration; systemic infection was associated with mortality. While GI colonization was found in compromised and in naive

Table 5. *In vivo* target systems of inhibition of candida adhesion and prevention of infection

I. Animal Models
A. Muco - Cutaneous Candidiasis
1. Vaginitis in:
a. Naive (untreated) mice
b. Mice with induced constant oestrus
c. Mice with induced diabetes
2. Oral Infection in:
a. Naive (untreated) mice
b. Mice treated with antibiotics
3. Cutaneous Infection in:
Naive (untreated) guinea pigs
B. Gastrointestinal (GI) Colonization in:
a. Naive (untreated) mice
b. Irradiated mice
c. Methotroxate (MTX) treated mice
d. 5-Fluorocytosine (5-FU) treated mice
C. Systemic Infection from GI Route in:
a. Irradiated mice
b. MTX treated mice
c. 5-FU treated mice
II. Human trials
Prevention of recurrence of cutaneous infection in susceptible individuals.

mice, although in the former quantitatively at a higher level, systemic infection developed, primarily, in the compromised animals (Sandovsky-Losica and Segal 1993).

In the mucosal or dermal infections assessment of efficacy of application of the inhibitor as preventive measure was carried out by pretreating the animals topically (intravaginally, on the oral mucosa, or on the skin) prior to their inoculation with *candida*, and shortly threrafter. In the GI-colonization and systemic infection models the inhibitor was administered orally (added to the drinking water) prior and post fungal inoculation.

The data obtained in the different models are summarized, again in a concize, not detailed manner, in Table 6. The data in this table refer to CSE, although the efficacy of chitin, both of that obtained from commercial source as well as of that isolated from *C.albicans*, and of NAG, has also been assesed in some models. Assessment of efficacy was determined by the rate of prevention of development of infection (mucosal, dermal, GI or systemic), amelioration of morbidity (clearance of infection, reduced fungal colonization in tissues [lower CFU]) and reduction of mortality rate (in systemic infections).

It can be seen, that vaginitis could be prevented in a significant number of the treated mice. In the case of the very severe type of infection, the infection cleared at a higher rate and/or faster than in the untreated animals. Amelioration of infection was also obtained in cutaneous candidiasis in guinea pigs and in oral candidiasis in mice, demonstrated primarily by lower *candida* CFU in tissue. Orally admministered CSE affected GI colonization in naive mice, and most
importantly, prevented development of systemic infection and the resulting mortality in a significant proportion of the MTX compromised animals. So these in vivo data indicate that by inhibiting the adhesion to GI mucosa it is possible to reduce the rate of development of systemic, lethal infection in debilitated hosts.

The inhibitory effect of CSE in vitro and in vivo, leading to prevention of infection was confirmed by other investigators (Ghanoum et al 1991). These investigators explored the effect on the in vitro adhesion of *C.albicans* to murine duodenal tissue discs. In vivo they

Table 6. Activity of a Chitin derivative (CSE) *in vivo*

A. Mucosal and Dermal Candidiasis				
Animal	Infection	Application of CSE	Prevention %	Clearance[a]/ decrease[b]
Mice-Naive	Vaginal	Topical	80	
Mice-constant estrus	Vaginal	Topical	45	
Mice-diabetic	Vaginal	Topical	30	[a]day 21-28(35)
Mice-pretreated-with antibiotics	Oral	Topical		[b]74-90%
Guinea pigs	Cutaneous	Topical		[b]75-90%
B. Gastronintestinal (GI) - Systemic Candidiasis				
Animal	Infection	Application of CSE	Prevention(%)[c]/	Mortality(%)[d]
Mice-Naive	GI	Systemic(oral)	49	
Mice compromised	Systemic	Systemic(oral)	60	60

Legend:
[a]Clearance of infection in comparison to control in parentheses.
[b]Decrease in fungal colonization (CFU).
[c]Prevention of infection.
[d]Decrease in mortality.

assessed GI colonization and dissemination to visceral organs, demonstrating the protective ability of the substance.

A limited human trial has been carried out thus far. The trial was confined to evaluation of prevention of recurrence of dermal candidiasis in susceptible individuals (suffering from recurences) by use of a topical preparation the inhibitor (following clearance of infection by conventional antifungal drugs). Promising results have been obtained in the double blind trial (placebo vs. inhibitor) indicating a low rate of reccurence among the patients receiving the preparation containing the inhibitor as compared to the patients receiving placebo. Additional studies are now in progress.

CONCLUDING REMARKS

- Adhesion of *candida* to host tissues plays a role in pathogenesis of candidiasis:

1. In mucosal - cutaneous entities;
2. In deep-seated and systemic infection; particularly that of the endogenous source, originating from the gastrointestinal tract.

- Adhesion can be competitively intercepted by analogs of the fungal adhesin(s).
- Blocking of adhesion can prevent candidiasis.
- Most importantly, anti-adhesive therapy could serve as a novel, effective approach to manage these infections; an approach of particular significance in compromised hosts.

REFERENCES

Bennet JE, 1995, Antifungal agents, In: Principles and Practice of Infectious Diseases, 4th edn., Eds. Mandell GL et al., pp. 401-410.

Calderone RE & Braun P, 1991, Adherence and receptor relationships of *candida* albicans, Microbiol Rev, 55: 1-20.

Calderone RE et al, 1994, Host cell-fungal cell interaction, J Med Vet Mycol, 32 (Suppl. 1): 151-168.

Critchley IA & Douglas LJ, 1987, Role of glycosides as epithelial cell receptors for *candida* albicans, J Gen Microbiol, 133: 637-643.

Douglas LJ, 1987, Adhesion of *candida* species to epithelial surfaces, CRC Crit Rev Microbiol, 15: 27-43.

Edwards JE, 1995, *candida* species, In: Principles and Practice of Infectious Diseaes, 4th edn., Eds. Mandell GL et al., pp. 2285-2305.

Ghannoum MA et al, 1991, Protection against *candida* albicans gastrointestinal colonization and dissemination by saccharides in experimental animals, Microbios, 67: 95-105.

Gottlieb S et al, 1991, Adhesion of *candida* albicans to epithelial cells - effect of polyoxin D, Mycopathologia, 115: 197-205.

Hostetter MK, 1994, Adhesins and ligands involved in interaction of *candida* sp. with epithelial and endothelial surfaces, Clin Microbiol Rev, 7: 29-42.

Kalo A et al, 1988, Interaction of *candida* albicans with genital mucosal surfaces: involvement of fibronectin in the adherence of the yeasts, J Inf Dis, 157: 1253-1256.

Kennedy MJ, 1988, Adhesion and association mechanisms of *candida* albicans, Curr Top Med Mycol, 2: 73-169.

Kennedy MJ et al, 1992, Molecular basis of *candida* albicans adhesion, J Med Vet Mycol, 30 (Suppl. 1): 95-122.

Klotz S et al, 1992, Effect of an arginine-glycine-aspartic acid containing peptide on hematogenous *candida*l infection in rabbits, Antimicrob Agents Chemother, 36: 132-136.

Lehrer N et al, 1983, In vitro and in vivo adherence of *candida* albicans to mucosal surfaces, Ann Microbiologie, 134: 293-306.

Lehrer N et al, 1986, Pathogenesis of vaginal candidiasis: studies with a mutant which has reduced ability to adhere in vitro, J Med Vet Mycol, 24: 127-131.

Li RK & Cutler JE, 1993, Chemical definition of an epitope/adhesin molecule on *candida* albicans, J Biol Chem, 268: 18293-18299.

Miyakawa Y et al, 1992, Role of specific determinants in mannan of *candida* albicans serotype A in adherence to human buccal epithelial cells, Inf Immun, 60: 2493-2499.

Sandovsky-Losica H & Segal E, 1990, Interaction of *candida* albicans with murine gastrointestinal mucosa from methotroxate and 5- Fluorouracyl treated animals: in vitro adhesion and prevention, J Med Vet Mycol, 28: 274-287.

Sandovsky-Losica H & Segal E, 1993, Effect of a chitin derivative on experimental candidiasis of gastrointestinal origin in compromised mice, Immunol & Inf Dis, 3: 155-159.

Segal E et al., 1984, Correlative relationship between adherence of *candida* albicans to human vaginal epithelial cells in vitro and *candida* vaginitis, J Med Vet Mycol, 22: 191-200.

Segal E et al, 1988, *candida*l vaginitis in hormone treated mice: prevention by a chitin extract, Mycopathologia, 102: 157-163.

Segal E & Josef-Lev A, 1995, Induction of *candida*l vaginitis in diabetic mice and attempts to prevent the infection, J Med Vet Mycol, 33: 1-8.

Segal E & Sandovsky-Losica H, 1996, Basis for *candida* albicans adhesion and penetration, In: Biomedical Mycology, 321-334 (In Press).

Sobel JD, 1993, Genital candidiasis, In: Candidiasis: Pathogenesis, Diagnosis and Treatment, 2nd edn, Ed. Bodey GP, pp. 225-247.

Srebrnik A & Segal E, 1990, Comparison of *candida albicans adherence to human corneocytes populations, Acta Derm Ven, 70: 459-462*

TRICHOMONAS VAGINALIS ADHESIN PROTEINS DISPLAY MOLECULAR MIMICRY TO METABOLIC ENZYMES

J. A. Engbring, J. L. O'Brien, and J. F. Alderete

Department of Microbiology*
The University of Texas Health Science Center
San Antonio, Texas 78284-7758

1. INTRODUCTION

Trichomonas vaginalis is a flagellated protozoan responsible for trichomonosis, one of the most prevalent sexually transmitted diseases. Women are mainly affected and have a broad range of symptomatology; minor irritation to severe inflammation of the vaginal epithelium, accompanied with discharge, itching and abdominal pain have been reported (Krieger *et al.*, 1990). Because of these varied symptoms and the lack of a good clinical test, diagnosis is less than adequate with approximately 50% of women misdiagnosed.

Recent findings indicate trichomonosis presents more of a problem than previously believed. Recommended treatment for this disease is metronidazole, but drug-resistant isolates have emerged (Müller *et al.*, 1980). During the current HIV epidemic, it has been observed that patients infected with *T. vaginalis* are more likely to seroconvert to HIV positive (Nzila *et al.*, 1991; Laga *et al.*, 1993). Also, women infected with *T. vaginalis* are at increased risk for preterm rupture of membranes, preterm delivery, and low birth weight infants (Cotch *et al.*, 1991; Minkoff *et al.*, 1984; Read and Klebanoff, 1993).

T. vaginalis is a non-self-limiting infection in women, illustrating the adaptation by the parasite for survival in the adverse vaginal environment. Recently several *T. vaginalis* virulence factors have been identified. Since the vaginal environment is nutrient-limiting, it is not surprising that trichomonads have receptors that specifically bind to host proteins as part of an elaborate nutrient-acquisition system. *T. vaginalis* parasites are unable to synthesize or modify lipids (Lund and Schorb, 1962; Roitman *et* al., 1978; Lindmark, 1983; Holz *et al.*, 1987). Thus, receptor-mediated binding of apoprotein CIII of lipoproteins (Peterson and Alderete, 1984a; b), coupled with adhesin binding and subsequent hemolysis of erythrocytes (Lehker *et al.*, 1990; Dailey *et al.*, 1990) represent excellent sources of lipids. It has been shown that purified lipoproteins derived from serum or lipids extracted from erythro-

* This work was supported by Public Health Service grant AI 18768 given by the National Institutes of Health. JAE was funded, in part, through NIH training grant AI 07271.

Toward Anti-Adhesion Therapy for Microbial Diseases, edited by Kahane and Ofek
Plenum Press, New York, 1996

cyte membranes can be used in lieu of serum in the complex growth medium (Lehker *et al.*, 1990).

A repertoire of low-iron-induced surface proteins have also been identified as the receptors for several iron-binding and iron-containing proteins (Lehker *et al.*, 1990). *T. vaginalis* binds lactoferrin, ferritin, cytochrome c and hemoglobin by specific receptors (Peterson and Alderete, 1984c). Among the best studied is the lactoferrin receptor (Peterson and Alderete, 1984c), and it has been shown that lactoferrin is an *in vivo* source of iron for the parasite (Masson *et al.*, 1966). Transferrin, another iron-binding protein of plasma, is not a source of iron for trichomonads (Peterson and Alderete, 1984c), illustrating the highly evolved nature for utilization of specific iron sources. The *in* vivo fluctuations in lactoferrin parallel the hormonal changes (Cohen *et al.*, 1987), and this has been suggested to represent an important *in vivo* environmental cue regulating a variety of parasite responses, such as growth rates and expression of immunogens and adhesins (Lehker *et al.*, 1991; Lehker and Alderete, 1992).

Up to twenty-three cysteine proteinase activities have been identified by two-dimensional substrate electrophoresis (Neale and Alderete, 1990). That all proteinases are expressed *in vivo* is exemplified by the detection of serum antibodies to most, if not all, cysteine proteinases in patients with trichomonosis (Alderete *et al.*, 1991a; b). Interestingly, evaluation of proteinase patterns of freshly isolated parasites showed only subsets of the entire repertoire (Neale and Alderete, 1990; Alderete *et al.*, 1991b), perhaps indicating environmental control of the expression of these genes. The proteinases, as well as anti-proteinase antibodies have been detected in vaginal washes of patients (Alderete *et al.*, 1991b).

The significance of so many cysteine proteinases within trichomonads may ultimately be determined through substrate specificity studies. Recent work has shown that some are involved in the degradation of immunoglobulins (Provenzano and Alderete, 1995). An iron-regulated proteinase has also been found to render otherwise highly susceptible trichomonads resistant to lysis by the C3 component of complement (Alderete *et al.*, 1995b). The ability to degrade these proteins is likely protective for trichomonads, and also has implications for immunity/vaccines against co-infecting STDs. Hemolysis and contact-dependent cytolysis have been shown to involve trichomonad cysteine proteinases (Dailey *et al.*, 1990; Arroyo and Alderete, 1995). The expression of substrate-specific proteinases, as mentioned above, is under the control of environmental cues at the site of infection (Alderete *et al.*, 1995b; Provenzano and Alderete, 1995).

Among ethnic populations, African Americans have higher rates of trichomonosis (Stevens-Simon *et al.*, 1994). Approximately 50% of *T. vaginalis* isolates contain a segmented double-stranded RNA virus (Khoshnan and Alderete, 1993); recent unpublished observations suggest a relationship between virus-harboring *T. vaginalis* isolates and infection of African American women. This may be of particular importance, since African American women appear at risk for adverse outcomes in pregnancy and low-birth-weight infants (Bramley, 1976).

Only isolates with the virus are capable of the interesting phenomenon of phenotypic variation, as shown for a highly immunogenic protein called P270 (Wang *et al.*, 1987; Alderete *et al*, 1986a; b). These isolates vary between surface and cytoplasmic expression of P270. Virus-minus trichomonal isolates synthesize P270, but are unable to surface express the immunogen. That this variation represents an immune evasion technique is evidenced by the fact that antibody in patient serum to P270 is parasiticidal (Alderete and Kasmala, 1986). Finally, a recent report documented viral up-regulation of P270 (Khoshnan and Alderete, 1994), and regulation by the virus of other trichomonad proteins also has been demonstrated (Provenzano and Alderete, unpublished results).

As mentioned above, cytoadherence is one of the early steps essential for colonization and persistence of *T. vaginalis* in the human urogenital tract. All mucosal pathogens must overcome host factors and responses in order to colonize the mucosa and establish infection. Some of these factors include the extensive mucus layer, nutrient-limiting conditions, immune surveillance, and the constant fluid flow of the vagina. The complexity of this host environment is further illustrated by the hormonal influences at the site of infection.

T. vaginalis has evolved adaptive responses, such as adherence to host cells, which aid in the establishment of infection. Overall, the emerging picture of trichomonal cytoadherence suggests a very complex series of steps, and in our opinion, this property is more accurately viewed as a cascade set of reactions rather than a simple ligand-receptor interaction. For instance, one or more cysteine proteinases is required for parasite cytoadherence (Arroyo and Alderete, 1989; 1995). It has also been shown that iron regulates adhesin synthesis at the transcriptional level (Lehker *et al.*, 1991). Furthermore, binding to host cells results in distinct signals received by the parasite (Arroyo *et al.*, 1993). First, motile trichomonads, upon contact with vaginal epithelial cells, undergo a morphologic transformation and convert to an ameboid form, possibly for enhanced association with host cell surfaces. Second, trichomonad adhesin synthesis is rapidly up-regulated following contact with the host cell. Third, other trichomonads are recruited to the site of contact on the vaginal epithelial cell. These observations reinforce the concept that *T. vaginalis* attachment is a complex process and shows the ability of the parasite to be responsive to the constantly changing vaginal environment.

It has been shown previously that *T. vaginalis* adherence to host cells is a highly specific event that indicates parasite-cell ligand-receptor type associations (Alderete and Garza, 1985; 1987; Alderete *et al.*, 1988). We established that cytoadherence is time, temperature, and pH dependent (Alderete and Garza, 1985; 1987). Also, *T. vaginalis* organisms demonstrated a tropism for epithelial cells (Alderete and Garza, 1985). Loss of the ability to cytoadhere after live organisms were treated with trypsin, and prevention of cytoadherence regeneration by inhibition of protein synthesis indicated that parasite surface proteins were involved in host cell recognition (Alderete and Garza, 1985).

The *T. vaginalis* adhesins were finally identified through the use of a ligand assay (Alderete and Garza, 1988; Lehker *et al.*, 1991; Arroyo *et al.*, 1992) in which a detergent extract of intrinsically or extrinsically-radiolabeled parasites was incubated with chemically-stabilized host cells. Avidly bound proteins were eluted and resolved on SDS-PAGE-fluorography or autoradiography. Four trichomonad proteins capable of binding to host cells were identified and designated AP65, AP51, AP33 and AP23 based on M_r (Fig. 1A). Various established criteria (Beachey, 1989), some of which are summarized in Table I, were satisfied to verify that these proteins are bona fide adhesins (Arroyo *et al.*, 1992).

Polyclonal antiserum and monoclonal antibody (mAb) were generated against each of these proteins (Fig. 1A). Screening of a phagemid expression library with these antibodies yielded several reactive clones; three for AP65, one for AP51 and six for AP33 (Arroyo *et al.*, 1995). The expression patterns of the recombinant proteins in *E. coli* are shown in figure 1B. The readily visible bands for the AP51 and AP33 clones, in comparison to control *E. coli* proteins, verify the overexpression of these recombinant proteins. In contrast, the recombinant AP65 proteins were not apparent by Coomassie brilliant blue staining, but could be detected by immunoblotting.

The functionality of each of the recombinant proteins was then readily established. The immunoblot in figure 2A shows that a representative of each of the recombinant proteins was capable of binding to host cells, as did the natural parasite adhesin. Importantly, as shown in figure 2B, the representative recombinant proteins inhibited binding of the respective native trichomonad adhesins (Arroyo *et al.*, 1995).

Cloning the genes encoding the adhesins, and expression of functional recombinant proteins are essential first steps toward understanding interrelationships between the adhesins, as well as the nature of the interactions with host cell receptors. The initial charac-

Table 1. Four *T. vaginalis* surface proteins function as adhesins

1. Four distinct, non-immuno-crossreactive surface proteins of *T. vaginalis* bind to host cells.
2. A direct relationship between amounts of the four surface-expressed proteins and cytoadherence levels was established.
3. The four purified proteins and antibody to each of the four proteins inhibits cytoadherence in a concentration-dependent fashion.
4. Host cells have a saturable number of receptors for the four proteins.
5. Recombinant proteins retain function and inhibit binding of the natural trichomonad proteins to the host cell surface.

terization of genes encoding three of the *T. vaginalis* adhesins is presented here. Each of these adhesins was found to be encoded by a family of genes that share homology with metabolic enzymes; evidence is presented strongly suggesting that expression and subsequent protein localization of the adhesins may be regulated by multiple mechanisms. This data illustrates an added level of complexity for these important biofunctional proteins.

2. MOLECULAR CHARACTERIZATION OF THE AP65 GENES

2.1. Multiple-Gene Family

2.1.1. The cDNAs Represent Three Unique Genes. Insert analysis revealed that two of the clones were of sufficient size to encode the full-length adhesin, while the third was truncated and missing approximately one-third of the gene. Surprisingly, restriction enzyme mapping showed that, although similar, each of the cDNAs had clear differences. Subsequent

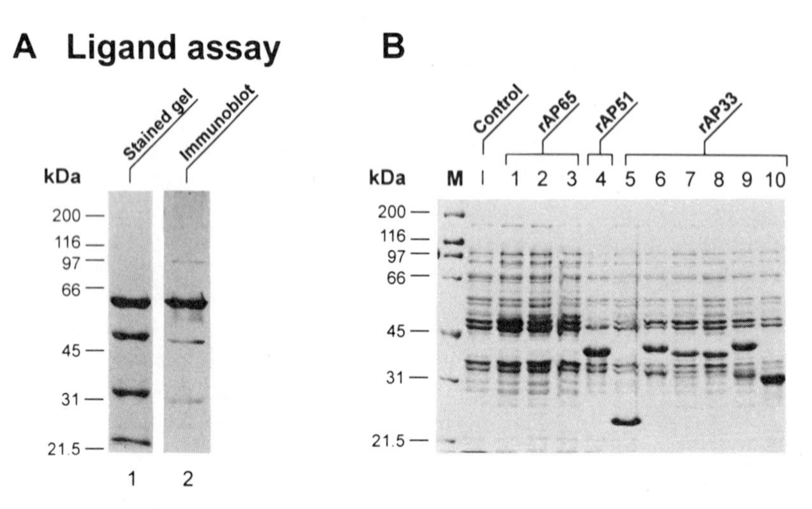

Figure 1. Analysis of trichomonad adhesins (A) and the recombinant proteins expressed in *E. coli* (B). (A) Trichomonad adhesins detected in stained gels after the ligand assay (lane 1) or after immunoblotting of adhesins with a mixture of antisera to the four adhesins. (B) Commassie brilliant blue-stained total proteins of control *E. coli* lysates (lane labeled control) were compared with proteins in lysates of recombinant *E. coli* (lanes 1 through 10). Intense recombinant protein bands specific for cDNA clones 4 through 10 had M_rs from ~23- to 42-kDa and were not seen in control *E. coli* lysates. Molecular weight markers (BioRad) are on the left in kilodaltons (kDa) (x1000). Printed with permission of Archives of Medical Research.

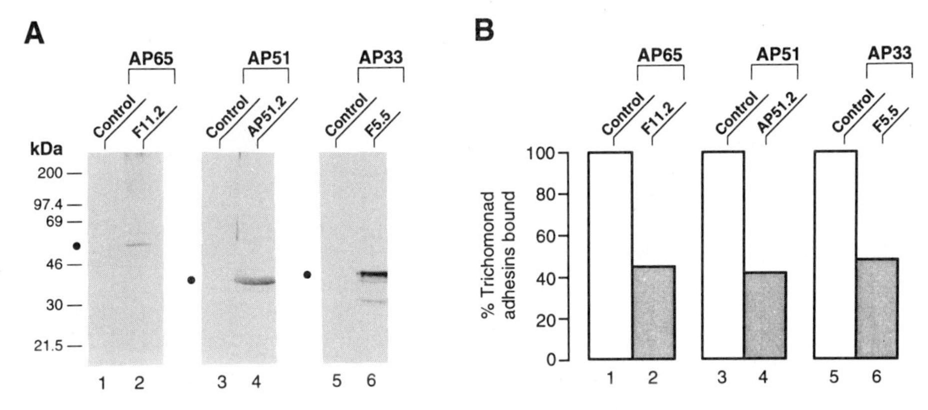

Figure 2. Representative experiments showing that recombinant proteins bind to HeLa cells (A) and compete with ^{35}S-labeled adhesins in *T. vaginalis* extract for HeLa cell binding (B). The recombinant *E. coli* lysate used for the experiments is designated above each lane. Control refers to *E. coli* harboring the vector without inserts used as negative control (lanes 1, 3 and 5). (A) Each recombinant protein bound and released from HeLa cell surfaces was electrophoresed and blotted onto nitrocellulose for detection with specific antiserum or mAb as for Figure 1. Control prebleed rabbit serum or myeloma culture supernatant failed to detect any of the bound recombinant proteins as for Figure 1. Dots designate the major recombinant adhesin bound to fixed HeLa cells. The numbers on the left show the size in kilodaltons (kDa) (x1000) of the prestained Rainbow molecular weight markers (Amersham Corp., Arlington, IL). (B) Fixed HeLa cells were interacted first with recombinant *E. coli* lysates and then with solubilized radiolabeled trichomonads in a ligand assay. Densitometric scanning analysis of the resulting fluorogram was used to determine the percentage of ^{35}S-labeled trichomonad adhesins bound to HeLa cells. The percentage of trichomonad adhesins binding after first interacting HeLa cells with control *E. coli* lysates represents 100% binding (lanes 1, 3 and 5). This experiment was repeated three times, and representative results are shown. Printed with permission of Archives of Medical Research.

sequencing of these cDNAs confirmed that each represented a unique gene, providing insight into the existence of a multigene family for this adhesin (Alderete *et al.*, 1995a; O'Brien *et al.*, 1996).

2.1.2. Obtaining and Comparing the Full-Length Sequences. Further analysis of the nucleotide sequences showed that the two largest cDNAs contained open reading frames (ORFs) that encoded for proteins of 543 amino acids, with predicted molecular masses of approximately 60,000 daltons (60-kDa). As the apparent size of the adhesin (Fig 1A) was not consistent with that predicted from the sequences, and all three cDNAs had a 3′ poly-A tail, this indicated that a portion at the 5′ end was missing from each. Since the cDNAs were similar but not identical, it was possible to identify unique oligonucleotide regions for the 5′ ampliFINDER RACE method to obtain the 5′ ends. Sequence analysis revealed a new translational start site, with each of the ORFs being identical in size. The complete sequences coded for proteins of 567 amino acids with predicted molecular masses of 63.3-kDa, 63.1-kDa, and 63.1-kDa, consistent with the native AP65 adhesin. Comparisons between the sequences showed 87%, 88% and 96% identity at the nucleotide level, and 91%, 89% and 96% identity for the proteins (Alderete *et al.*, 1995a; O'Brien *et al.*, 1996).

2.1.3. Each Gene Is Multi-Copy. The 5′-end PCR products described above and the highly specific oligonucleotide used to generate the 5′-end sequence for the truncated, third cDNA were shown to hybridize only with the corresponding cDNA from which it was derived. This gave us gene-specific probes for Southern analysis of genomic DNA of individual isolates, as well as of agar clones derived from single trichomonads. The DNA was restricted with enzymes that do not cut within the AP65 genes. Unexpectedly, each of the probes generated multiple bands, indicating that the three genes are present in the *T. vaginalis* genome and likely in multiple copies (Arroyo *et al.*, 1995; Alderete *et al.*, 1995a; O'Brien *et al.*, 1996). The data show that the AP65 adhesin is a member of a multigene family, each of which is present in the genome of all trichomonads.

2.1.4. Each of the Recombinant AP65s has Properties Similar to the Native Adhesin. Each of the recombinant proteins was immunoreactive with polyclonal antibodies and mAbs to the natural AP65 obtained from the ligand assay. Furthermore, each recombinant AP65 was tested for functionality and found to be capable of binding to host cells. The binding of native AP65 was inhibited by each recombinant protein, indicating that each of the cDNAs encoded a functional, yet distinct adhesin protein (Arroyo *et al.*, 1995; Alderete *et al.*, 1995a; O'Brien *et al.*, 1996).

2.2. AP65 Has Homology to Malic Enzyme

2.2.1. Database Search Reveals Homology to Malic Enzyme. Database analysis found significant similarity between the AP65 clones and the decarboxylating malic enzymes. Malic enzyme is active in metabolic pathways and present in the trichomonad hydrogenosome, the anaerobic equivalent of mitochondria (Lahti *et al.*, 1992). All AP65s showed ~54% identity at the nucleotide level and 38% identity at the amino acid level when compared to human malic enzyme. However, there were four regions in which the AP65s were almost identical, and also showed a high degree of identity, up to 78%, with malic enzyme. It may be important that the amino acid positions of all these regions are similar for the AP65 clones and the malic enzymes (Alderete *et al.*, 1995a; O'Brien *et al.*, 1996).

2.2.2. The Malic Enzyme Conserved Domains. Three of the four conserved regions correspond to dinucleotide-binding sites identified in several enzymes, and the other is a divalent cation-binding site of malic enzyme. The first region is the putative NADP-binding site of human and murine malic enzyme, goose fatty-acid synthetase, and human glyceraldehyde 3-phosphate dehydrogenase (Loeber *et al.*, 1991; Bagchi *et al.*, 1987; Poulouse and Kolattukudy, 1983). The AP65s show homology to 7 of the 9 malic enzyme amino acids spanning this domain. A cysteine residue believed to be the malate-binding site (Satterlee and Hsu, 1991) immediately follows this domain, and is conserved among all malic enzymes, as well as the AP65s. It is likely that this cysteine is important in binding of L-malate, but not NADPH, since modification of the residue only affects malate binding (Satterlee and Hsu, 1991).

The second conserved domain is the ADP-binding βαβ fold (Wierenga *et al.*, 1985). A consensus sequence (Wierenga *et al.*, 1985) consisting of hydrophilic and hydrophobic residues, as well as a highly conserved six amino acid motif has been established from sequence analysis of various enzymes, including human and duck malic enzymes, horse alcohol dehydrogenase, dogfish lactate dehydrogenase, and lobster glyceraldehyde phosphate dehydrogenase (Loeber *et al.* 1991; Hsu *et al.*, 1992; Wierenga *et al.*, 1985). This motif, GXGXXG, is essential for NAD-binding (Scrutton *et al.*, 1990). The AP65s follow the motif exactly, and also show a high degree of homology to the rest of the consensus sequence (Alderete *et al.*, 1995a; O'Brien *et al.*, 1996).

The third conserved dinucleotide-binding fold includes the GXGXXA sequence necessary for the binding of NADP in human and duck malic enzyme, human adrenodoxin reductase, *Agrobacterium* octopine synthase, and *E. coli* glutathione reductase (Loeber *et al.*, 1991; Hsu *et al.*, 1992; Scrutton *et al.*, 1990). The AP65s all conform perfectly to this motif as well as following the general pattern surrounding it (Alderete *et al.*, 1995a; O'Brien *et al*,.,1996).

Finally, alignment of the AP65s corresponded well with the putative Mn^{2+}-binding site of malic enzymes (Wei *et al.*, 1994), including pigeon, human, duck, murine, *Ascaris suum*, and maize (Wei *et al.*, 1994; Loeber *et al.*, 1991; Hsu *et al.*, 1992; Bagchi *et al.*, 1987; Kulkarni *et al.*, 1993; and Rothermel and Nelson, 1989). This domain includes a highly conserved aspartic acid believed to be one of the Mn^{2+} ligand sites of malic enzyme.

2.2.3. Malic Enzyme Does Not Exhibit Adhesive Function. It was of interest to determine if malic enzyme had properties similar to the AP65 adhesins. A monoclonal antibody generated against AP65 was found to be immuno-crossreactive with commercially-purchased malic enzyme, indicating recognition of a common epitope. However, malic enzyme was not able to inhibit binding of native AP65 or the recombinant AP65s to host cells in a ligand assay, even at high concentrations. Likewise, malic enzyme did not compete with live trichomonads for host cell binding sites in a cytoadherence competition experiment, suggesting that the four conserved domains are insufficient for adherence to host cells (Alderete *et al.*, 1995a).

2.3. Regulation of Expression/Localization

2.3.1. Iron Regulation. Earlier work had shown that iron regulated cytoadherence and synthesis of the *T. vaginalis* adhesins (Lehker *et al.*, 1991). The isolation of three different and independent clones indicated *a priori* that each of the three AP65 genes was transcribed. Northern analysis of RNA isolated from trichomonads showed that the specific probes hybridized to transcripts of ~1.8 kb, a size consistent with the AP65 genes. Interestingly, transcripts were only detected in total RNA from high-iron-grown trichomonads. Detection of transcripts in RNA derived from low-iron-grown organisms required prolonged exposure of X-ray film, indicating that trichomonads maintain basal levels of each transcript. Therefore, iron plays a role in upregulating transcriptional expression of each AP65 gene. Transcripts of all three genes were seen in the parental population as well as all clones, confirming the presence of each gene in all trichomonads, and suggesting that each gene is expressed within individual trichomonads (Alderete *et al.*, 1995a; O'Brien *et al.*, 1996).

2.3.2. Promoter Regions. The 5′ untranslated regions of the AP65 clones were fairly short at 13 to 17 nucleotides, and did not contain any of the typical eukaryotic promoter elements, such as the TATA box. Sequence analysis 5′ to the start site showed similarity to the promoter elements of other reported *T. vaginalis* protein-coding genes, including ferredoxin, β succinyl-CoA synthetase, α succinyl-CoA synthetase, α-tubulin, β-tubulin, 70-kDa heat-shock protein, and P-glycoprotein 1 (Quon *et al.*, 1994). Based on these genes, a 13 bp consensus sequence for *T. vaginalis* promoters, TCAYTWYTCATTA, has been defined (Quon *et al.*, 1994). Alignment of the AP65s with this sequence showed that between 8-10 bases of each AP65 5′ end conformed to the pattern. Of particular interest was the observation that each of the AP65 promoters have differences at the -9 and -10 positions (Alderete *et al.*, 1995a; O'Brien *et al.*, 1996). These gene-specific bases may affect RNA polymerase recognition of the promoters, and therefore play a role in transcriptional regulation of the AP65 adhesins.

2.3.3. 3′ Untranslated Regions (UTRs). The AP65 genes show little or no homology in the 3′-UTRs, the sequence from the stop codon to the poly-A tail. It is interesting that two of the transcripts for the AP65 genes have similar features putatively involved in destabilizing mRNA. Both have short 15 bp poly-A tails, easily degraded relative to the third transcript, which has a 55 nucleotide poly-A tail. Both also contain the sequence motif ATTTA in the cDNA, or AUUUA in mRNA, found in AU-rich elements (AREs); these elements have been shown to confer instability on mRNAs (McCarthy and Kollmuss, 1995). A separate report suggested that additional surrounding bases were also important for destabilization, lengthening the sequence to UUAUUUAU (Zubiaga *et al.*, 1995). This more complete destabilizing sequence was found in the 3′-UTR of one of the AP65 genes (Alderete *et al.*, 1995a; O'Brien *et al.*, 1996). Therefore, it is possible that each of the AP65 transcripts are subject to translational regulation through differential degradation of select AP65 transcripts.

2.3.4. Leader Sequences. From N-terminal amino acid sequencing it was determined that the mature AP65 polypeptides were missing twelve amino acids at the 5′ ends, thus identifying potential signal sequences (Alderete *et al.*, 1995a). A comparison of the signal peptides to other known leaders revealed similarities to *T. vaginalis* hydrogenosomal protein signal sequences and mitochondrial leaders (Alderete *et al.*, 1995a; O'Brien *et al.*, 1996; Lahti *et al.*, 1992; von Heijne *et al.*, 1989). Consistent with previous reports, all of the sequences begin with Met-Leu and have an arginine at the -2 position relative to the cleavage site (Lahti *et al.*, 1992). It is possible that the differences in the signal peptides, albeit slight, selectively localize each of the AP65s to distinct cellular sites.

3. INITIAL CHARACTERIZATION OF AP51 AND AP33

This section is based on preliminary, unpublished data.

3.1. Multi-Gene Families

3.1.1. Three Similar Yet Unique Genes Exist for AP51 and AP33. Restriction enzyme analysis and sequence data confirmed that the six clones for AP33 represent 3 similar, yet distinct, genes. The clones contain inserts that encode near full-length proteins for each, and procedures similar to those used for AP65, described above were used to get complete gene sequences. Homology between these AP33 genes is 91.8%, 92.7% and 95.9%, and between the proteins is 97%, 97.3% and 99.3% respectively, indicating the highly conserved nature of this adhesin.

Only one antibody-reactive cDNA clone was isolated for AP51, and insert analysis, along with sequence information indicated approximately one-third of the 5′ end was missing. While trying to generate the 5′ end for this gene, evidence was obtained for the existence of two additional AP51 genes. This data demonstrates that the three *T. vaginalis* adhesin proteins so far studied are encoded by multigene gene families and, importantly, each family exists within individual trichomonads.

3.1.2. The Genes Encode Functional Adhesin Proteins. We had already shown that the AP51 clone and one of the AP33 clones bound to host cells and efficiently inhibited the corresponding native adhesin from binding (Arroyo *et al.*, 1995). The two additional AP33 proteins were tested for these activities and were found to behave similarly.

3.2. Homology to Succinyl-CoA Synthetase. Database searches of reported DNA and protein sequences revealed that the AP33 genes and the AP51 gene have homology to the α and β subunits of succinyl-CoA synthetase, respectively. As with AP65, these homologous proteins are believed to be active in a metabolic pathway and have also been found in the *T. vaginalis* hydrogenosome (Lahti *et al.*, 1992; Lahti *et al.*, 1994). Proteins encoded by the three AP33 genes have 53.8%, 54.5%, and 54.9% identity with the α subunit of *E. coli* succinyl-CoA synthetase (Buck *et al.*, 1985). The AP51 clone shows 44.2% identity to the β subunit at the amino acid level (Buck *et al.*, 1985). There are three conserved domains in the α subunit which may contribute to nucleotide binding. Each AP33 gene is highly homologous to the enzyme subunit in these domains. There is also a conserved region surrounding a histidine residue, which, when phosphorylated, gives rise to the active form of the enzyme (Lou and Nishimura, 1991; Majumdar *et al.*, 1991), and this residue is present in the AP33 proteins.

3.3. Regulation of Expression/Localization

3.3.1. Iron Regulation. Although unique probes for each of the AP33 and AP51 genes have not yet been generated, Northern analysis revealed that transcripts for AP33 and AP51 were only detected in RNA isolated from parasites grown in iron-replete medium (Arroyo, *et al.*, 1995). This suggests that, as for AP65, transcription of each of these genes is regulated by iron.

3.3.2. Promoter Sequences. Although each AP33 promoter is unique, the sequences follow that reported for other *T. vaginalis* genes (Quon *et al.*, 1994). To date, the promoter for only one of the AP51 genes has been identified, and interestingly, it is identical to one of the AP33 promoters, but different from the AP65 promoters. It is not unlikely that the subunits of an enzyme would have the same promoter element, however, it is unknown at this point whether AP33 and AP51 form a complex during the adherence event. Each protein/subunit is capable of adhering to host cells and inhibiting native adhesin binding, suggesting that such a complex is not necessary for adhesin activity. However, using the same promoter may ensure generation of equal amounts of transcripts for at least one of the AP33 and AP51 proteins.

3.3.3. 3′ UTRs. Major differences between each of the AP33 3′ UTRs confirms the presence of three distinct genes. This is also the case for the three AP51 genes, confirming that both of these adhesins are encoded by multigene families. Although none of the AP33 genes contains the destabilizing AREs, one of the AP51 genes does. One gene for AP33 and one for AP51 have the same promoter element, and this implies that these genes are coordinately regulated at the transcriptional level. This raises the possibility that these genes may also be coordinately regulated at the translational level. Therefore, more genes may exist for either AP33, or both AP51 and AP33 if the proteins function as subunits in a complex and share similar regulatory mechanisms.

3.3.4. Leader Sequences. As for the recombinant AP65s, leader sequences were identified for each of the AP33 proteins, and these peptides were similar to other known leader sequences (Alderete *et al.*, 1995a; O'Brien *et al.*, 1996; Lahti *et al.*, 1992; von Heijne *et al.*, 1989). Two of these signal peptides were identical, indicating a possibility for similar localization of two AP33 proteins within the organism, while the third may be trafficked to other areas. The 5′ ends of the AP51 genes, including sequences encoding potential leader peptides, are still being generated.

4. CONCLUSIONS AND QUESTIONS

This overview of ongoing research activities summarizes the molecular characterization of *T. vaginalis* proteins involved in adherence to host cells. Results show that multiple-gene families encode three of four trichomonad adhesins, referred to as AP65, AP51, and AP33. That we have discovered three members, or isoforms, of each multigene family and that each of those genes is present in multiple copies reaffirms the importance and significance of these proteins to the overall biology of this parasite in general, and the property of cytoadherence in particular.

Iron is an *in vivo* signal to *T. vaginalis* (Lehker and Alderete, 1991; 1992), and iron-limited organisms provided with iron increase transcription and synthesis of adhesins (Lehker and Alderete, 1991). Adherence to host cells represents another signal; immediately after contact, the parasites undergo a dramatic morphological transformation concomitant with the rapid synthesis of adhesins (Arroyo *et al.*, 1993). That each of the adhesins is encoded by more than one gene may insure for the expression and synthesis of adequate amounts of adhesins within a short period of time following contact. Also possible is that the different copies of each adhesin gene, although coordinately regulated by iron, respond to other yet undefined environmental signals. It is conceivable, for example, that only one or two of the genes for each adhesin are directly involved in transcription and expression of the adhesin following contact with the host cell surface (Arroyo *et al.*, 1993). The other gene(s) may be responsive at times other than contact, for instance when amounts and types of vaginal iron sources fluctuate, such as during menstruation. Thus, it is essential that the arrangement and regulation of each member of the adhesin gene family be examined further.

Unexpectedly, each of the three *T. vaginalis* adhesins show significant homology to metabolic enzymes known to exist in the trichomonad hydrogenosome. It has not yet been determined whether the trichomonad adhesins express enzymatic activity. Nevertheless, in inhibition experiments using commercially available malic enzyme as an analog for AP65, the enzyme did not interfere with *T. vaginalis* cytoadherence to host cells, showing that the receptor-binding epitope is a sequence unique from the regions conserved among malic enzymes (Alderete *et al.*, 1995a).

Although percent identities between the gene members of each adhesin family are high, as much as 99% identity between the predicted amino acid sequences of two AP33 genes, the resulting variation in amino acids may result in differences in function. In the case of the *fimH* gene of *Escherichia coli*, gene variants displaying more than 98% homology and encoding proteins that differ by as little as one amino acid confer distinct adhesive phenotypes (Sokurenko *et al.*, 1994). Thus, the minor sequence variations among members of a *T. vaginalis* adhesin gene family may confer enzymatic activity for one protein and adhesive capacity for another. Alternatively, the divergence in sequence may lead to differences in receptor specificity. Functional assays of the recombinant proteins will clarify the effects of sequence variations.

We would be remiss to exclude the possibility that one of the members of each adhesin family represents the corresponding trichomonad enzyme. Comparison of the adhesin gene sequences with the reported enzyme sequences reveals that a number of the genes are

Table 2. Enzymes on mammalian cells and functions outside the metabolic pathways

α-enolase	plasminogen receptor
Lens proteins/crystallins	structural lens proteins
1. lactate dehydrogenase	
2. alcohol dehydrogenase	
3. hydroxyacyl CoA dehydrogenase	
4. NADPH-dependent reductases	
5. α-enolase	
6. glutathione S-transferase	
7. argininosuccinate lyase	

References cited in text.

identical. Recent reports present the sequences of *T. vaginalis* malic enzyme (Hrdy and Müller, 1995), and the α (Lahti *et al.*, 1994) and β (Lahti *et al.*, 1992) subunits of succinyl-CoA synthetase. However, transcripts from these genomic copies have not been shown to yield proteins with enzymatic activity. It is critical to determine whether the hydrogenosomal enzymes and the trichomonad adhesins are, in fact, encoded by identical genes, a phenomenon known as gene sharing (Piatigorsky and Wistow, 1989). The adhesin proteins may possess bifunctionality, having both the catalytic function of metabolic enzymes and adhesive properties. Although this very well may be the first report of a protozoan using surface-expressed metabolic enzymes as adhesins, this phenomenon is not without precedence among other microbial pathogens, as detailed below.

Some of the earliest reports of enzymes having important roles beyond those in metabolism (Table 2) were the crystallins, the structural proteins of the lens, which were revealed to be metabolic enzymes (Piatigorsky and Wistow, 1989; Wistow, 1993; Wistow and Piatigorsky, 1987). One of the crystallins, α-enolase, has also been implicated as a plasminogen receptor when surface-expressed (Miles *et al.*, 1991). A number of reports have described metabolic enzymes on the surfaces of microbial pathogens, as listed in Table 3 (Camara *et al.*, 1994; Joe *et al.*, 1994; Lottenberg *et al.*, 1992; Pancholi and Fischetti, 1992; Vacca-Smith *et al.*, 1994; Goudot-Crozel, 1989). These surface enzymes possessed multiple functions, including adherence (Lottenberg *et al.*, 1992; Pancholi and Fischetti, 1992; 1993; Vacca-Smith *et al.*, 1994). Interestingly, glyceraldehyde-3-phosphate dehydrogenase (GAPDH) has been found on the surface of Group A Streptococci (Lottenberg *et al.*, 1992; Pancholi and Fischetti, 1992) and on the outer membrane of *Schistosoma mansoni* (Goudot-Crozel *et al.*, 1989). On Group A Streptococci GAPDH functions as a plasmin receptor (Lottenberg *et al.*, 1992; Pancholi and Fischetti, 1992), however, a number of additional functions have been described including uracil DNA glycosylase activity (Meyer-Siegler *et al.*, 1991), protein kinase activity (Kawamoto and Caswell, 1986), bundling of microtubules

Table 3. Enzymes on microbial pathogens and parasite surfaces

Streptococcus pneumoniae	neuraminidase
Porphyromonas gingivalis	glutamate dehydrogenase
Group A Streptococci	glyceraldehyde-3-phosphate dehydrogenase
Streptococcus gordonii	glucosyltransferase
Schistosoma mansoni	glyceraldehyde-3-phosphate dehydrogenase

References cited in text.

(Huitorel and Pantaloni, 1985), and binding of fibronectin, lysozyme, and cytoskeletal proteins (Pancholi and Fischetti, 1992). Another surface-localized enzyme, glucosyltransferase, has been demonstrated to mediate the adhesion of *Streptococcus gordonii* to human endothelial cells (Vacca-Smith *et al.*, 1994).

It is equally plausible that a new structural role was developed following gene duplication and subsequent separation of function (Piatogorsky and Wistow, 1989; Wistow, 1993). Moreover, features acquired after gene duplication may contribute to the evolution of the distinct protein functions. For example, since function might be determined by location, different signal peptides may direct the proteins to distinct cellular sites. We cannot exclude the possibility that the slight differences in the leader sequences selectively localize each molecule to distinct cellular locations. Alternatively, a recent report demonstrates that mRNAs contain site-directing determinants (zip-codes) for translation and expression of proteins in specific regions of the cell, a phenomenon documented for the transcripts that encode actin, tubulin and vimentin (Kislauskis and Singer, 1992). Like the AP65, AP51, and AP33 adhesins, actin is a family of almost identical proteins that differ in the 3′-UTRs. It has been proposed that the mRNAs for the different actin isoforms dictate different compartmentalization for the synthesis of each isoform (Kislauskis and Singer, 1992). The use of antibodies specific for the recombinant proteins of each adhesin in immunogold labeling of *T. vaginalis* could identify the cellular location(s) of the adhesins.

We have shown numerous times that the adhesin genes are regulated by iron (O'Brien *et al.*, 1996; Alderete *et al.*, 1995a; Lehker *et al.*, 1991; Lehker and Alderete, 1992). If, in fact, any of the adhesin proteins do represent the corresponding *T. vaginalis* enzymes, these findings show that these and possibly other hydrogenosomal enzymes are up-regulated by iron. To our knowledge, this would be the first time that the expression of trichomonad hydrogenosomal enzymes has been shown to be regulated by iron. A recent report demonstrates that changes in culture conditions result in changes in the activities of *T. vaginalis* metabolic enzymes (Ter Kuile, 1994). Thus, it is critical to consider certain factors when studying the regulation of *T. vaginalis* enzymes and, therefore, of detection of trichomonad enzymatic pathways.

Upon contact with VECs or HeLa cells, trichomonads synthesize greater amounts of all four adhesins (Arroyo *et al.*, 1993); we hypothesized that the increase in adhesin synthesis was due to utilization of internal iron pools (Arroyo *et al.*, 1993). Although speculative, it may be important to consider that adhesin synthesis is regulated by the rate of transcript turnover. Recent reports, for example, demonstrate that mRNA turnover involves specific sequences and regions that contribute to the instability of transcripts (Peltz and Jacobson, 1992; Zubiaga *et al.*, 1995; McCarthy and Kollmus, 1995; Chen and Shyu, 1994). It may be significant that the AU-rich element (ARE), a sequence element involved in RNA destabilization (Chen and Shyu, 1994; Zubiaga *et al.*, 1995), is present in the 3′-UTRs of some of the adhesin genes. The presence of these destabilizing elements in select adhesin mRNAs may suggest differential degradation of the transcripts. The signalling event described earlier may lead to a decrease in mRNA turnover, especially of the transcripts containing AREs, providing for a mechanism of rapid expression of the individual adhesin proteins.

Another possible device for the expeditious expression of all four adhesins following signalling is the packaging of the proteins into vacuoles. Considering that the four *T. vaginalis* adhesins are coordinately expressed, one might envision that all four proteins are placed within a vesicle and exported in this fashion. This would allow for rapid simultaneous surface localization of the four adhesins. Although the precise interrelationship between the adhesins remains unknown, some form of packaging within organelles for subsequent coordinated expression would be consistent with the fact that each adhesin is essential for cytoadherence (Alderete and Garza, 1985; 1988; Arroyo *et al.*, 1992). The exact interrelationship of the adhesins once they reach the parasite surface also remains a mystery. While the four adhesins may exist on the surface separately, it is equally plausible that, particularly among the three found to have homology to enzymes, the proteins act as subunits forming adhesive complexes.

In view of the highly conserved nature of metabolic enzymes and the metabolic enzyme sequences in the adhesins, the placement of host-like proteins on the surface of this sexually transmitted pathogen may be significant and play a role in immune evasion. This would exemplify a type of molecular mimicry, something that has received increasing attention as an important mechanism by which parasites escape recognition by the host immune system (Damian, 1989). We have already reported that the adhesins are immunorecessive in nature, and preliminary results suggest that adhesin antibody, if present at all, is at low levels in vaginal wash or human serum (Alderete *et al.*, 1991b; unpublished observations). This tactic used by the parasite raises important concerns for vaccine development using biofunctional trichomonad molecules, such as the adhesins, which mimic host proteins. Localization of the receptor-binding epitope and the extent of similarity of this epitope with host sequences will be crucial in the consideration of these proteins as vaccine candidates and development of diagnostics.

The results to date prompt the following questions:

1. What is the interrelationship between the adhesins? How are they arranged on the parasite surface?
2. Are the receptors on the host-cell surface the same for all four adhesins, unique for each adhesin, or even unique for each isoform within each adhesin family?
3. Does the increase in adhesin expression following attachment to host cells result from the utilization of internal iron pools, or are there other yet undefined environmental signals? Does the signalling event lead to stabilization of otherwise degradable adhesin transcripts and hence lead to increased adhesin synthesis?
4. Are all isoforms of an adhesin family surface-expressed? If not, what signals regulate the differential expression of these proteins?
5. Which of the adhesin genes discussed here encode genuine hydrogenosomal enzymes?

REFERENCES

Alderete, J.F., and Garza, G.E., 1985, Specific nature of *Trichomonas vaginalis* parasitism of host cell surfaces, *Infect. Immun.* 50:701-708.

Alderete, J.F., and Garza, G.E., 1987, *Trichomonas vaginalis* attachment to host cell surfaces and role of cytoadherence in cytotoxicity, *Acta Universitatis Carolinae - Biologica.* 10:373- 380.

Alderete, J.F., and Garza, G.E., 1988, Identification and properties of *Trichomonas vaginalis* proteins involved in cytoadherence, *Infect. Immun.* 56:28-33.

Alderete, J.F., and Kasmala, L., 1986, Monoclonal antibody to a major glycoprotein immunogen mediates differential complement-independent lysis of *Trichomonas vaginalis*, *Infect. Immun.* 53:697-699.

Alderete, J.F., Demeš, P., Gombošova, A., Valent, M., Fabušova, M., Janoška, A., Stefanovic', J., and Arroyo, R., 1988, Specific parasitism of purified vaginal epithelial cells by *Trichomonas vaginalis*, *Infect. Immun.* 56:2558-2562.

Alderete, J.F., Kasmala, L., Metcalfe, E., and Garza, G.E., 1986a, Phenotypic variation and diversity among *Trichomonas vaginalis* isolates and correlation of phenotype with trichomonal virulence determinants, *Infect. Immun.* 53:285-293.

Alderete, J.F., Newton, E., Dennis, C., and Neale, K.A., 1991a, Antibody in sera of patients infected with *Trichomonas vaginalis* is to trichomonad proteinases, *Genitourin. Med.* 67:331-334.

Alderete, J.F., Newton, E., Dennis, C., and Neale, K.A., 1991b, The vagina of women infected with *Trichomonas vaginalis* has numerous proteinases and antibody to trichomonad proteinases, *Genitourin. Med.* 67:469-474.

Alderete, J.F., O'Brien, J.., Arroyo, R., Engbring, J.A., Musatovova, O., Lopez, O., Lauriano, C., and Nguyen, J., 1995a, Cloning and molecular characterization of two genes encoding adhesion proteins involved in *Trichomonas vaginalis*, *Mol. Microbiol.* 17:69-83.

Alderete, J.F., Provenzano, D., and Lehker, M.W., 1995b, Iron mediates *Trichomonas vaginalis* resistance to complement lysis, *Microbial Pathogen.* 19:93-103.

Alderete, J.F., Suprun-Brown, L., and Kasmala, L., 1986b, Monoclonal antibody to a major surface glycoprotein immunogen differentiates isolates and subpopulations of *Trichomonas vaginalis, Infect. Immun.* 52:70-75.

Arroyo, R., and Alderete, J.F., 1989, *Trichomonas vaginalis* surface proteinase activity is necessary for parasite adherence to epithelial cells, *Infect. Immun.* 57:2991-2997.

Arroyo, R., and Alderete, J.F., 1995, Two *Trichomonas vaginalis* surface proteinases bind to host epithelial cells and are related to levels of cytoadherence and cytotoxicity, *Arch. Med. Res.* 26:279-285.

Arroyo, R., Engbring, J., and Alderete, J.F., 1992, Molecular basis of host epithelial cell recognition by *Trichomonas vaginalis*, *Mol. Microbiol.* 6:853-862.

Arroyo, R., Engbring, J., Nguyen, J., Musatovova, O., Lopez, O., Lauriano, C., and Alderete, J.F., 1995, Characterization of cDNAs encoding adhesion proteins involved in *Trichomonas vaginalis* cytoadherence, *Arch. Med. Res.* 26:361-369.

Arroyo, R., González-Robles, A., Martinez-Palomo, A., and Alderete, J.F., 1993, Signalling of *Trichomonas vaginalis* for amoeboid transformation and adhesion synthesis follows cytoadherence, *Mol. Microbiol.* 7:299-309.

Bagchi, S., Wise, L.S., Brown, M.L., Bregman, D., Su, J.S., and Rubin, C.S., 1987, Structure and expression of murine malic enzyme mRNA. Differentiation-dependent accumulation of two forms of malic enzyme mRNA in 3T3-L1 cells, *J. Biol. Chem.* 262:1558-1565

Beachey, E.H., 1989, Bacterial adherence, in *Molecular Mechanisms of Microbial Adhesion*, (Switalski, L., Hook, M., and Beachey, E.H., eds.), pp.1-4, Springer-Verlag, New York.

Bramley, M., 1976, Study of female babies of women entering confinement with vaginal trichomoiasis, *Brit. J. Vener. Dis.* 52:58-62.

Buck, D., Spencer, M.E., and Guest, J.R., 1985, Primary structure of the succinyl-CoA synthetase of *Escherichia coli, Biochemistry* 24:6245-6252.

Camara, M., Boulnois, G.T., Andrew, P.W., and Mitchell, T.J., 1994, A neuramindase from *streptococcus pneumoniae* has the features of a surface protein, *Infect. Immun.* 62:3688- 3695.

Chen, C.-Y.A., and Shyu, A.-B., 1994, Selective degradation of early-response-gene mRNAs: functional analyses of sequence features of the AU-rich elements, *Mol. Cell. Biol.* 14:8471-8482.

Cohen, M.S., Britigan, B.E., French, M., and Bean, K., 1987, Preliminary observations on lactoferrin secretion in human vaginal mucus: variation during menstrual cycle, evidence of hormonal regulation and implications for infection with *Neisseria gonorrhoeae, Am. J. Obstet. Gynecol.* 157:1122-1125.

Cotch, M.F., Pastorek II, J.G., Nugent, R.P., Yerg, D.E., Martin, D.H., and Eschenbach, D.A., 1991, Demographic and behavioral predictors of *Trichomonas vaginalis* infection among pregnant women, *Obstet. Gynecol.* 78:1087-1092.

Dailey, D.C., Chang, T., and Alderete, J.F., 1990, Characterization of *Trichomonas vaginalis* haemolysis, *Parasitol.* 101:171-175.

Damian, R.T., 1989, Molecular mimicry: parasite evasion and host defense, *Curr. Top. Microbiol. Immunol.* 145:101-115.

Goudot-Crozel, V., Caillol, D., Djabali, M., and Dessein, A.J., 1989, The major parasite surface antigen associated with human resistance to schistosomiasis is a 37-kD glyceraldehyde- 3P-dehydrogenase, *J. Exp. Med.* 170:2065-2080.

Holz, G.G., Lindmark, D.G., Beach, D.J., Neale, K.A., Singh, B.N., 1987, Lipids and lipid metabolism of trichomonads, symposium on trichomonads and trichomoniasis, *Acta Unversitatis Carolinae-Biologica,* Praque, Czechoslovakia.

Hrdy, I., and Müller, M., 1995, Primary structure of the hydrogenosomal malic enzyme of *Trichomonas vaginalis* and its relationship to homologous enzymes, *J. Euk. Micro.* 42:593-603.

Hsu, R.Y., Glynias, M.J., Satterlee, J., Feeney, R., Clarke, A.R., Emery, D.C., Roe, B.A., Wilson, R.K., Goodridge, A.G., and Holbrooks, J.J., 1992, Duck liver malic enzyme expression in *Escherichia coli* and characterization of the wild-type enzyme and site-directed mutants, *Biochem. J.* 284:869-876.

Huitorel, P., and Pantaloni, D., 1985, Bundling of microtubules by glyceraldehyde-3-phosphate dehydrogenase and its modulation by ATP, *Eur. J. Biochem.* 150:265-269.

Joe, A., Murray, C.S., and McBride, B.C., 1994, Nucleotide sequence of a *porphyromonas gingivalis* gene encoding a surface-associated glutamate dehydrogenase and construction of a glutamate dehydrogenase- deficient isogenic mutant, *Infect. Immun.* 62:1358-1368.

Kawamoto, R.M., and Caswell, A.H., 1986, Autophosphorylation of glyceraldehyde phosphate dehydrogenase and phosphorylation of protein from skeletal muscle microsomes, *Biochemistry.* 25:656-661.

Khoshnan, A., and Alderete, J.F., 1993, Multiple double-stranded RNA segments are associated with virus particles infecting *Trichomonas vaginalis, J. Virol.* 67:6950-6955.

Khoshnan, A., and Alderete, J.F., 1994, *Trichomonas vaginalis* with a double-stranded RNA virus has upregulated levels of phenotypically variable immunogen mRNA, *J. Virol.* 68:4035-4038.

Kislauskis, E.H., and Singer, R.H., 1992, Determinants of mRNA localization, *Curr. Opin. Cell. Biol.* 4:975-978.

Krieger, J.N., Wolner-Hanssen, P., Steven, C., and Holmes, K.K., 1990, Characteristics of *Trichomonas vaginalis* isolates from women with and without colpitis macularis, *J. Infect. Dis.* 161:307-311.

Kulkarni, G., Cook, P.F., and Harris, B.G., 1993, Cloning and nucleotide sequence of a full- length cDNA encoding *Ascaris suum* malic enzyme, *Arch. Biochem. Biophys.* 300:231- 237.

Laga, M.A., Manoka, A., Kivuvu, M., Malele, B., Tuliza, M., Nzila, N., Goeman, J., Behets, F., Batter, V., Alary, M., Heyward, W.L., Ryder, R.W., Piot, P., 1993, Non-ulcerative sexually transmitted diseases as risk factors for HIV-1 transmission in women: results from a cohort study, *AIDS.* 7:95-102.

Lahti, C.J., Bradley, P.J., and Johnson, P.J., 1994, Molecular characterization of the α-subunit of *Trichomonas vaginalis* hydrogenosomal succinyl CoA synthetase, *Mol. Biochem. Parasitol.* 66:309-318.

Lahti, C.J., d'Oliveira, C.E., and Johnson, P.J., 1992, Beta-succinyl coenzyme A synthetase from *Trichomonas vaginalis* is a soluble hydrogenosomal protein with an amino terminal sequence that resembles motochondrial presequences, *J. Bacteriol.* 174:6822-6830.

Lehker, M., and Alderete, J.F., 1992, Iron regulates growth of *Trichomonas vaginalis* and the expression of immunogenic proteins, *Mol. Microbiol.* 6:123-132.

Lehker, M., Arroyo, R., and Alderete, J.F., 1991, The regulation by iron of the synthesis of adhesins and cytoadherence levels in the protozoan *Trichomonas vaginalis*, *J. Exp. Med.* 174:311-318.

Lehker, M.L., Chang, T.H., Dailey, D.C., and Alderete, J.F., 1990, Specific erythrocyte binding is an additional nutrient acquisition system for *Trichomonas vaginalis, J. Exp. Med.* 171:2165-2170.

Lehker, M.W., and Alderete, J.F., 1991, The regulation by iron of the synthesis of adhesins and cytoadherence levels in the protozoan *Trichomonas vaginalis, J. Exp. Med.* 174:311-318.

Lindmark, D.G., 1983, Failure of trichomonads to convert or retroconvert long chain fatty acids or cholesterol, *J. Protozool.* 30:5A.

Loeber, G., Infante, A.A., Maurer-Fogy, I., Krystek, E., and Dworkin, M.B., 1991, Human NAD^+-dependent mitochondrial malic enzyme, *J. Biol. Chem.* 266:3016-3021.

Lottenberg, R., Broder, C.C., Boyle, M.D.P., Kain, S.J., Schroeder, B.L., and Curtis III, R., 1992, Cloning sequence analysis and expression in *Escherichia coli* of a streptococcal plasmin receptor, *J. Bacteriol* 174:5204-5210.

Lou, G.X., and Nishimura, J.S., 1991, Site-directed mutagenesis of *Escherichia coli* succinyl- CoA synthetase, *J. Biol. Chem.* 266:20781-20785.

Lund, P.G., and Shorb, M.S., 1962, Steroid requirement of trichomonads, *J. Protozool.* 9:151- 154.

Majumdar, R., Guest, J.R., and Bridger, W.A., 1991, Functional consequences of substitution of the active site (phospho) histidine residue of *Escherichia coli* succinyl-CoA synthetase, *Biochim. Biophys. Acta.* 1076:86-90.

Masson, P.L., Heremans, J.F., and Dive, C.H., 1966, An iron-binding protein common to many external excretions, *Clin. Chim. Acta.* 14:735-739.

McCarthy, J.E.G., and Kollmuss, H., 1995, Cytoplasmic mRNA-protein interactions in eukaryotic gene expression, *Trends Biochem. Sci.* 20:191-197.

Meyer-Siegler, K., Mauro, D.J., Seal, G., Wurzer, J., DeRiel, J.K., and Sirover, M.M., 1991, A human nuclear uracil DNA glycosylase is the 37 kDa subunit of glyceraldehyde-3- phosphate dehydrogenase *Proc. Natl. Acad. Sci. USA* 88:8460-8464.

Miles, L.A., Dahlberg, C.M., Plescia, J., Felez, J., Kato, K., and Plow, E.F., 1991, Role of cell- surface lysines in plasminogen binding to cells: identification of α-enalase as a candidate plasminogen receptor, *Biochemistry.* 30:1682-1691.

Minkoff, H., Grunebaum, A.N., Schwarz, R.H., Feldman, J., Cummings, M., Crobleholme, W., Clark, L., Pringle, G., and McCormack, W.M., 1984, Risk Factors for prematurity and premature rupture of membranes: a prospective study of the vaginal flora in pregnancy, *Am. J. Obstet. Gynecol.* 150:965-972.

Müller, M., Meingassner, J.G., Miller, M.A., and Ledger, W.J., 1980, Three metronidazole- resistant strains of *Trichomonas vaginalis* from the U.S.A., *Am. J. Obstet. Gynecol.* 138:808-812.

Neale, K.A., and Alderete, J.F., 1990, Analysis of the proteinases of representative *Trichomonas vaginalis* isolates, *Infect. Immun.* 58:157-162.

Nzila, N., Laga, M., Thiam, M.A., Mayimona, K., Edidi, B., Van Dyck, E., Behets, F., Hassig, S., Nelson, A., Mokwa, K., Ashley, R.L., Piot, P., and Ryder, R.W., 1991, HIV and other sexually transmitted diseases among female prostitutes in Kinshasa, *AIDS.* 5:715-721.

O'Brien, J.L., Lauriano, C.M., and Alderete, J.F., 1996, Molecular characterization of a third AP65 adhesin gene of *Trichomonas vaginalis, Microbial Pathogen.* In press.

Pancholi, V., and Fischetti, V.A., 1992, A major surface protein on group A streptococci is a glyceraldehyde-3-phosphate-dehydrogenase with multiple binding activity, *J. Exp. Med.* 176:415-426.

Pancholi, V., and Fischetti, V.A., 1993, Glyceraldehyde-3-phosphate dehydrogenase on the surface of group A streptococci is also an ADP-ribosylating enzyme, *Proc. Natl. Acad. Sci. USA* 90:8154-8158.

Peltz, S.W., and Jacobson, A., 1992, mRNA stability: in *trans*-it, *Curr. Opin. Cell. Biol.* 4:979- 983.

Peterson, K.M., and Alderete, J.F., 1984a, Selective acquisition of plasma proteins by *Trichomonas vaginalis* and human lipoproteins as a growth requirement by this species, *Mol. Biochem. Parasitol.* 12:37-48.

Peterson, K.M., and Alderete, J.F., 1984b, *Trichomonas vaginalis* is dependent on uptake and degradation of human low density lipoproteins, *J. Exp. Med.* 160:1261-1271.

Peterson, K.M., and Alderete, J.F., 1984c, Iron uptake and increased intracellular enzyme activity follows host lactoferrin binding by *Trichomonas vaginalis* receptors, *J. Exp. Med.* 160:398-410.

Piatigorsky, J., and Wistow, G.J., 1989, Enzyme/crystallins: gene sharing as an evolutionary strategy, *Cell* 57:197-199.

Poulouse, A.J., and Kolattukudy, P.E., 1983, Sequence of a tryptic peptide from the NADPH binding site of the enoyl reductase domain of fatty acid synthase, *Arch. Biochem. Biophys.* 220:652-656.

Provenzano, D., and Alderete, J.F., 1995, Analysis of human immunoglobulin-degrading cysteine proteinases of *Trichomonas vaginalis, Infect. Immun.* 63:3388-3395.

Quon, D.V.K., Delgadillo, M.G., Khachi, A., Smale, S.T., and Johnson, P.J., 1994, Similarity between a ubiquitous promoter element in an ancient eukaryote and mammalian initiator elements, *Proc. Natl. Acad. Sci.* 91:4579-4583.

Read, J.S., and Klebanoff, M.A., 1993, Sexual intercourse during pregnancy and preterm delivery: effects of vaginal microorganisms, *Am. J. Obstet. Gynecol.* 168:514-519.

Roitman, I., Heyworth, P.G., and Gutteridge, W.E., 1978, Lipid synthesis by *Trichomonas vaginalis*, *Ann. Trop. Med. Parasitol.* 72:583-585.

Rothermel, B.A., and Nelson, T., 1989, Primary structure of the maize NADP-dependent malic enzyme, *J. Biol. Chem.* 264:19587-19592.

Satterlee, J., and Hsu, R.Y., 1991, Duck liver malic enzyme: sequence of a tryptic peptide containing one cysteine residue labelled by the substrate analog bromopyruvate, *Biochim. Biophys. Acta.* 1079:247-252.

Scrutton, N.S., Berry, A., and Perham, R.N., 1990, Redesign of the coenzyme specificity of a dehydrogenase by protein engineering, *Nature* 343:38-43.

Sokurenko, E.V., Courtney, H.S., Ohman, D.E., Klemm, P., and Hasty, D.L., 1994, FimH family of type 1 fimbrial adhesins: functional heterogeneity due to minor sequence variations among *fimH* genes, *J. Bacteriol.* 176:748-755.

Stevens-Simon, C., Jamison, J., McGregor, J.A, and Douglas, J.M., 1994, Racial variation in vaginal pH among healthy sexually active adolescents, *Sex. Trans. Dis.* 21:168-172.

Ter Kuile, B.H., 1994, Adaptation of the carbon metabolism of *Trichomonas vaginalis* to the nature and availability of carbon source, *Microbiol.* 140:2503-2510.

Vacca-Smith, A.M., Jones, C.A., Levine, M.J., and Stinson, M.W., 1994, Glucosyltransferase mediates adhesion of *streptococcus gordonii* to human endothelial cells *in vitro, Infect. Immun.* 62:2187-2194.

von Heijne, G., Steppuhn, J., and Herrmann, R.G., 1989, Domain structure of mitochondrial and chloroplast targeting peptides, *Eur. J. Biochem.* 180:535-545.

Wang, A., Wang, C.C., and Alderete, J.F., 1987, *Trichomonas vaginalis* phenotypic variation occurs only among trichomonads infected with the double-stranded RNA Virus, *J. Exp. Med.* 166:142-150.

Wei, C.H., Chou, W.Y., Huang, S.H., Lin, C.C., and Chang, G.G., 1994, Affinity cleavage at the putative metal-binding site of pigeon liver malic enzyme by the Fe^{2+}-ascorbate system, *Biochemistry* 33:7931-7936.

Wierenga, R.K., DeMaeyer, M.C.H., and Hol, W.G.J., 1985, Interaction of pyrophosphate moities with α-helixes in dinucleotide-binding proteins, *Biochemistry* 24:1346-1357.

Wistow, G., 1993, Lens crystallins: gene recruitment and evolutionary dynamism, *Trends Biochem. Sci.* 18:301-306.

Wistow, G., and Piatigorsky, J., 1987, Recruitment of enzymes as lens structural proteins, *Science* 236:1554-1555.

Zubiaga, A.M., Belasco, J.G., and Greenberg, M.E., 1995, The nonamer UUAUUUAUU is the key AU-rich sequence motif that mediates mRNA degradation, *Mol. Cell. Biol.* 15:2219- 2230.

26

CELL SURFACE MOLECULES OF PATHOGENIC AND NONPATHOGENIC *ENTAMOEBA HISTOLYTICA* AND THEIR RELATION TO VIRULENCE

David Mirelman, Sue Moody-Haupt, Steven Becker, Yael Nuchamowitz, Rivka Bracha, and Rinat Alon

Department of Membrane Research and Biophysics
The Weizmann Institute of Science
Rehovot 76100, Israel

In recent years it has become increasingly clear that human infections with *Entamoeba histolytica* are caused by two types of morphologically indistinguishable trophozoites (McKerrow, 1992; Mirelman, 1992). Symptomatic disease is always associated with the pathogenic type (P), whereas the nonpathogenic (NP) ameba or *Entamoeba dispar* as many prefer to name it, usually behaves more as a commensal that does not cause symptoms. The correlation, however, between the clinical picture of the patient and the identity of the type of parasite with which he is infected, does not always fit. This complication may be due to the fact that both P and NP types of amebae can sometimes be present in the same stool, as has been recently reported for quite a number of cases (Acuna-Soto, et al. 1993; Walderich et al. 1995).

Considerable efforts are being made in many laboratories around the world to elucidate the molecular components which differ between the P and NP types of parasites so as to better understand those which are responsible for its virulence. A variety of molecular markers such as monoclonal antibodies (mAbs) and DNA probes have been developed so that nowadays the two types of amebae can be rapidly determined (Garfinkel et al. 1989; Clark & Diamond, 1991). Numerous studies have concentrated on identifying differences on surface molecules of the trophozoites. The best characterized cell surface component of the ameba is the galactose-binding lectin, discovered 15 years ago by Ravdin, whose function is to bind to receptors on the surface of mammalian or bacterial cells (Ravdin et al. 1988). Sequence comparisons of the Gal-lectin genes of P and NP *E. histolytica* revealed some differences (10%) but these apparently do not result in any significant defect in the lectin binding activity of NP trophozoites (Petri et al. 1990).

Structural differences between other cell surface antigens of P and NP amebae have also been studied. A mAb which specifically recognizes a 30 kDa, lysine-rich antigen on cell membranes of NP trophozoites and does not react with P amebae, was prepared in our lab.

Toward Anti-Adhesion Therapy for Microbial Diseases, edited by Kahane and Ofek
Plenum Press, New York, 1996

A very similar 30 kDa Lys-rich antigen with 95% sequence homology was detected on surfaces of P amebae. Expression of the two gene products revealed that only epitopes on the NP antigen react with the mAb. Unexpectedly, a P-type ameba was isolated not long ago from a human patient which positively reacted with the NP-specific mAb. Using specific oligonucleotide primers and PCR, we established that this particular P-strain was most likely a natural hybrid that expressed both the P- and the NP-specific 30 kDa antigen (Bracha et al. 1995). Using a novel recombinant vector for stable transformation of amebae (Grodberg et al. 1990; Vines et al. 1995), we have recently managed to introduce the NP copy of the 30 kDa Lys-rich antigen into a P strain and detected, with the mAb, its expression on the trophozoite cell surface. These preliminary findings support the notion that genetic transfer between NP and P amebae may occur also under natural conditions , perhaps during simultaneous infections with both P and NP strains .

Another family of surface antigens which appear to differ between strains of different virulence are the lipophosphoglycans (LPG) (Bhattacharya et al. 1992). Similar types of molecules have been extensively characterized in another protozoan parasite, the *Leishmania sp.* The *E. histolytica* molecules, which appear to differ in certain structural elements from those of *Leishmania,* were first identified in a pathogenic isolate using a procedure for the isolation of bacterial lipopolysaccharides. We have used solvent as well as hydrophobic and anion exchange chromatography to purify two distinct lipid anchored glycolipids whose composition was indicative of an LPG and a lipophosphopeptidoglycan (LPPG) (Moody et al. 1995). A direct correlation was observed between the relative abundance of these molecules in different amebic isolates and their virulence. The glycolipids extracted from avirulent *E. histolytica* contained extremely low levels of LPG molecules. A similar observation was made in trophozoites that had lost their virulence following long term *in vitro* cultivation in cholesterol-poor medium. Upon restoration of virulence by passaging of the ameba in a hamster liver, or by cultivation in high cholesterol, there was a detectable reappearance of LPG molecules on the surface of the parasite. The detailed chemical structures of the LPG and LPPG molecules of various strains are currently being investigated.

The ability of parasites to modify surface molecules is one of their inherent strategies for survival in the host. Better understanding of the regulatory mechanisms and structure variations that are involved in the modulation of *E. histolytica* cell surface antigens and glycoconjugates is very important for any rational design of a therapeutic approach and control of this pathogen.

ACKNOWLEDGMENTS

This investigation was supported in part by a grant from the Commission of the European Communities, Avicenne Program.

REFERENCES

Acuna-Soto, R., Samuelson, J., Girolami, P.D., Zarate, L., Millan-Velasco, F., Schoolnick, G. and Wirth, D. (1993) Application of the polymerase chain reaction to the epidemiology of pathogenic and nonpathogenic *Entamoeba histolytica*. Am. J. Trop. Med. Hyg. 48, 58-70.

Bhattacharya, A., Prasad, R. and Sacks, D.L. (1992) Identification and partial characterization of a lipophosphoglycan from a pathogenic strain of *Entamoeba histolytica*. Mol Biochem Parasitol 56, 161-168.

Bracha, R., Nuchamowitz, Y. and Mirelman, D. (1995) Molecular cloning of a 30-kilodalton lysine-rich surface antigen from a nonpathogenic *Entamoeba histolytica* strain and its expression in a pathogenic strain. Infect. Immun. 63, 917-925.

Clark, C.G. and Diamond, L.S. (1991) Ribosomal RNA genes of "pathogenic" and "nonpathogenic" *Entamoeba histolytica* are distinct. Mol. Biochem. Parasitol. 49, 297-302.

Garfinkel, L.I., Giladi, M., Huber, M., Gitler, C., Mirelman, D., Revel, M. and Rozenblatt, S. (1989) DNA probes specific for *Entamoeba histolytica* having pathogenic and non pathogenic zymodemes. Infect. Immun. 57, 926-931.

Grodberg, J., Salazar, N., Oren, R. and Mirelman, D. (1990) Autonomous replication sequences in an extrachromosomal element of a pathogenic Entamoeba histolytica. Nucl. Acids Res. 18, 5515-5519.

McKerrow, J.H. (1992) Pathogenesis in amebiasis: Is it genetic or acquired? Infect. Agents and Dis. 1, 11-14.

Mirelman, M. (1992) Pathogenic versus Nonpathogenic *Entamoeba histolytica*. Infect. Agents and Dis. 1, 15-18.

Moody, S., Becker, S., Nuchamowitz, Y. and Mirelman, D. (1995) Virulent and avirulent Entamoeba histolytica differ in their cell surface phosphorylated glycolipids. The First Meeting of the F.I.S.E.B Eilat, Oct. 1995.

Petri, W.A., Jackson, T.F.H.G., V, G., Kress, K., Saffer, L.D., Snodgrass, T.L., Chapman, M.D., Keren, Z. and Mirelman, D. (1990) Pathogenic and nonpathogenic strains of *Entamoeba histolytica* can be differentiated by monoclonal antibodies to the galactose-specific adherence lectin. Infect. Immun. 58, 1802-1806.

Ravdin, J.I., Petri, W.A. and Mirelman, D. (1988) Mechanisms of adherence by *Entamoeba histolytica.* In: Amebiasis: Human Infection by *Entamoeba histolytica.* Ravdin, J.I. (ed.). John Wiley & Sons, N.Y. (publ), pp. 205-218.

Vines, R.R., Purdy, J.E., Ragland, B.D., Samuelson, J., Mann, B.J. and Petri, W.A. (1995) Stable episomal transfection of *Entamoeba histolytica*. Mol. Biochem. Parasitol. 71, 265-267.

Walderich, B., Muller, Burchard, G.D. and Mirelman, D. (1995) Patients infected with *Entamoeba histolytica* and *Entamoeba dispar* as detected by isoenzymes and two different PCR systems. European Conference on Tropical Medicine. Hamburg, Oct. 1995 Abs. L119, 150.

27

THE ROLE OF LECTINS IN RECOGNITION AND ADHESION OF THE MYCOPARASITIC FUNGUS *TRICHODERMA* spp. TO ITS HOST

Jacob Inbar and Ilan Chet

The Otto Warburg Center for Agricultural Biotechnology
The Hebrew University
Faculty of Agriculture
Rehovot 76100, Israel

The destructive parasitic mode in *Trichoderma*, a natural antagonist to other fungi and well-known biocontrol agent of plant pathogenic fungi, appears to be a complex process, made up from several successive steps. Upon contact with the host *Trichoderma* coils around or grows along the host hyphae and forms hook-like structures that aid in penetrating the host's cell wall (Chet, 1987). In *Trichoderma*, this reaction has been found to be rather specific (Dennis and Webster, 1971), and lectin-carbohydrate interactions were assumed to mediate the attachment and recognition between *Trichoderma* and soil-borne plant pathogenic fungi (Barak *et al.*, 1985).

In an attempt to test this hypothesis, we developed a biomimetic system based on the binding of lectins to the surface of nylon fibers. This system simulates the host hyphae and enabled us to examine the role of lectins in mycoparasitism (Inbar and Chet, 1992). The first direct evidence for this role was provided by the mycoparasitic biocontrol fungus *T. harzianum*, which, when allowed to grow on nylon fibers treated with concanavalin A or crude *Sclerotium rolfsii* agglutinin, coiled around the nylon fibers and produced hooks in a pattern similar to that observed with the real host hyphae.

Recently, we have isolated and purified a novel lectin from the culture filtrate of the soilborne plant pathogenic fungus *S. rolfsii* (Inbar and Chet, 1994). Agglutination of *Escherichia coli* cells by the purified lectin (a protein with a molecular mass of ca. 45 kDa) could be inhibited by the glycoproteins mucin and asialomucin. Proteases as well as β-1,3-glucanase, were found to be totally destructive to agglutination activity, indicating that both protein and β-1,3-glucan are necessary for agglutination. Using the biomimetic system, it was apparent that the presence of the purified agglutinin on the surface of the fibers specifically induces mycoparasitic behaviour in *T. harzianum*. *T. harzianum* formed tightly adhering coils, which were significantly more frequent with the purified agglutinin-treated fibers than with the untreated ones, or with those treated with non-agglutinating extracellular proteins from *S. rolfsii*. Secretion of adhesive material aiding in the establishment of the interaction of *Trichoderma* with the lectin-treated fibers was also observed.

Toward Anti-Adhesion Therapy for Microbial Diseases, edited by Kahane and Ofek
Plenum Press, New York, 1996

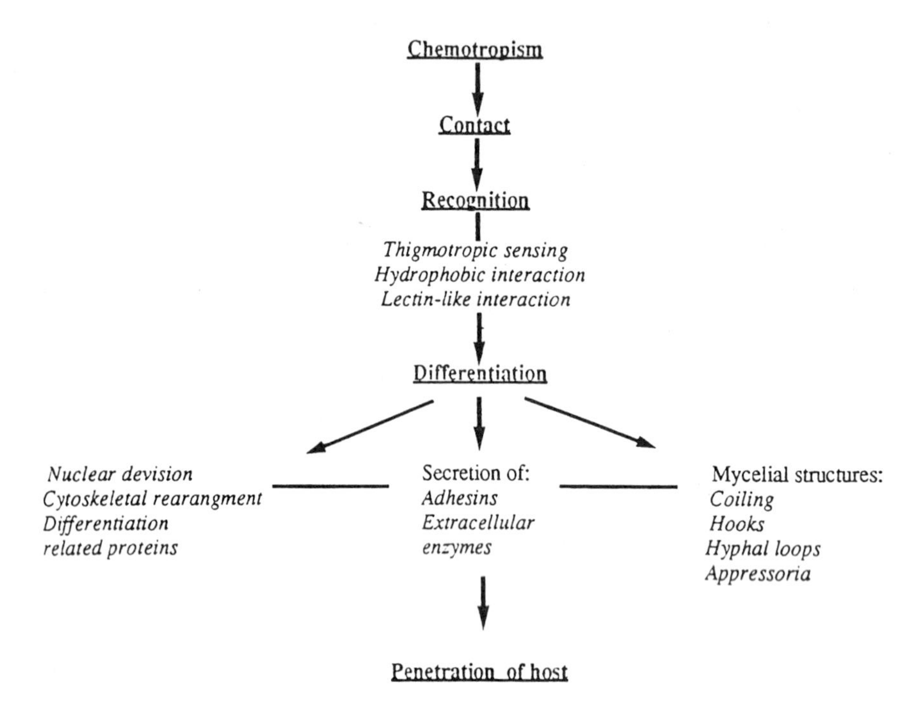

Figure 1. Possible interaction mechanisms in host -parasite systems (cf. Tunlid *et al.*, Mycol. Res. 96:401-412, 1992)

The induction of chitinolytic enzymes in the biocontrol agent *T. harzianum* during parasitism on *S. rolfsii*, and the role of fungal-fungal recognition in this process were studied (Inbar and Chet, 1995). A change in the chitinolytic enzyme profile was detected during the interaction between the fungi, grown in dual culture on synthetic medium. An increase in the activity of CHIT 102 (one of the *Trichoderma* chitinases) was detected on the biomimetic system, suggesting that the induction of chitinolytic enzymes in *Trichoderma* is an early event which is elicited by the recognition signal (i.e. lectin-carbohydrate interactions).

Models proposed to explain the interaction mechanism in fungal parasitism have suggested several putative events following recognition and attachment (Chet, 1987, 1990; Tunlid *et al.*, 1992). Tunlid *et al.* (1992) (chart 1) suggested that as a consequence of these first steps, differentiation processes take place leading to the formation of various infection structures and are accompanied by nuclear division, cytoskeletal rearrangement and synthesis of differentiation-related proteins. Simultaneously, adhesins are secreted to consolidate the attachment, together with extracellular enzymes, resulting in penetration of the host. This model is partially confirmed in our work: when the purified surface agglutinin from the host fungus *S. rolfsii* was bound to nylon fibers, differentiation processes in *T. harzianum* leading to formation of infection structures (coils, hooks, hyphal loops and appressoria) were observed. Induction and secretion of chitinolytic enzyme and adhesive material(s) aiding in establishing this interaction and penetration of the host cell wall were also observed. It is postulated that recognition is the first step in a cascade of antagonistic events which triggers the parasitic response in *Trichoderma*.

REFERENCES

Barak, R., Elad, Y., Mirelman, D. and Chet, I. (1985). Lectins: a possible basis for specific recognition in *Trichoderma-Sclerotium rolfsii* interaction. Phytopathology 75:458-462.

Chet, I. (1987). *Trichoderma* - application, mode of action, and potential as biocontrol agent of soilborne plant pathogneic fungi. pp. 137-160. In: Chet, I. (ed.). Innovative Approaches to Plant Disease Control, John Wiley & Sons, N.Y.

Dennis, C. and Webster, J. (1971). Antagonistic properties of species-groups of *Trichoderma*. III. Hyphal interaction. Trans. Br. Mycol. Soc. 57: 363-369.

Inbar, J. and Chet, I. 1992. Biomimics of fungal cell-cell recognition by use of lectin-coated nylon fibers. J. Bacteriol. 174: 1055-1059.

Inbar, J. and Chet, I. 1994. A newly isolated lectin from the plant-pathogenic fungus *Sclerotium rolfsii:* purification, characterization and role in mycoparasitism. Microbiology 140: 651-657.

Inbar, J. and Chet, I. (1995). The role of recognition in the induction of specific chitinases during mycoparasitism by *Trichoderma harzianum*. Microbiology 141:2823-2829.

Tunlid, A., Jansson, H.B. and Nordbring-Hertz, B. (1992). Fungal attachment to nematodes. Mycol. Res. 96: 401-412.

28

MICROBIAL COAGGREGATION IN THE ORAL CAVITY

Ervin I. Weiss, Blanka Shenitzki, and Roni Leibusor

The Maurice and Gabriela Goldschleger School of Dental Medicine
Tel Aviv University
Tel Aviv, Israel

PERIODONTAL DISEASES

Gingivitis and periodontitis are the most common human diseases with bacterial etiology. Almost all adults worldwide experience gingivitis and some degree of periodontitis. Gingivitis can persist for prolonged periods without significant progression or can serve as a precursor to periodontitis in susceptible individuals. Periodontitis lesions are virtually irreversible and exhibit destruction of the connective tissue attachment, apical migration of the junctional epithelium and resorption of the adjacent alveolar bone. Clinically, it results in deep periodontal pockets, exposure of the root cementum to the oral environment, and eventual loss of dentition. The natural occurrence of initiation and subsequent progression of periodontal disease has been followed in a homogeneous population in Sri Lanka over a 20 year period (Loe, 1986). This study has shown that in the complete absence of oral hygiene and oral health care, all participants exhibited visible plaque, gingivitis and supra- and subgingival calculus in all teeth. But despite the remarkable homogeneity of the group, the severity of the disease among individuals varied greatly. On the basis of the rate of periodontal attachment loss, this population could be divided into three groups: (i) those who showed rapidly progressing periodontitis (approximately 8%), (ii) those who showed moderately progressing periodontitis (80%), and (iii) a small but significant group who exhibited no progression of the disease beyond chronic gingivitis. It is the observation of this last group which challenges the understanding of the mechanism(s) of how and what turns on the progression of the periodontal lesion, and how it is turned off. This is still unresolved.

DENTAL PLAQUE AND ORAL HYGIENE

Dental plaque contains the principal etiologic components for initiation and propagation of periodontal diseases. It consists primarily of proliferating microorganisms along with a scattering of epithelial cells, leukocytes, and macrophages, in an adherent intercellular matrix firmly attached to teeth. Bacteria, which make up over 80% of this material, co-exist

Toward Anti-Adhesion Therapy for Microbial Diseases, edited by Kahane and Ofek
Plenum Press, New York, 1996

in an extremely complex arrangement. There may be as many as 200 different species in one site with some species not currently identified and others impossible to culture. Among 51,000 bacterial isolates from the gingival crevice, over 500 different taxa were detected (Moore and Moore, 1994). Dental plaque may also contain mycoplasma, fungi, protozoa, and viruses. The intermicrobial matrix may contain numerous inflammation-inducing substances, such as proteolytic enzymes, antigenic substances, endotoxin, and low-molecular-weight toxic metabolites. Salivary composition and flow rate, diet, oral hygiene, host response, genetic predisposition, environmental conditions, habitual function and dysfunction of the dentition and teeth alignment, all separately and/or in combination, could contribute to the circumstances that lead to the uncontrolled outbreak of these complex mixed infections.

The relationships among improper oral hygiene, formation of dental deposits, and development of dental diseases have been recognized for centuries. In a letter to the Royal Society of London (Sept 17, 1683), Anton Van Leeuwenhoek described the bacteria of dental plaque and oral hygiene. He stated that in order to keep periodontal health "I am in the habit of rubbing my teeth with salt...... and then rinsing my mouth with water; and often after eating to clean my back teeth with a toothpick, as well as rubbing them hard with a cloth...."

Since then, substantial scientific evidence implicates dental plaque as the most important factor in the development of periodontal diseases.

- Epidemiologic studies reveal an almost linear relationship between severity of periodontal disease and lack of oral hygiene.
- Gingivitis was experimentally induced in humans when oral hygiene measures were abolished. When oral hygiene was reinstituted, restoration of gingival health occurred within a week.
- The transition from gingivitis to periodontitis has not been demonstrated experimentally in humans. However, in dogs, plaque accumulation leads to gingivitis which, if allowed to continue, causes destruction of tooth-supporting tissues, similar to human periodontal disease.
- Studies in humans has shown that the progression of periodontal disease has been retarded by improved oral hygiene and with the use of anti-microbials, such as chlorhexidine.

SPECIFICITY OF BACTERIAL ETIOLOGY

Adult periodontitis lesions contain high proportions of anaerobes (90%), gram-negative organisms (75%), and spirochetes (30%). The composition of the periodontopathic microflora differs markedly from patient to patient and from pocket to pocket in a given patient. However, a limited number of species appears to be responsible for or associated with the conversion of gingivitis to a progressive periodontitis lesion. Because of the unavailability of an appropriate animal model, Koch's postulates to delineate etiologic significance have not been fulfilled for periodontal microorganisms.

Based on the criteria of association with active disease, remission following elimination, immune response, virulence factors and effects in animal studies, *P. gingivalis, P. intermedia, A. actinomycetemcomitans, B. forsythus, Campylobacter rectus, Eikanella corrodens, Peptostrepto-coccus micros, Fusobacterium* sp., *Treponema* sp., *Selenomonas* sp., and *Eubacterium* sp., have been implicated as possible pathogens in adult periodontitis. Certain subgingival microbial complexes show a particularly strong relationship to active periodontitis, underscoring the mixed infectious nature of the disease (Haffajjee and Socransky, 1994).

Moore and Moore (1994) claim that the favored subjects included in many clinical studies, conducted to determine the cause of periodontal disease, were those with the most advanced cases of periodontitis. Samples taken after the disease was established contained bacterial species most associated with existing tissue damage, but did not reveal which species initiated the periodontal lesion or which species proliferated *post factum*. The responsible species must be a relatively large part of the flora in order to play an important role in the initiation of periodontal disease. The results of their extensive studies, based on thorough microbiological sampling, identification and statistical analysis, indicate that the bacterial microbiota follows a predictable progression in the gingival crevice during development of gingivitis and periodontitis. *F. nucleatum* was shown to be the principal and most frequent cause of gingival inflammation. It is also the most common predominant pathogen in subsequent periodontal destruction. Additionally 28 bacterial, 5 treponemal and one mycoplasmal species may contribute to further tissue destruction. The diversity of bacterial species in the periodontal flora, the variation in the composition of the floras among individuals, and the variation in host response are the major reasons that specific etiology of periodontal diseases has not been established.

MANIPULATION OF THE BACTERIAL ECOLOGY

Despite the advances in understanding the pathogenesis of periodontitis and the development of new periodontal diagnostics, few changes in therapeutic strategy have occurred. Periodontal diseases, though widely recognized as having a specific bacterial etiology, continue to be principally treated by mechanical procedures alone. Non-specific mechanical and chemical control for complete suppression of supragingival plaque formation is still recommended by professionals as standard prevention and treatment protocols. However, since the state of the art oral hygiene techniques are time consuming and often require high dexterity, most patients have inadequate motivation to maintain sufficient oral hygiene habits conducive to proper periodontal health.

To better control these mixed infections, it is essential to control the organisms that cause them. The difficulty arises from the complex array of reservoirs where these organisms can persist. In addition to the supra- and subgingival environments which are the principal reservoirs, several secondary reservoirs exist: (i) the dentinal tubules and the opposing periodontal soft tissue exposed to the disease site are invaded by bacteria and may contain viable periodontal pathogens (Adriaene *et al.*, 1988); (ii) *P. gingivalis, P. intermedia* and *A. actinomycetemcomitans* could be isolated from the dorsum of the tongue, the tonsillar area and the cheek mucosa of periodontitis patients (van Winkellhoff *et al.*, 1988); and (iii) studies indicate that the putative periodontitis pathogens may be transmitted from parents to their children or between spouses (Alaluusua *et al.*, 1991; van Steenbergen *et al.*, 1993). The various sources for recolonization and reinfection cause a considerable problem in devising means to effectively deliver therapeutic agents to these locations.

Manipulation of the oral bacterial ecology will likely represent the most effective strategy for long-term control of periodontal diseases. There are three main environmental compartments that communicate with the periodontium and through which the presence of persisting bacteria can be manipulated or therapeutic agents can be administered. (1) The supragingival region that can be reached by routine mechanical oral hygiene measures, such as a toothbrush, dental floss, proximal brushes and toothpicks, as well as mouthrinse or toothpaste. (2) The subgingival area, access to which is more difficult and often needs professional help. Subgingival irrigation, with antibacterial agents or antibiotics using specially designed local delivered systems or slow release devices, were recently introduced as part of periodontal disease treatment and control. (3) The systemic environment that can serve for *per os* medication using antibiotic therapy and for immunization once a safe, effective and feasible periodontal vaccine is available.

Antibiotics can be delivered through all three environmental compartments. Nevertheless, antibacterial therapy can be considered only as a limited set of tools for that purpose. Periodontal diseases are chronic in their nature and prolonged usage of antibiotics, locally delivered or systemically ingested, will possibly raise resistant strains or cause hypersensitivty reactions, thus further complicating the treatment of the disease. Anti-adherence therapy on the other hand, has the potential to irradicate the causative microorganisms without the side effects of antibiotics.

INHIBITION OF COAGGREGATION

Dental plaque formation involves two major processes: (1) The binding of bacteria to the pellicle-coated tooth surface (van Houte, 1980). Oral streptococci, and to some extent actinomyces, are the prominent early colonizers of tooth surfaces (Nyvad and Kilian, 1990) and apparently attach to macromolecules selectively adsorbed to the enamel through adhesin-receptor recognition (Gibbons *et al.*, 1991) or via hydrophobic interactions (Rosenberg *et al.*, 1983). The attached bacteria proliferate giving rise to a variety of microcolonies. (2) Further aggregation and accretion of bacteria to the temporarily stable bacterial population, via cell-to-cell interactions. Microscopic observations of *in situ* plaque samples reveal that intimate contact occurs among the vast majority of habitates of this biomass. Interbacterial adhesion is probably the most important interaction to stabilize microorganisms to withstand the mechanical forces and the salivary flow that tend to dislodge the plaque from the oral cavity. This process of interaction and adhesion between different bacteria is termed coaggregation and can be demonstrated *in vitro* (Kolenbrander *et al.*, 1993). Mixing bacteria suspensions from two different genera often results in the formation of large aggregates that settle to the bottom of the tube within seconds after mixing, leaving the supernatant transparent. This test has served in quick screening and identification of partners for coaggregation and isolation of mutants that lack adhesins and/or receptors (Weiss *et al.*, 1987a,b). McIntire *et al.* (1978) first showed that coaggregation is mediated by a lectin-like protein, which specifically recognizes a sugar receptor upon the partner bacterium. Kolenbrander *et al.* (1993) have screened hundreds of oral bacteria strains and found that almost all strains of *Streptococci* and *Actinomyces,* as well as many other gram-negative pathogenic oral bacteria, participate in coaggregation processes. Most coaggegations occur among strains belonging to different genera and these processes are mediated by lectin-like adhesins and can be inhibited by lactose and other galactosides.

It is only natural therefore, that lactose and galactosides become prime candidates as anti-adhesion therapeutic agents. Unfortunately, the use of these simple sugars, clinically, is problematic for the following reasons: 1) Lactose and galactose are readily fermented by oral microorganisms specially by cariogenic bacteria resulting in acid production and the potential for developing caries lesions. 2) Preliminary *in vitro* results indicate that these simple sugars *per se* are ineffective in plaque dispersal and/or removal. Bauman (1991) demonstrated that supragingival dental plaque collected *in situ*, washed and mechanically dispersed, is capable of spontaneous reaggregation. The rate of reaggregation can be followed turbidometrically, which was found to increase by mild aggitation, and to almost completely stop by pre-heating the plaque suspension at 80°C for 30 min. None of the simple sugars reported to be effective as a coaggregation inhibitor when tested separately on coaggregating pairs, were capable of affecting the *in vitro* reaggregation. 3) Galactoside containing more complex receptor analogs which potentially could be effective *in vivo* are at present unavailable.

Because cranberry juice was found to contain inhibitors of bacterial adhesion to host cells, and because recent studies suggest that the cranberry juice inhibitor(s) are clinically effective in preventing infections, the effect of cranberry juice on the coaggragation of several bacterial pairs was examined. Cranberry juice reversed the coaggregation between *F. nucleatum* and *A. naeslundii*, *S. morbillorum* and *P. gingivalis* but had no effect on the coaggregation *with C. ochracea* (Table 1). With the exception of the coaggregation between *F. nucleatum* PK1594 and *P. gingivalis* PK1924, all coaggregation interactions were unaffected by lactose or related sugars (data not shown). These results indicate that the natural inhibitor component in cranberries affects some of the coaggregation interactions to which no other inhibitor has been found so far. Furthermore, the ability of the cranberry inhibitor to reverse already formed coaggregates underscores its potential as a natural anti-adhesion agent. It is yet unclear whether the inhibitor for one pair is the same or different from that for another pair or from that with the anti-adhesive activity toward *E. coli* adhesin. Although the molecular structure of the inhibitor has not been described and the mechanism of action has not been elucidated, the promising *in vitro* data, and the natural inexpensive source of the active ingredient turns it into a potential substance for clinical usage.

Cranberry juice inhibitor, sugar components, receptor analogs and other inhibitory agents are envisioned to be effective in controlling bacterial accumulation by access to the supragingival region and to a much lesser extent to the subgingival area. Creating an immune response in the host through vaccination will specifically modulate the attachment and colonization of certain bacterial species and will be effective, primarily on the subgingival compartment and other hard to reach niches, in controlling plaque maturation and disease prevention. Adhesin molecules are ideal for vaccines designed to irradicate the respective periodontal pathogens from their ecological reservoirs. We hypothesis that the various types of adhesin molecules that mediate attachment and coaggregation of respective bacteria are limited in number.

The adhesins of certain gram-negative microorganisms which may have an important role in plaque maturation and disease initiation were recently identified and characterized in our laboratory. Among the species studied, capnocytophagae exhibited defined intergeneric coaggregations with certain gram-positive human oral bacteria (Kolenbrander and Andersen, 1984, Kolenbrander and London, 1993). *Capnocytophaga ochracea* ATCC 33596 coaggregates with the widest variety of gram-positive strains including members of the general *Actinomyces, Rothia and Streptococcus*, while *C. sputigena* ATCC 33612 recognizes a subset of these that includes only strains of *Actinomyces* and *R. dentocariosa*. *C. gingivalis* ATCC 33624 coaggregates only with two strains of *A. israelii* which are also common coaggregation partners of *C. ochracea* and *C. sputigena*. In all of these coaggregation reactions, the adhesin appears to be part of the outer membranes of the capnocytophagae

Table 1. The effect of cranberry juice on coaggregation interactions

Bacterial pairs tested		Coaggregation scores in the presence of cranberry inhibitor (mg/ml)[a]		
A	B	2.5	0.6	Control
F. nucleatum PK1594	*C. ochracea* ATCC 33506	4	4	4
	P. gingivalis PK1924	0	3	4
F. nucleatum PK1909	A. naesundii PK29	0	0	4
	S. morbillorum PK509	0	0	4
	P. gingivalis PK1924	0	1	4

[a] Coaggregates were scored visually on a scale of 0 - no visible coaggregation to 4 - maximum coaggregation

(Kagermeier and London, 1986; Weiss *et al.*, 1990), while the carbohydrate-bearing receptors are on the surfaces of the gram-positive coaggregation partners (Cassels *et al.*, 1990). Three types of coaggregation-defective (Cog$^-$) mutants of *C. ochracea* ATCC 33596 have been isolated (Weiss *et al.*, 1987b). When tested against three selected gram-positive partners, the mutants exhibited a successive loss of coaggregation activities: type I coaggregated with two of the partners, type II coaggregated with just one, and type III coaggregated with none. A model was proposed in which the three adhesin sites operated in a functionally interdependent fashion on the surface of *C. ochracea* (Weiss *et al.*, 1987b). The sugars that inhibited the coaggregations of the three species of *Capnocytophaga* with their respective partners, correlated precisely with the sugars that inhibited the coaggregations of the two types of coaggregation defective mutants with these partners (Table 2). Furthermore, *C. ochracea* cell surface specific polyclonal antibodies inhibited coaggregation with all gram-positive partners, also cross reacted with *C. sputigena* and *C. gingivalis* to inhibit coaggregation interaction with their respective partners. These findings of functionally, and probably immunogenically conserved adhesins expressed on the surfaces of genetically diverse bacteria, support the notion of the existence of a finit number of adhesin genes in the genetic pool of the periodontal microbiota. To identify and characterize the adhesins, monoclonal antibodies (MAbs) were prepared by using inhibition of coaggregation as the assay for screening positive hybridoma cell lines (Weiss *et al.*, 1990). The Mabs served as probes in immunoblot analysis and recognized a polypeptide of 155 kD molecular weight present in the wild type organism but absent in a coaggregation defective mutant. Inhibition of biological activity (i.e., coaggregation) was previously used by us to obtain adhesin specific MAbs against two distinct fimbria-associated adhesins of *P. loeschei* PK1295 (Weiss *et al.*, 1988). The possibility to obtain these adhesion blocking MAbs also indicates that the adhesin molecules contain immunogenic domains located in close proximity to the active sites which makes these domains excellent candidates for engineering vaccine molecules.

Another periopathogenic organism studied in our laboratory *F. nucleatum,* was found to coaggregate with over 600 partners representing 11 genera of human oral bacteria, more than any other group of oral bacteria tested. Different strains express different adhesins (Kolenbrander and Andersen, 1989). Three types of adhesins have been described: (1) mediate coaggregation inhibited by galactose and related sugars, such as lactose and n-acetyl galactosamine (Kaufman and DiRienzo, 1988; Kolenbrander and Andersen, 1989; Kinder and Holt, 1993); (2) mediate coaggregation inhibited by arginine (Takemoto *et al.*, 1993); and (3) mediate coaggregation for which an inhibitor has not yet been found (Kaufman and DiRienzo, 1988; Kolenbrander *et al.*, 1989; Takemoto *et al.*, 1993). *F. nucleatum* does not

Table 2. Phenotypic similarity between *C. ochracea* Cog$^-$ mutants and two species of *Capnocytophaga*

	Coaggregation score[a]				
	C. ochracea				
Coaggregation partner strains	Parent	Cog$^-$ type I	Cog$^-$ type II	*C. sputigena*	C. gingivalis
S. oralis H1[b]	3	0	0	0	0
A. naeslundii PK947[b]	4	2	0	2	0
A. israelii PK16	4	3	3	4	3

[a]Coaggregates were scored visually

[b]All coaggregations were inhibited by 80 mM L-rhamnose or D-fucose final concentrations.

express fimbriae, so adhesins are most probably found on its outer membrane. We characterized the adhesins of *F. nucleatum* PK1594, which mediate its coaggregation and adhesion to exposed surfaces in the oral cavity. MAb specific for the two types of adhesins on *F. nucleatum* PK1595 which inhibit the coaggregation with *P. gingivalis* and *A. israelii*, respectively, has recently been prepared in our laboratory. None of the MAbs obtained reacted with *F. nucleatum* polypeptide in immunoblot experiments.

The idea of anti-adherence therapy, as part of the strategy for preventing oral diseases, has been used in dentistry for many decades. Most dentifrices and mouthwashes contain detergents, amphipathic molecules, or surface tension reducing agents, all with the rationale of dislodging adherent microorganisms and preventing their reattachment. Despite the tremendous popularity gained by mouthwashes and dentifrices, the success of most commercial formulations in preventing disease is marginal. The anti-adhesion strategy for dental plaque control has merit if it focus on inhibition and/or reversal of specific adhesion interactions, crucial in the development and maturation of dental plaque. Plaque control can be achieved by either using receptor analogs, such as carbohydrate-containing substances, or by using adhesin binding agents, such as adhesion blocking antibodies produced via immunization. Specific intervention in the maturation process of the plaque, by inhibiting the transition to a potentially pathogenic microbiota, requires careful analysis of the molecular mechanisms involved in these adhesion processes.

BIBLIOGRAPHY

Adriaene PA, De Boever A, Loesche WJ. Bacterial invasion in root cementum and radicular dentin of periodontally diseased teeth in humans. A reservoir of periodontopathic bacteria. J Periodontol 1988;59:222-230.

Alaluusua S, Asikainen S, Lai C-H. Intrafamilial transmission of *Actino-bacillus actinomycetemcomitans*. J Periodontol 1991;62:207-210.

Bauman A. Reaggregation of dental plaque. DMD Thesis, Tel Aviv University, Tel Aviv, Israel 1991.

Cassels FJ, London J. Isolation of a coaggregation-inhibiting cell wall poly-saccharide from *Streptococcus sanguis* H1. J Bacteriol 1989;171:4019-4025.

Cassels FJ, Fales HM, London J, Carlson RW, van Halbeek H. Structure of a streptococcal adhesin carbohydrate receptor. J Biol Chem 1990;265:14127-14135.

Cisar JOS, Kolenbrander PE, McIntire FC. The specificity of coaggregation reactions between human oral streptococci and strains of *Actinomyces viscosus* or *Actinomyces naeslundii*. Infect Immun 1979;24:742-752.

Gibbons RJ, Nygaard M. Interbacterial aggregation of plaque bacteria. Arch Oral Biol 1970;15:1397-1400.

Gibbons RJ, Hay DI, Schlesinger DH. Delineation of a segment of adsorbed salivary proline-rich proteins which promotes adhesion of *Streptococcus gordonii* to apatitic surfaces. Infect Immun 1991;59:2948-2954.

Haffajjee AD, Socransky SS. Microbial etiological agents of destructive periodontal diseases. In: Socransky SS, Haffajjee AD (eds.), Microbiology and immunology of periodontal diseases. Periodontol 2000 1994;5:78-111.

Kagermeier A, London J. Identification and preliminary characterization of a lectin like protein from *Capnocytophaga gingivalis* (emended). Infect Immun 1986;51:490-494.

Kaufman J, DiRienzo JM. Evidence for the existence of two classes of corncob (coaggregation) recetor in *Fusobacterium nucleatum*. Oral Microbiol Immunol 1988;3:145-152.

Kinder SA, Holt SC. Localization of the *Fusobacterium nucleatum*. T18 adhesin activity mediating coaggregation with *Porphyromonas gingivalis* T22. J Bacteriol 1993;175:840-850.

Kolenbrander PE, Andersen RN. Cell to cell interactions of *Capnocytophaga* and *Bacteroides* species with other oral bacteria and their potential role in development of plaque. J Periodontal Res 1984;19:564-569.

Kolenbrander PE, Andersen RN. Inhibition of coaggregation between *Fuso-bacterium nucleatum* and *Porphyromonas (Bacteroides) gingivalis* by lactose and related sugars. Infect Immun 1989;57:3204-3209.

Kolenbrander PE, London J. Adhere today, here tomorrow: Oral bacterial adherence. J Bacteriol 1993;175:3247-3252.

Kolenbrander PE, Andersen RN, Moore LVH. Coaggregation of *Fuso-bacterium nucleatum, Selenomonas flueggei, Selenomonas infelix, Selenomonas noxia*, and *Selenomonas sputigena* with strains from 11 genera of oral bacteria. Infect Immun 1989;57:3194-3204.

Kolenbrander PE, Ganeshkumar N, Carrels FJ, Hughes CV. Coaggregation: Specific adherence among human and plaque bacteria. FASEB J 1993;7: 406-413.

Loe H. Progression of natural untreated periodontal disease in man. In: Lhener T, Gimasoni G (eds.), The Borderland Between Caries and Periodontal Disease III. Editions Medicine et Hygiene, Geneva, 1986:11-29.

McIntireFC, Vatter AE, Baros J, Arnold J. Mechanism of coaggregation between *Actinomyces viscosus* T14V and *Streptococcus sanguis* 34. Infect Immun 1978;21:978-988.

Moore WEC, Moore LVH. In: Socransky SS, Haffajjee AD (eds.) Micro-biology and immunology of periodontal diseases. Periodontol 2000 1994;5: 78-111.

Nyvad B, Kilian M. Comparison of the initial streptococcal microflora on dental enamel in caries-active and in caries-inactive individuals. Caries Res 1990;24:267-272.

Rosenberg M, Judes H, Weiss E. Cell surface hydrophobicity of dental plaque microorganisms *in situ*. Infect Immun 1983;42:831-834.

Takemoto T, Ozaki M, Shirakawa M, Hino T, Okamoto H. Purification of arginine-sensitive hemagglutinin from *Fusobacterium nucleatum* and its role in coaggregation. J Periodontal Res 1993;28:21-26.

van Houte J. Bacterial speificity in the etiology of dental caries. Int Dent J 1980;30:305-326.

van Steenbergen TJJM, Petit MDA, Scholte LHM, van der Velden UU, de Graaff. Transmission of *Porphyromonas gingivalis* between spouses. J Clin Periodontol 1993;20: 340-345.

van Winkellhoff AJ, van der Velden U, Clement M, de Graaff J. Intraoral distribution of black-pigmented *Bacteroides* species in periodontitis patients. Oral Microbiol Immunol 1988;3:83-85.

Weiss EI, Eli I, Shenitzki B, Smorodinsky N. Identification of the rhamnose-sensitive adhesin of *Capnocytophaga ochracea* ATCC 33596. Arch Oral Biol 1990;35 (Suppl):127S-130S.

Weiss EI, London J, Kolenbrander PE, Andersen RN, Fischler C, Siraganian R. Characterization of monoclonal antibodies to fimbria-associated adhesins of *Bacteroides loeschei* PK1295. Infect Immun 1988;56:219-224.

Weiss EI, Kolenbrander PE, London J, Hand AR, Andersen RN. Fimbriae-associated proteins of *Bacteroides loeschei* PK1295 mediate intergeneric coaggregations. J Bacteriol 1987a;169:4215-4222.

Weiss EI, London J, Kolenbrander PE, Kagermeier AS, Andersen RN. Characterization of lectin like surface components on *Capnocytophaga ochracea* ATCC 33596 that mediate coaggregation with gram-positive oral bacteria. Infect Immun 1987b;55:1198-1202.

29

HYDROPHOBIC INTERACTIONS AS A BASIS FOR INTERFERING WITH MICROBIAL ADHESION

Mel Rosenberg, Ronit Bar-Ness Greenstein, Mira Barki, and Sarit Goldberg

The Maurice and Gabriela Goldschleger School of Dental Medicine
Department of Human Microbiology
Sackler Faculty of Medicine
Tel-Aviv University
Ramat-Aviv 69978, Israel

INTRODUCTION

Hydrophobic interactions are considered important in a variety of microbial adhesion phenomena in both the open environment, as well as the host (for a review on hydrophobicity in the context of microbial adhesion, see Rosenberg and Doyle, 1990). Examples of medically-related adhesion phenomena that appear to be influenced by hydrophobic interactions include adhesion to teeth, fatty tissues, insect cuticle, catheters, intrauterine contraceptive devices, bioprostheses, contact lenses, and other biomaterials, as well as a variety of epithelial and phagocytic cell types. In 1980, we proposed a simple technique for measuring bacterial (later including eucaryotic microorganisms) adhesion to liquid hydrocarbons based on simply vortexing washed cell suspensions with different test hydrocarbons, with concomitant adhesion at the oil:water interface. The observation that adhesion could be reversed by isopropanol was one indication that hydrophobic interactions were involved, leading us to propose this test as a simple technique for studying cell surface hydrophobicity (Rosenberg *et al.*, 1980). Many tests exist for studying hydrophobic surface properties of microorganisms, including some recently proposed techniques (Lin *et al.*, 1995), but their discussion is beyond the scope of this chapter (for reviews, see Rosenberg and Doyle, 1990; Rosenberg, 1991; Rosenberg *et al.*, 1991a; Van der Mei *et al.*, 1991). The microbial adhesion to hydrocarbon (MATH) test subsequently led us into various avenues of investigation related to identification of cell surface components which promote or interfere with adhesion at the oil:water interface, as well as soluble agents which inhibit or promote adhesion. Furthermore, the observation that many oral microorganisms adhere at the oil:water interface led us to develop a novel two-phase microbe-desorbing mouthrinse, which has been sold successfully in Israel since 1992.

Toward Anti-Adhesion Therapy for Microbial Diseases, edited by Kahane and Ofek
Plenum Press, New York, 1996

HYDROPHOBINS AND HYDROPHILINS

In 1986, Rosenberg and Kjelleberg coined the term "hydrophobin" to denote hydrophobic surface components that promote cell surface hydrophobicity and adhesion. Conversely, hydrophilins were defined as cell surface components that reduced hydrophobic surface properties of microbial cells. In recent years the term hydrophobin has been extended to include hydrophobic surface components of fungal cells that may be involved in adhesion processes. Studies on hydrophobins and hydrophilins of several microbial species are discussed below.

ACINETOBACTER CALCOACETICUS

Acinetobacter calcoaceticus species, although often found in the environment, can inhabit the human body and are sometimes associated with nosocomial infections. Stationary phase cells of the oil-degrading *A. calcoaceticus* RAG-1 exhibit a high affinity for the oil:water interface. Mutant MR-481, selected for its inability to adhere to oil droplets, was devoid of thin fimbriae present on wild type cells. Revertants bearing thin fimbriae regained partial adhesion characteristics. These data established a role for the thin fimbriae as a hydrophobin of RAG-1 cells (Rosenberg *et al.*, 1982). However, another RAG-1 mutant was found to be able to adhere at the oil:water interface, although it lacked thin fimbriae. Thus, other RAG-1 hydrophobin(s) remain to be elucidated.

Emulsan, a capsular anionic heteropolysaccharide of *A. calcoaceticus* RAG-1 acts as a hydrophilin when present on RAG-1 cells. When sloughed off in solution, emulsan is a potent inhibitor of microbial adhesion (see further). Similarly, the rhamnose-rich capsule of *Acinetobacter calcoaceticus* BD 4, although completely different in composition, also acts as a hydrophilin. As the capsule is enzymatically degraded, adhesion to oil droplets rises from zero to almost 100% (Rosenberg *et al.*, 1983b).

STREPTOCOCCUS PYOGENES

Streptococcus pyogenes is a common cause of throat infections, and other, sometimes serious, diseases. Cells of *S. pyogenes* often adhere avidly to hexadecane and other oil droplets (Ofek *et al.*, 1983). LTA and M protein have both been considered as contributing to the hydrophobic surface properties. Conversely, hyaluronate capsule appears to act as a hydrophilin, and the hydrophobic/hydrophilic conversion of cells may be critical for successful colonization.

SERRATIA MARCESCENS

Serratia marcescens is a ubiquitous microbial species which can be a troublesome source of nosocomial infections, particularly in neonate wards. It has been known for some time that prodigiosin, the red, oil-soluble pigment characteristic of pigmented strains imparts hydrophobic surface characteristics, enabling cells to partition at air:water interfaces. However, most clinical isolates are nonpigmented. Nevertheless, nonpigmented strains are also capable of adhering (albeit, less avidly) to hydrocarbons and polystyrene (Rosenberg *et al.*, 1986). Furthermore, growth of cells at 37°C leads to loss of pigment, whereas cells are still adherent. Finally, we have isolated pigmented mutants which lost their surface hydrophobicity as well as nonpigmented mutants that retained partial hydrophobic surface characteristics (Rosenberg *et al.*, 1984a). In 1989, Bar Ness and Rosenberg showed that a 70 kDa *S. marcescens* (wild type) surface protein, serraphobin, binds at the oil:water interface and is associated with cell surface hydrophobicity (Bar Ness-Greenstein, 1992). Transformation of *Escherichia coli* with DNA from *S. marcescens* resulted in a transformant clone exhibiting an increase in hydrophobic surface properties, concomitant with appearance of *S. marcescens* antigen/s (as yet unidentified) on the surface of transformant cells. Rabbit antiserum raised against whole cells of wild type hydrophobic strain of *S. marcescens* (and adsorbed on the appropriate control cells), reacted with the transformant in colony immuno-blotting. The insert size of the hydophobic transformant was 5.4 kb, as revealed by restriction mapping (BarNess-Greenstein *et al.*, 1995).

Another amphipathic cell surface component, the cyclic aminolipid, serratamolide, turned out to act as a hydrophilin. Mutants lacking this surfactant molecule exhibited higher adhesion levels than isogenic wild type cells (Bar-Ness *et al.*, 1988).

CANDIDA ALBICANS

Candida albicans is an opportunistic yeast-like microorganism which may give rise to superficial and systemic infections. Although hydrophobic surface properties of *C. albicans* were found in several investigations (Calderone & Braun, 1991; Hazen *et al.*, 1991; Tronchin *et al.*, 1988) the cell surface components involved are unclear. In order to look for surface hydrophobin(s) of this microorganism, a genomic library of *C. albicans* in a nonadherent strain of *Saccharomyces cerevisiae* was constructed. Selection for adhesion to polystyrene yielded a transformant which exhibited enhanced adhesion to treated and untreated polystyrene, as well as pronounced autoaggregation, as compared to appropriate control cells bearing the vector alone. Analysis of this clone revealed an insert of ca. 4.5 kb from *C. albicans*. Curing of the plasmid resulted in loss of adhesion and autoaggregation properties (Barki *et al.*, 1993). A subclone bearing a reduced insert of 3.3 kb retained the ability to autoaggregate, to bind to treated and untreated polystyrene, and to adhere to buccal epithelial cells, as compared to appropriate controls. Evidence was subsequently presented that the adhesion/autoaggregation phenotype observed in *S. cerevisiae* cells transformed with the candidal DNA fragment is due to expression of a *C. albicans* surface antigen (Barki *et al.*, 1994). Rabbit antiserum, raised against transformant *S. cerevisiae* cells, was adsorbed on *S. cerevisiae* bearing the vector alone. Immunofluorescence micrography showed that the adsorbed antisera bound to the surface of transformant *S. cerevisiae* cells as well as to *C. albicans* cells (especially to the invasion-associated mycelial phase), but only marginally to the *S. cerevisiae* control. Furthermore, adsorbed antisera specifically inhibited autoaggregation of transformant cells. Autoaggregative activity was also abolished following treatment by trypsin and proteinase K. Western immunoblot analysis yielded an antigen of ca. 30 kDa molecular weight, present in transformant cell surface extracts, but absent in the control. The extent to which this antigen acts as an adhesin/hydrophobin on *C. albicans* cells is currently under investigation.

SOLUBLE AGENTS THAT AFFECT MICROBIAL ADHESION VIA PROMOTION OF OR INTERFERENCE WITH HYDROPHOBIC INTERACTIONS

A wide variety of amphipathic substances may inhibit adhesion to animate and inanimate surfaces (for a review see Rosenberg and Doyle 1990). Some examples are detergents (e.g., sodium dodecyl sulfate, Tween 80, Tween 20, Triton X-100), chaotropes (e.g., thiocyanate anion), components obtained from body fluids (e.g., bovine serum albumin [particularly defatted], normal human serum, salivary components, caseinoglycopeptides) and small organic molecules (e.g., p-nitrophenol, tetramethyl urea, propanol, fatty acids).

Amphipathic components in saliva appears to act as a barrier, preventing free microorganisms from adhering to oral surfaces. Washed microorganisms obtained directly from dental plaque were capable of adhering to buccal epithelial cells in vitro, but this was largely inhibited when saliva was added (Rosenberg *et al.*, 1989). Saliva prevented microbial adhesion to polystyrene, but when the saliva was extracted with hexadecane, it lost its inhibitory ability (Eli *et al.*, 1989). Babu and coworkers demonstrated in 1986, inhibition of adhesion of *Streptococcus sanguis* to hexadecane by specific hydrophobic proteins of saliva.

Adhesion of microorganisms to hexadecane and polystyrene is promoted in the presence of ammonium sulphate (Rosenberg *et al.*, 1984b). As discussed further below, amphipathic cationic molecules (e.g., cetylpyridinium chloride, chlorhexidine, lysozyme, poly L-lysine, chitosan) may promote adhesion by "hydrophobizing" cells. This presumably occurs by binding of such molecules to negatively charged cell surface residues, thereby simultaneously decreasing surface charge and adding hydrophobic moieties to the cell outer surface (Goldberg *et al.*, 1990a and 1990b). Covalent blocking of negatively-charged outer surface components similarly results in increased hydrophobicity and adhesion.

One interesting aspect is the inhibition by amphipathic molecules of adhesion apparently mediated by stereospecific, lectin-like interactions. Adhesion of *E. coli* to epithelial cells, while sensitive to mannose, can also be blocked by unrelated amphipathic compounds, such as p-nitrophenol (Falkowski *et al.*, 1986). *Actinomyces viscosus* cells preferentially adhered to *Porphyromonas gingivalis* cells, rather than to hexadecane, yet this ostensibly specific interaction was inhibited by defatted bovine serum albumin (Rosenberg *et al.*, 1991b).

The therapeutic use of anti-adhesion agents in controlling microbial infections must consider that in most cases, we become aware of the intruding microorganisms after they have already adhered. In other words, for most medical purposes, anti-adhesion agents must be able to desorb already attached microorganisms, as well as to inhibit adhesion. Unfortunately, prevention of initial attachment is easy, as compared with detaching microorganisms after they have already adhered, interacted with neighbouring bacteria, elaborated a capsular glycocalyx, and so on.

Relatively few studies have dealt with microbial desorption. In one such investigation, E. Rosenberg and coworkers were able to show that when the capsular anionic heteropolysaccharide of *A. calcoaceticus* RAG-1 ("emulsan") is sloughed off into the aqueous phase, it is not only capable of inhibiting adhesion to buccal epithelial cells, but also desorbs a substantial fraction of the indigenous attached microbiota (Rosenberg *et al.*, 1983a).

Another consideration facing anti-adhesion therapy is the observation that bacteria will eventually adhere to and foul practically every available interface. As Kevin C. Marshall has succinctly remarked in this context, "The Bacteria Always Win." In the following section, we describe an alternative anti-adhesion strategy, i.e., the BAWA ("beat adhesion with adhesion") approach. According to this strategy, adhering bacteria can be desorbed from one interface by preferential adhesion to a second interface (Rosenberg *et al.*, 1983c). We have successfully used oil droplets to remove bound bacteria from oral surfaces in this manner (Goldberg and Rosenberg, 1991). It is important to emphasize that in such an instance, the mechanisms governing the initial adhesion are probably of secondary importance, as compared with the affinity of the bound cells for the displacing interface.

FROM OIL POLLUTION TO ORAL POLLUTION: RESEARCH AND DEVELOPMENT OF A BACTERIA-DESORBING MOUTHWASH

Over the past thirteen years, much of our interest in anti-adhesion therapy has focused on the potential of two-phase, oil:water mixtures to remove oral bacteria and to combat bad breath. In 1982, Nesbitt and coworkers showed that *Streptococcus sanguis* adheres to toluene droplets. In the same year, Weiss *et al.* reported that most oral microbial isolates retrieved from extracted teeth adhered to hexadecane. In 1983, Rosenberg and coworkers subsequently demonstrated that washed, dispersed suspensions of microorganisms obtained directly from supragingival dental plaque adhered in high proportions to droplets of liquid hydrocarbon (Rosenberg *et al.,* 1983d). These initial observations led to the notion of using two-phase oil:water mouthwashes for physical desorption of oral microbiota. Indeed, other studies have demonstrated that a wide variety of laboratory strains of oral microorganisms, including *Actinobacillus actinomycetemcomitans, Actinomyces viscosus and Actinomyces naeslundii, Porphyromonas gingivalis, Streptococcus salivarius*, *Streptococcus sanguis*, and certain *Streptococcus mutans* strains, adhered with high affinity to hydrocarbons (for a review, see Doyle *et al.,* 1990). Initial experiments showed that microorganisms which adhere to hydrocarbons also bind to other nontoxic oils (Rosenberg *et al.,* 1983c). Furthermore, in vitro experiments showed that water:oil combinations could desorb bacteria from plastic surfaces. Desorption could also be demonstrated *in vivo*, based on removal of erythrosine-stained bacteria and debris from the mouth following rinsing with a two phase rinse consisting of saline and olive oil (Weiss *et al.*, 1985). Similar combinations were also able to remove bound cells from polystyrene cuvettes (Goldberg and Rosenberg, 1991).

Despite these initial desorption experiments, the simple aqueous:oil combinations did not prove particularly effective in improving oral hygiene over time. Furthermore, the two immiscible oil and water phases posed technical difficulties. In 1989, Goldberg and coworkers found that adhesion of microorganisms and oral debris to oil was promoted by low concentrations of cetylpyridinium chloride (CPC), an amphipathic cationic agent which is commonly added to mouthwashes. This was surprising, since amphipathic agents were generally considered to inhibit adhesion at the oil:water interface. Since CPC is an antibacterial agent, its addition had several other advantages, including inhibition of microbial activity in the mouth, and increasing shelf life of the product by preventing microbial outgrowth. Furthermore, addition of an amphipathic agent such as CPC to oil:water mixtures results in a temporary emulsion following shaking. This in turn enabled the mixture to be supplied in a single bottle.

Prototype formulations, consisting of an aqueous phase containing CPC, and an oil phase comprising a mixture of vegetable and essential oils, were subsequently developed and tested. Such formulations maintained dramatic *in vitro* bacterial desorption properties, removing over 90% of a bacterial film of *A. calcoaceticus* RAG-1 on polystyrene cuvettes, as compared to 15-50% removal for a series of commercial mouthwashes (Goldberg and Rosenberg, 1991).

In an initial clinical study, the prototype two-phase rinse was compared with (i) a placebo rinse; and (ii) a mouthrinse containing 0.2% chlorhexidine gluconate (Corsodyl™, ICI). Subjects performed oral rinsing prior to bedtime and in the morning of the following day. Measurements were taken in the late afternoon (e.g. at least eight hours following rinsing) of the second day, and were compared to those taken the previous afternoon prior to rinsing. The data demonstrated that rinsing with the two-phase mouthwash resulted in daylong reduction in microbial levels, as well as in oral malodor (Rosenberg *et al.,* 1992). In another study employing the same two phase formulation, Yaegaki and Sanada showed dramatic reductions in volatile sulphide levels 3.5 hours after use, as compared to a commercial mouthwash (Skoal™). These authors suggested that physical desorption of the particulate oral debris by the mouthrinse was an important factor in malodor reduction (Yaegaki and Sanada, 1992).

In 1992, the prototype two-phase oil-water mouthwash was reformulated and introduced in the Israeli market. A clinical study (Kozlovsky *et al.*, in press) was performed to compare the efficacy of the commercial version of the two-phase mouthwash with Listerine™, a mouthrinse which has been previously shown to be effective in reducing levels of odor-related microorganisms, as well as plaque and gingivitis (Pitts *et al.*, 1983; DePaola *et al.*, 1989). In the study, fifty subjects rinsed with one of the two mouthwashes for 30 seconds twice a day over six weeks, while continuing their regular oral hygiene habits. In both mouthwash groups, dramatic improvements were observed in parameters associated with oral microbial levels, malodor, periodontal health and plaque accumulation. For example, as compared to time zero scores, whole mouth odor as well as tongue dorsum anterior and posterior odors decreased continuously over time, attaining 80%, 79% and 70% reductions, respectively, following 6 weeks, in the two-phase mouthwash (Assuta™) group, vs. 70%, 77% and 59% for the Listerine™ group. For whole mouth and tongue dorsum posterior, the reductions observed in the two-phase mouthwash group were significantly greater than those obtained with Listerine™ ($p=0.026$ and $p=0.025$, respectively), suggesting that the two-phase mouthwash is superior to Listerine™ in long-term reduction of oral malodor.

CONCLUDING REMARKS

Hydrophobic interactions appear to be involved in numerous examples of microbial adhesion. Whereas many amphipathic agents (detergents, chaotropes, etc.) can inhibit adhesion which is mediated via hydrophobic interactions, desorption of already-bound cells (e.g., from dental plaque) is a more difficult prospect. Emulsan, a capsular polysaccharide from *A. calcoaceticus* RAG-1 is one example of a bacteria-desorbing polymer. Unfortunately, this potential, demonstrated many years ago, has not been further exploited. One approach which works for the oral cavity is the "beat adhesion with adhesion" approach, in which attached bacteria are simultaneously hydrophobized (by cetylpyridinium chloride) and desorbed onto oil droplets. Further research is necessary to determine whether similar techniques can be used to desorb bacterial films from other surfaces of medical interest.

REFERENCES

Babu, J.P., E.H. Beachey and W.A. Simpson. Inhibition of the interaction of *Streptococcus sanguis* with hexadecane droplets by 55- and 66- kilodalton hydrophobic proteins of human saliva. Infect. Immun. 1986. 53:278-284.

Bar Ness-Greenstein R., M. Rosenberg, R.J. Doyle, and N. Kaplan. DNA from *Serratia marcescens* confers a hydrophobic character in *Escherichia coli.* FEMS Microbiol. Lett. 1995. 125:71-76.

Bar-Ness Greenstein, R. Isolation and characterization of the hydrophobicity factor/s in *Serratia marcescens.* Ph.D. thesis, Tel Aviv University. 1992.

Bar-Ness R. and M. Rosenberg. Putative role of a 70 kD surface protein in mediating cell surface hydrophobicity of *Serratia marcescens.* J. Gen. Microbiol. 1989. 135(8):2277-2281.

Bar-Ness R., N. Avrahamy, T. Matsuyama and M. Rosenberg. Increased cell surface hydrophobicity of a *Serratia marcescens* NS 38 mutant lacking wetting activity. J. Bacteriol. 1988. 170:4361-4364.

Barki M., Y. Koltin, M. Yanko, A. Tamarkin and M. Rosenberg. Isolation and characterization of a *Candida albicans* DNA sequence conferring adhesion and aggregation on *Saccharmoyces cerevisiae.* J. Bacteriol. 1993. 175:5683-5689.

Barki M.,Y. Koltin, M. van Wetter and M. Rosenberg. A *Candida albicans* surface antigen mediating adhesion and autoaggregation in *Saccharomyces cerevisiae.* Infect. Immun. 1994. 62:4107-4111.

Calderone, R.A., and P.C. Braun. Adherence and receptor relationships of *Candida albicans*. Microbiol. Rev. 1991. 55:1-20.

DePaola, LG., C.D. Overholser, T. F. Meiller, G.E. Minah and C. Niehaus. Chemotherapeutic inhibition of supragingival dental plaque and gingivitis development. J. Clin. Periodontol. 1989. 16:311-315.

Doyle R.J., M. Rosenberg and D. Drake. 1990. Hydrophobicity of oral bacteria. In: R.J. Doyle, M. Rosenberg (eds). Microbial Cell Surface Hydrophobicity. ASM Publications, Washington, D.C. pp. 387-419.

Eli I., H. Judes and M. Rosenberg. Saliva-mediated inhibition and promotion of bacterial adhesion to polystyrene. Biofouling 1989. 1:203-211.

Falkowski, W., M. Edwards and A.J. Schaeffer. Inhibitory effect of substituted aromatic hydrocarbons on adherence of Escherichia coli to human epithelial cells. Infect. Immun. 1986. 52:863-866.

Goldberg S. and M. Rosenberg. Bacterial desorption by commercial mouthwashes vs two-phase oil:water formulations. Biofouling 1991. 3:193-198.

Goldberg S., Y. Konis and M. Rosenberg. Effect of cetylpyridinium chloride on microbial adhesion to hexadecane. Appl. Environ. Microbiol. 1990a. 56:1678-1682.

Goldberg S., R.J. Doyle and M. Rosenberg. Mechanism of enhancement of microbial cell hydrophobicity by cationic polymers. J. Bacteriol. 1990b. 172:5650-5654.

Hazen, K.C., D.L. Brawner, M.H. Riesselman, M.A. Jutila and J.E. Cutler. Differential adherence of hydrophobic and hydrophilic *Candida albicans* yeast cells to mouse tissues. Infect. Immun. 1991. 59:907-912.

Kozlovsky A., S. Goldberg, I. Natour, A. Rogatky, I. Gelernter and M. Rosenberg. Efficacy of Assuta™ mouthwash in controlling mouth odor, gingivitis and plaque, in comparison with Listerine™. J. Periodontol. (in press).

Lin L., M. Rosenberg, K.G. Taylor, and R.J. Doyl. Kinetic analysis of ammonium sulfate dependent aggregation of bacteria. Colloids and Surfaces B: Biointerfaces 1995. 5:124-127.

Nesbitt W.E., R.J. Doyle and K.G. Taylor. Hydrophobic interactions and the adherence of *Streptococcus sanguis* to hydroxylapatite. Infect. Immun. 1982. 38:637-644.

Ofek, I., Whitnack, E., and Beachy, E.H. Hydrophobic Interactions of group A streptococci with hexadecane droplets. J. Bacteriol. 1983. 154: 139-145.

Pitts, G., C. Brogdon, L. Hu, T. Masurat, R. Pianotti and P. Schumann. Mechanism of action of an antiseptic, anti-odor mouthwash. J. Dent. Res. 1983. 62:738-742.

Rosenberg M., I. Gelernter, M. Barki and R. Bar-Ness. Daylong reduction of oral malodor by a two-phase oil:water mouthrinse, as compared to chlorhexidine and placebo rinses. J. Periodontol. 1992. 63:39-43.

Rosenberg, M. Basic and applied aspects of microbial adhesion at the oil:water interface CRC Crit. Rev. Microbiol. 1991. 18:159-173.

Rosenberg M., M. Barki, R. Bar-Ness, S. Goldberg and R.J. Doyle. Microbial adhesion to hydrocarbons (MATH). Biofouling 1991a. 4:121-128.

Rosenberg M., I.A. Buivids and R.P. Ellen. Adhesion of *Actinomyces viscosus* to *Porphyromonas (Bacteroides) gingivalis*-coated hexadecane droplets. J. Bacteriol. 1991b. 173:2581-2589.

Rosenberg M. and R.J. Doyle. 1990. Microbial cell surface hydrophobicity: History, measurement and significance. In: R.J. Doyle, M. Rosenberg (eds). Microbial Cell Surface Hydrophobicity. ASM Publications, Washington, D.C. pp. 1-37.

Rosenberg M., H. Tal, E. Weiss and S. Guendelman. Adhesion of non-coccal dental plaque microorganisms to buccal epithelial cells: inhibition by saliva and amphipathic agents. Microbial Ecology in Health and Disease. 1989. 2:197-202.

Rosenberg M. and S. Kjelleberg. Hydrophobic interactions: Role in microbial adhesion. Adv. Microbiol. Ecol. 1986. 9:353-393.

Rosenberg M., Y. Blumberger, H. Judes, R. Bar-Ness, E. Rubenstein and Y. Mazor. Cell surface hydrophobicity of pigmented and nonpigmented clinical *Serratia marcescens* strains. Infect. Immun. 1986. 51:932-935.

Rosenberg M. Isolation of pigmented and nonpigmented mutants of *Serratia marcescens* deficient in cell surface hydrophobicity. J. Bacteriol. 1984a. 160:480-482.

Rosenberg M. Ammonium sulphate enhances adherence of *Escherichia coli* J-5 to hydrocarbon and polystyrene. FEMS Microbiol. Lett. 1984b. 25:41-45.

Rosenberg E., A. Gottlieb and M. Rosenberg. Inhibition of bacterial adherence to epithelial cells and hydrocarbons by emulsan. Infect. Immun. 1983a. 39:1024-1028.

Rosenberg E., N. Kaplan, O. Pines, M. Rosenberg and D. Gutnick. Capsular polysaccharides interfere with adherence of *Acinetobacter calcoaceticus* to hydrocarbons. FEMS Microbiol. Lett. 1983b. 17:157-160.

Rosenberg M., H. Judes and E. Weiss. Desorption of adherent bacteria from a solid hydrophobic surface by oil. J. Microbiol. Meth. 1983c. 1:239-244.

Rosenberg M., H. Judes and E. Weiss. Cell surface hydrophobicity of dental plaque microorganisms *in situ* . Infect. Immun. 1983d. 42:831-834.

Rosenberg M., E.A. Bayer, J. Delarea and E. Rosenberg. Role of thin fimbriae in adherence and growth of *Acinetobacter calcoaceticus* on hexadecane. Appl. Environ. Microbiol. 1982. 44:929-937.

Rosenberg, M., D. Gutnick and E. Rosenberg. Adherence of bacteria to hydrocarbons: a simple method for measuring cell surface hydrophobicity. FEMS Microbiol. Lett. 1980. 9:29-33.

Tronchin, G., J.P. Bouchara, R. Robert and J.M. Senet. Adherence of *Candida albicans* to plastic: ultrastructural and molecular studies of fibrillar adhesins. Infect. Immun. 1988. 56:1987-1993.

Van der Mei H.C., M. Rosenberg and H.J. Busscher. 1991. Assessment of microbial cell surface hydrophobicity. In: N. Mozes, P.S. Handley, H.J. Busscher, P.G. Rouxhet (eds). Microbial Cell Surface Analysis: Structural and Physiochemical Methods. VCH Publishers, Inc., New York. pp. 263-287.

Weiss E., M. Rosenberg and H. Judes."Dental and Oral Preparation." 1985. U.S. Patent No. 4,525,342; Israel Patent No. 68027.

Weiss E., M. Rosenberg, H. Judes and E. Rosenberg. Cell surface hydrophobicity of adherent oral bacteria. Curr. Microbiol. 1982. 7:125-128.

Yaegaki, K. and K. Sanada. Effects of a tow-phase oil-water mouthwash on halitosis. Clin. Preven. Dent. 1992. 14: 5-9.

30

ANTI-ADHESINS OF *STREPTOCOCCUS SOBRINUS*

Qi Wang,[1] Sujan Singh,[2] K. G. Taylor,[2] and R. J. Doyle[1]

[1] Department of Microbiology and Immunology
University of Louisville
Louisville, Kentucky 40292
[2] Department of Chemistry
University of Louisville
Louisville, Kentucky 40292

INTRODUCTION

Several members of the genus *Streptococcus* seem to be able to elaborate surface lectins. The oral streptococci have been reported to bind sialic acid (*S. sanguis*) and α-glucans (Gibbons and Fitzgerald 1969). The organisms capable of binding α-glucans include *S. cricetus* and *S. sobrinus*. A few members of the highly cariogenic *S. mutans* also can complex with glucans. Most oral streptococci are also capable of interacting with α-glucans via lectin-independent sites (Robyt, 1995; Mooser, 1992). In the lectin-independent interactions, glucan synthases (dextransucrases or glucosyltransferases, GTFs) use glucans as acceptors for chain extension. The glucan-binding sites on these enzymes are distinct from the catalytic sites acting on sucrose, the sole and natural substrate of the enzymes (Mooser, 1992; Mooser and Wong, 1988; Robyt, 1982). In the presence of sucrose, streptococci are able to adhere to and colonize enamel surfaces leading to dental decay. Adhesion is a first step leading to dental caries. There have been numerous studies in the past few decades describing adhesion mechanisms of oral streptococci. If suitable anti-adhesins were available, it follows that bacterial colonization could be prevented. There are several factors which modify oral microbial adhesion, including antibodies, lysozyme, salivary amylase, salivary proteins (statherin and proline-rich proteins, PRPs), and resident microbiota capable of producing proteases and glycosidases. Some of the above named factors promote adhesion, whereas others inhibit streptococcal binding to enamel. In this paper, results are presented which show that an anti-adhesin of *S. sobrinus*, periodate-oxidized α-1,6 glucan, is a potent anti-caries agent. The periodate-oxidized glucan not only inhibits glucan-binding lectin (GBL) activities, but also is an inhibitor of streptococcal GTFs.

Toward Anti-Adhesion Therapy for Microbial Diseases, edited by Kahane and Ofek
Plenum Press, New York, 1996

FACTORS INFLUENCING GLUCAN-BINDING LECTIN ACTIVITY IN *STREPTOCOCCUS SOBRINUS*

The GBL of *S. sobrinus* seems to be well-regulated. Exponential cells have the highest GBL activities. There is a requirement for Mn^{2+} for expression of lectin activity in *S. cricetus* and *S. sobrinus* (Drake *et al.*, 1988b). Growth of the organisms in citrate or lactate causes a loss of GBL activity (Lü-Lü *et al.*, 1992), presumably because the citrate or lactate can sequester Mn^{2+}. Cells growing in high concentrations of fermentable carbohydrates can generate enough organic acid to deplete Mn^{2+} resulting in low GBL levels. Interestingly, citrate or other chelating agents can directly inhibit the GBLs (Lü-Lü *et al* ., 1992, Bauer *et al.*, 1993). The inhibition of the streptococcal GBLs by chelating agents is reversible, as washing the cells in fresh buffer restores lectin activity. There is no evidence the GBLs are manganoproteins. When culture supernatants of *S. sobrinus* are added to Sephadex G-75, several proteins are bound to the crosslinked dextran but chelating agents do not remove any of the proteins (Ma *et al.*, 1996).

Growth of the oral streptococci in subinhibitory concentrations of fluoride (1-7 mM) also causes a loss of the ability of the cells to be aggregated by α-1,6 glucan (Cox *et al.*, unpublished). The basis for the fluoride effect is unknown. In preliminary experiments, it has been shown that fluoride causes the production of proteins akin to heat shock proteins and the loss of other proteins, including the GBL and Mn-dependent superoxide dismutase (SOD). It may be that fluoride acts on a regulatory protein involved in GBL, SOD and other proteins. When extracts or culture supernatants of fluoride-grown *S. sobrinus* are subjected to SDS-PAGE, a 60 kDa GBL band is missing or decreased in intensity. Mutants deficient in GBL activity are also devoid of the 60 kDa protein, but not other Sephadex-binding proteins (Ma *et al.*, 1996). In addition to fluoride, it has been observed that some antibiotics enhance GBL activities, whereas others decrease the GBLs (Wu *et al.*, 1995). When GBL activity is increased by growth in the presence of an antibiotic, cell surface hydrophobicity is decreased. In contrast, antibiotics which decrease GBL activities tend to increase hydrophobicities. It thus appears that GBL activity may be modulated by several compounds. It would not be surprising that the compounds are acting at a common regulatory site.

SPECIFICITIES OF THE GLUCAN-BINDING ADHESINS

High molecular weight α-1,6 glucans are able to aggregate some oral streptococci (Gibbons and Fitzgerald, 1969; Drake *et al.*, 1988a). The glucan molecular weight must be at least 40,000 in order to promote aggregation. Most experiments employ glucan T-2000 (Pharmacia), a linear α-1,6 glucan of 2,000,000 molecular weight, to elicit rapid and reproducible aggregation of suspensions of the streptococci. When the suspension opacity decreases, the rate can be determined by spectrophotometric readings (Drake *et al.,* 1988a). Polysaccharides, such as α-1,4-linked glucans (amylose), glycogen, lichenan, laminarin, pullulan, mannans, arabinans, fucans, galactans and arabinogalactans are unable to cause aggregation. The aggregation can be inhibited only by oligomers of α-1,6 glucose. Short oligomers, such as isomaltose, isomaltotriose and isomaltotetraose are incapable of preventing the agglutination of the streptococci by glucan T-2000. Even concentrations of isomaltotriose sufficiently high to create a syrup are non-inhibitors. The superior inhibitors were found to contain 8-10 α-1,6-linked glucose units. The results clearly show that the GBL recognizes stretches of glucose, but not terminal non-reducing glucose residues. The GBLs of the streptococci are unlike plant glucan-binding lectins where non-reducing terminii are the preferred receptors. Interestingly, the 2-C hydroxyls of the α-1,6 glucans appear to be important contact sites as methylation of these hydroxyls results in loss of recognition by the GBLs (Wang *et al.*, 1995). The apparent absolute specificity of the GBLs of the oral streptococci may have an evolutionary advantage for the bacteria. The lectins recognize only the products of glucosyltransferases acting on sucrose. Dietary polysaccharides, such as starches and glycogens, are not products of GTFs acting on sucrose. These polysaccharides are not GBL receptors and are therefore unable to promote or inhibit oral streptococcal adhesion.

LECTIN AND GLUCOSYLTRANSFERASE ADHESINS OF *STREPTOCOCCUS SOBRINUS*

It is now clear that culture supernatants of *S. sobrinus, S. cricetus* and *S. mutans* contain proteins which are capable of binding to crosslinked dextrans (Sephadex). Although most strains of *S. mutans* do not aggregate with soluble dextrans (α-1,6 rich glucans), the bacteria secrete glucan-binding proteins. One of the proteins is highly immunogenic (Smith *et al.*, 1994) and may be a potential anti-caries vaccine. The other proteins are GTFs. The non-GTF glucan-binding protein of *S. mutans* does not appear to be antigenically similar to glucan-binding proteins of *S. sobrinus* and *S. cricetus,* however (Ma *et al.*, 1996).

When cell extracts of *S. sobrinus* are poured over Sephadex G-75, at least five distinct proteins bind to the polysaccharide gel, only to be eluted by low molecular weight α-1,6 glucan (glucan T-10) or by chaotropes (Ma *et al.*, 1996). The proteins range in size from 16 to 145 kDa. When SDS-PAGE gels are incubated with Triton X-100 and sucrose, the GTF bands swell and can be readily identified as 135 and 145 kDa proteins. No fructosyltransferase activities have been detected in *S. sobrinus* although the enzyme is found in *S. mutans* strains. Whole cell extracts appear to contain the same proteins found in culture supernatants. Secretion of GTFs and/or GBLs into the growth medium may be one way for the bacteria to adhere. For example, soluble GTFs from oral streptococci are able to bind hydroxylapatite (Schilling and Bowen, 1992). When the GTFs are subjected to sucrose (Figure 1), glucans (Figure 2) are synthesized and serve as receptors for oral streptococci. It thus makes evolutionary sense for oral streptococci to secrete GTFs. No studies have been performed to determine whether GBLs may serve as streptococcal receptors *in situ*.

It seems unusual that GTFs from oral streptococcal species are able to bind well to Sephadex, yet the bacteria bearing the GTFs on their cell surfaces are frequently unable to be aggregated by glucans. These same bacteria, refractory to glucans, are able to adhere to various kinds of surfaces (teeth, steel, glass, plastics, etc.) when incubated in the presence of sucrose. In addition, these same bacteria can be prevented from adhering when in the presence of GBL inhibitors, such as low molecular weight α-1,6 glucan (Doyle and Taylor, 1994). The GBL, leading to cellular aggregation by high molecular weight glucan, is therefore not absolutely required for adhesion.

In *S. sobrinus* 6715, the lectin is found in the growth medium and cell-bound. Protoplasts of *S. sobrinus* do not possess the lectin, showing that the lectin is cell wall-as-

Figure 1. Structure of sucrose (β-D-fructofuranosyl-α-D-glucopyranoside. The main features of the structure have been described by Adams and Lerner (1992), Bock and Lemieux (1982), Brown and Levy (1973) and Singh *et al.* (1983). The structure does not possess a reducing group. The glycosidic bond energy derived from sucrose is used for glucose polymerization by GTFs.

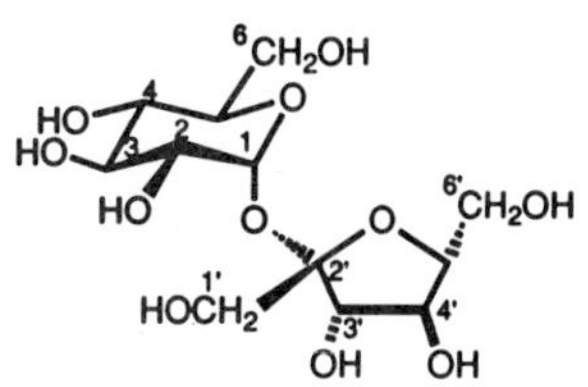

$^{6}CH_2OH$ HO HO HO O O O $^{6'}CH_2OH$ $HOCH_2$ OH OH 1 2 3 4 1' 2' 3' 4'

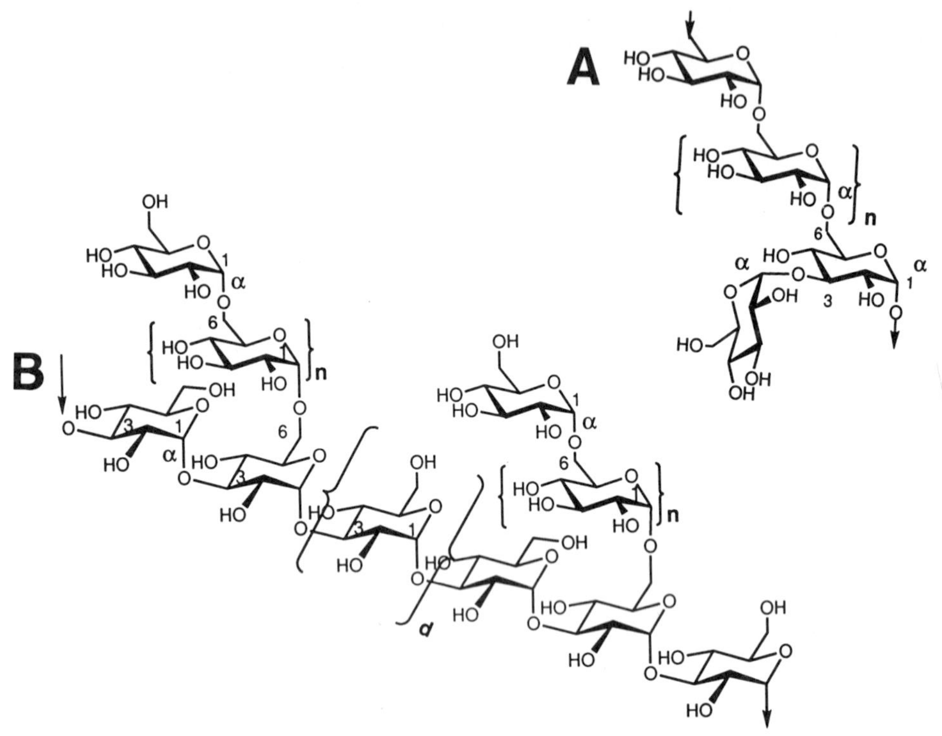

Figure 2. Representative structures of dextrans produced by oral streptococci. A. Water-soluble glucan rich in α-1,6 glucose residues. This polymer is recognized by glucan-binding lectins of *S. sobrinus* and *S. cricetus*. The α-1,6-linked backbone may have short α-1,3 branches. The extent of branching depends on the bacterium and its culture conditions. B. A representative water-insoluble dextran produced by oral streptococci. The α-1,3 linked backbone is not recognized by glucan-binding lectins. Branching usually occurs between α-1,6 sites, but some α-1,3 linkages may be found in some bacteria. Adapted from Cheetham *et al.* (1991), Davis *et al.* (1986), Shimamura *et al.* (1989), and Seymour *et al.* (1980).

sociated. The lectin is not a proteolytic end product of GTF(s) (Ma *et al.,* 1996). The lectin of *S. sobrinus* is N-terminally blocked and has an amino acid composition distinct from a 60 kDa glucan-binding protein of *S. mutans* (Banas *et al*., 1990). Table 1 summarizes some of the properties of glucan-binding proteins (GBPs) of oral streptococci, emphasizing the GBLs and GTFs of *S. sobrinus and S. mutans.* The proteins are inactivated by proteases, but soluble α-1,6 glucan affords some protection to the GBL of *S. sobrinus*. Similarly, glucan partially protects the GBL(s) against chaotropes, such as 6 M urea and sodium dodecyl sulfate (1.0 mg/ml).

Table 2 reviews some of the agents known to act as sucrose-dependent anti-adhesins of the oral streptococci. These agents range from sucrose analogues (6-aldehydo-sucrose), low molecular weight α-1,6 glucans, fluoride, to chelating agents, including citrate and lactate. Sucrose-independent adhesion was measured by determining binding of radioactive cells to hydroxylapatite or saliva-coated hydroxylapatite. Details on the methods for adhesion assays have been published elsewhere (Nesbitt *et al.,* 1982; Schilling and Doyle, 1995). Sucrose-dependent adhesion was determined by the binding of cell suspensions to glass tubes, washing the tubes and then staining with gentian violet. This assay is only semi-quantitative. Although the table does not describe individual strains or species, some interesting

Table 1. Comparison of properties between glucan-binding proteins of oral streptococci

	GBL(s) and GBP(s)	GTF(s)
Molecular weight(s) (kDa)	87,60,58,42,35,15,7.5	170,145,135,90
Glucan-binding domain(s)	+	+
Inhibition by fluoride	+	-
Inhibition by EDTA, citrate, lactate	+	-
Inhibition by periodate-oxidized α-1,6 glucan	+	+
Requires Mn^{2+} for phenotypic expression	+	+
Inducible by sucrose or glucan	-	-
Cell bound	some strains	most strains
Protease susceptible	protected by α-1,6 glucan	not protected by α-1,6 glucan
Denatured by 6 M urea	only slowly	more rapidly than GBL
Response to glucan T-10	inhibits	stimulates
Response to periodate-oxidized glucan T-10	inhibits	inhibits
Inhibition by α-methyl glucosides	-	+
Catalytic activity on sucrose	-	+
Contributes to surface hydrophobicity	+ (but glucan T-10 reduces hydrophobicity)	-
Role in sucrose-dependent adhesion	+	+

Taken from Bauer *et al.* (1993), Ciardi (1983), Doyle and Taylor (1994), Drake *et al.* (1988a), Landale and McCabe (1987), Liang *et al.* (1989), Loesche (1986), Lü-Lü *et al.* (1992), Ma *et al.* (1994, 1996), McCabe and Hamelik (1978); McCabe and Smith (1975); McCabe *et al.* (1977), Mooser (1992), and Robyt (1983, 1995). GBP = glucan-binding proteins but not required for glucan-dependent cellular aggregation.

results were noted. For example, *S. mutans* 10449, which does not possess a cell surface GBL, adheres well to glass in the presence of sucrose. Inhibitors of GBL, but not GTFs, such as citrate, fluoride and glucan T-10, tend to prevent sucrose-dependent adhesion. How this occurs is unknown, although it may be that adherent cells, unlike planktonic cells, are able to express GBL-like proteins or GTFs may act as GBLs. In unpublished research, it has been shown by our groups that washed and chloramphenicol- treated *S. mutans* 10449 can adhere to glass or hydroxylapatite in the presence of sucrose. This rules out the possibility the bacteria are synthesizing and expressing new proteins upon adhesion. For various reasons, including expense of preparation (as for the substituted sucroses), none (with the exception noted below) of the anti-adhesins listed in Table 2 are appropriate anti-caries agents. The last section shows that an oxidized derivative of glucan T-10 is a good anti-caries agent when mixed with cariogenic diets. The oxidized glucan is easily prepared and retains inhibitory activities in compounded diets.

PERIODATE-OXIDIZED GLUCAN T-10, A NEW ANTI-CARIES AGENT

An earlier report showed that periodate-oxidized α-1,6 glucans were effective inhibitors of streptococcal GTFs (Ono *et al.*, 1981). Because the oxidized dextrans (Ox-T10) are inhibitors of GTFs, it was considered they may be useful anti-caries agents because they also appear to be inhibitory against GBLs (Doyle and Taylor, 1994). A food supplement capable of preventing adhesion, inhibiting GBLs and also inhibiting GTFs is an attractive prospect for preventing sucrose-promoted caries. Several experiments were conducted

Table 2. Anti-adhesins of oral streptococci

Addition	Effect on GBL	Effect on GTF	Sucrose-indepen -dent adhesion	Sucrose -dependent adhesion	Disaggregation of sucrose dependent adherent cells
6-Aldehydo-α-methylglucoside	NI	I	NI	I	-
6-Aldehydosucrose	NI	I	NI	I	-
6-Aminosucrose	NI	I	NI	I	-
6′-Aminosucrose	NI	I	NI	I	-
6′6′-Aminosucrose	NI	I	NI	I	-
Citrate (50 mM)	I	NI	NI	I	+
Fluoride (200 mM)	I	NI	NI	I	+
Glucan T-10	I	NI	NI	I	+
Isomaltose	NI	I	NI	I	-
3-Ketosucrose	NI	I	NI	I	-
Lactate (50 mM)	I	NI	NI	I	+
Lipoteichoic acid(100 μg/ml)	NI	NI	NI	NI	-
Maltose	NI	I	NI	I	-
Nigerose	NI	I	NI	I	-
Periodate-oxidized glucan T-10	I	I	NI	I	+
Pyridine analogues	NI	I	NI	I	-

Taken from Cox *et al.* (1994), Doyle and Taylor (1994), Inoue and Smith (1980), Lü-Lü *et al.* (1992), Ma *et al.* (1994, 1996), McAlister *et al.* (1986,1989a,b), and Thaniyavarn *et al.* (1981, 1982, 1983). GBL activities were measured according to Drake *et al.* (1988a). GTF was measured according to Germaine *et al.* (1974). For GBL, the concentrations of carbohydrate inhibitors were 50-500 μg/ml. Sucrose-dependent adhesion required incubation of *S. sobrinus* in glass vials with 20 mg/ml sucrose at 37° for 16-20 hrs. Adherent cells and polysaccharides were visualized by staining with gentian violet. Disaggregation of sucrose-dependent adherent cells was determined by washing the sucrose-incubated mixtures twice with distilled water and then adding the prospective inhibitor in neutral phosphate buffer. NI, non-inhibitor; I, inhibitor; -, no observed effect on disaggregation; +, capable of disaggregating adherent cells from glass.

employing Ox-T10 (Figure 3) to study its effects on GBLs and on animal caries. A separate paper describes the preparation and purification of the Ox-T10 (Singh *et al*,. submitted). Approximately 4 kilograms of the Ox-T10 were purified and found to be free of unreactive periodate. When the Ox-T10 was used to inhibit glucan T-2000 promoted aggregation of *S. sobrinus* 6715, it was observed the derivative was a superior inhibitor, compared to the unmodified glucan (Figure 4). In these experiments, suspensions of *S. sobrinus* 6715 were subjected to oxidized or unmodified glucan T-10 for 30 min, then glucan T-2000 was added to promote cellular aggregation. The rates of the glucan T-2000 dependent aggregation of *S. sobrinus* were then determined (Drake *et al*., 1988a). When percent inhibition was plotted against log glucan concentration, it was possible to estimate the amounts needed for 50% inhibition of aggregation. For glucan T-10, the amount required for 50% inhibition was about 120 μg/ml, whereas for the oxidized derivative only approximately 35 μg/ml was needed for the same extent of inhibition. Clearly, in these experiments, the oxidized how molecular weight glucan was a superior inhibitor, compared to the unmodified glucan.

In theory, an aldehyde function can condense with an ammonium group to form a Schiff's base. Such bases are only slightly reversible. A test of Schiff's base participation in the oxidized glucan-GBL interaction was made by initially mixing the glucan with cells and then washing away any non-reactive polysaccharide. The washed cells were then mixed with

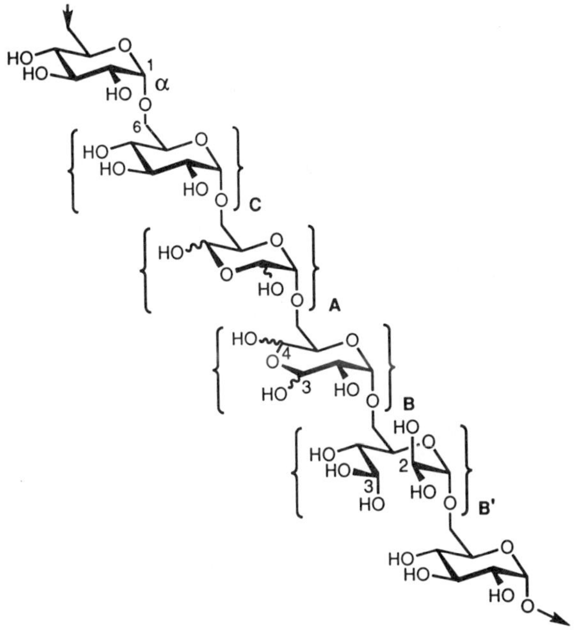

Figure 3. Structure of periodate-oxidized glucan T-10. The glucan (one gram) (degree of polymerization ave = 54) was oxidized with 220 mg sodium periodate. A = sites of double oxidation; B and/or B′ = sites of single oxidation; C= unoxidized sites. Sites A,B and B′ are not necessarily contiguous, but are randomly interspersed with C along the chain. Of the 54 glucose residues, 45 remain unoxidized whereas A = 5 and B (and/or B′) = 4. Unpublished results of Singh and Taylor.

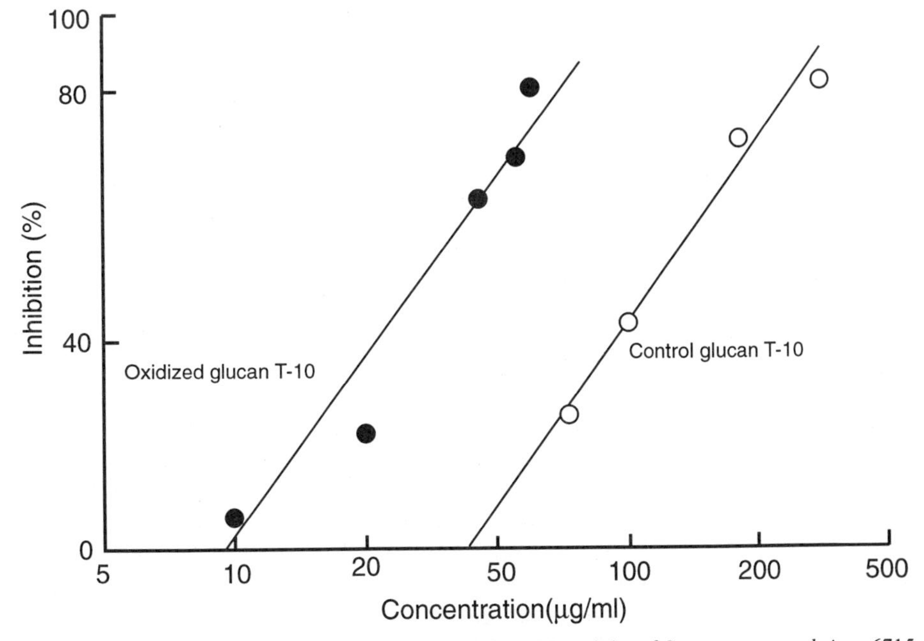

Figure 4. Oxidized glucan T-10 and glucan T-10 inhibit GBL activity of *Streptococcus sobrinus* 6715.

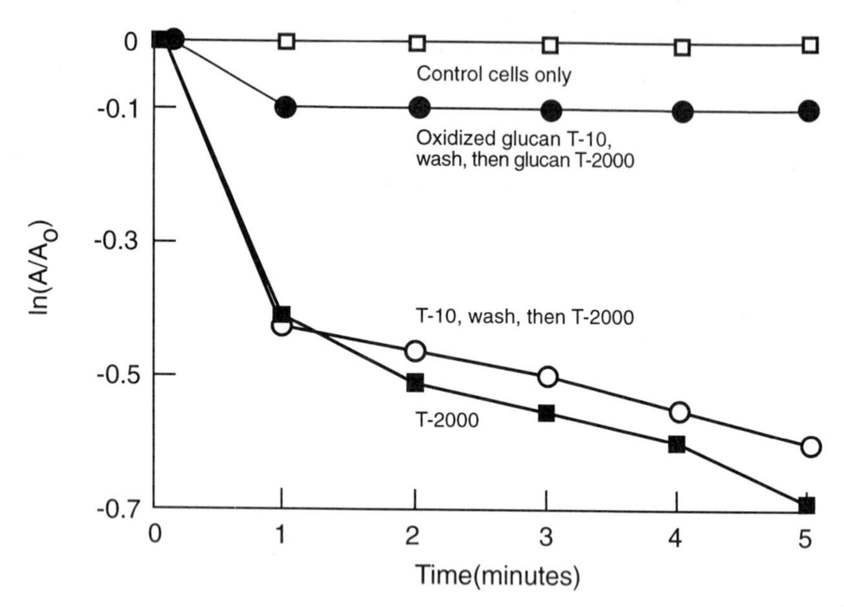

Figure 5. *Reversibility of the effects of oxidized glucan T-10 on the GBL of S. sobrinus* 6715. Cells were incubated for 30 min in the glucan T-10 (or oxidized glucan T-10) prior to washings. Solid circle: oxidized glucan T-10, followed by two washes in PBS and the addition of glucan T-2000; open circle: glucan T-10, followed by two washes in PBS and the addition of glucan T-2000; solid square: control, addition of glucan T-2000 only; square: no T-2000, control cells in PBS only.

glucan T-2000 in order to elicit aggregation. The results, shown in Figure 5, are clear that the oxidized glucan irreversibly (within the time frame of the experiment) inhibits the GBL. In contrast to the oxidized glucan, unmodified glucan T-10 could be washed away from the cells without causing any loss in GBL activity. In the absence of washing, both forms of glucan T-10 completely inhibited the glucan T-2000 dependent aggregation (not shown in a plot).

Sometimes aldehydes may be oxidized to acids. If this is the case for the aldehydo-glucans (Figure 3), it would seem likely the polysaccharide would lose activity upon storage. Solutions of glucan T-10 (and oxidized glucan T-10) were prepared in distilled water and stored at 4°C for up to 10 weeks. Samples were removed at intervals and used to inhibit the GBL of *S. sobrinus*. In Figure 6, it is shown that neither the oxidized glucan nor the unmodified glucan lost any inhibitory power for up to 10 weeks.

Reduced glucans can be derived by use of borohydride. The borohydride forms a single alcohol function at a reducing end of a polysaccharide, whereas for oxidized polysaccharides, there should be an alcohol function formed from each aldehyde function. The reduced derivative was prepared and tested for its effects on the GBL of *S. sobrinus*. Reduction of the oxidized glucan T-10 results in a loss of inhibitory power.

It is known that when cell suspensions of *S. sobrinus* are incubated with sucrose, adhesion and accumulation of the bacteria occur on vessel surfaces. With this in mind, it was considered that oxidized T-10 may be a good inhibitor of the sucrose-dependent adhesion. In Figure 7, it is shown that in the absence of sucrose, very little adhesion occurs. Supplementation of the cells with sucrose results in significant adhesion, as measured by dye uptake by the *in vitro* plaque. Oxidized glucan T-10 was a potent inhibitor of adhesion,

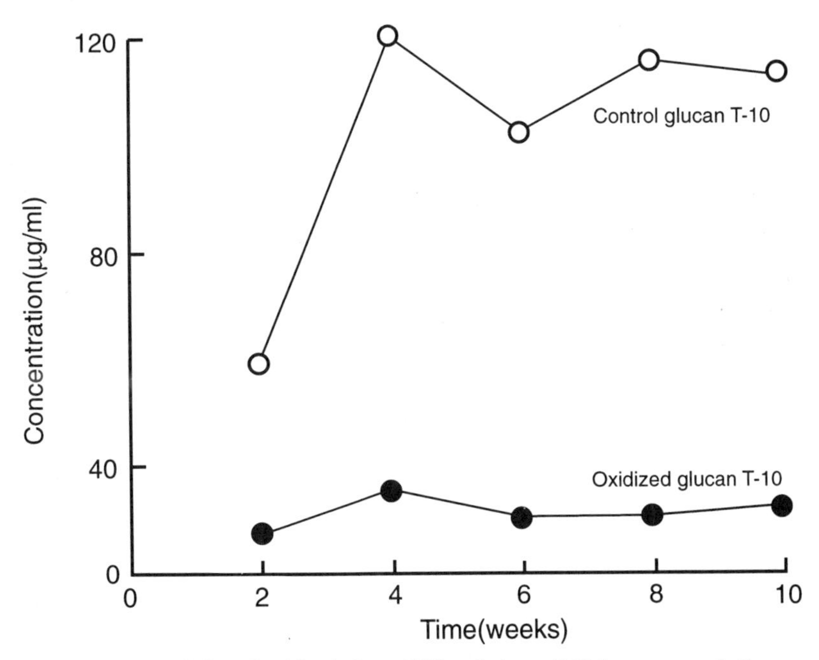

Figure 6. Stability of oxidized glucan T-10 and glucan T-10 in aqueous solution.

whereas glucan T-10, reduced glucan T-10, and oxidized-reduced glucan T-10 were poor inhibitors.

In Schiff's base formation, the aldehyde condenses better with free amino groups than with protonated amino groups, so it would be predicted that the oxidized glucan would inhibit GBL activity better at pH 7-9 then at pH<6. Some results are shown in Figure 8 where it is revealed that at pHs 7,8 and 9, the oxidized glucan T-10 was a much better inhibitor than at pHs or 6. At pH 5, approximately 870 μg/ml of oxidized T-10 was required for 50% inhibition. In contrast, at pH 9, the amount needed was about 25 μg/ml for 50% inhibition. For the unmodified glucan T-10, the amounts needed at the foregoing pHs were 300 μg/ml and 200 μg/ml, respectively. The studies established that oxidized α-glucans were effective inhibitors of GBL and streptococcal adhesion. When the oxidized glucan T-10 was used as a food supplement, it was observed that the rate of caries was lower in experimental rats (Table 3). In these experiments, rats were given a rich diet *ad libitum*. Sucrose was

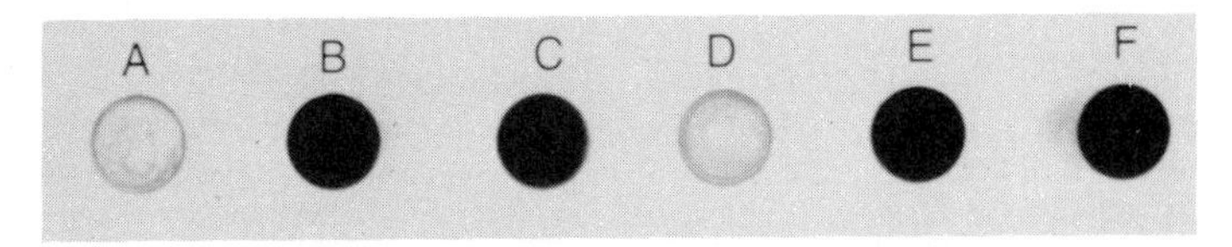

Figure 7. Different kinds of glucan T-10 affect sucrose-dependent accumulation of *S. sobrinus* onto glass surfaces. A washed cell suspension of *S. sobrinus* was mixed with sucrose (20 mg/ml final concentration) or sucrose plus oxidized glucan T-10 (200 μg/ml) or sucrose plus modified glucan T-10 (200 μg/ml final concentrations). A, control with no sucrose; B, sucrose - supplemented; C, sucrose plus glucan T-10; D, sucrose plus oxidized glucan T-10; E, sucrose plus reduced glucan T-10; F, sucrose plus oxidized, reduced glucan T-10.

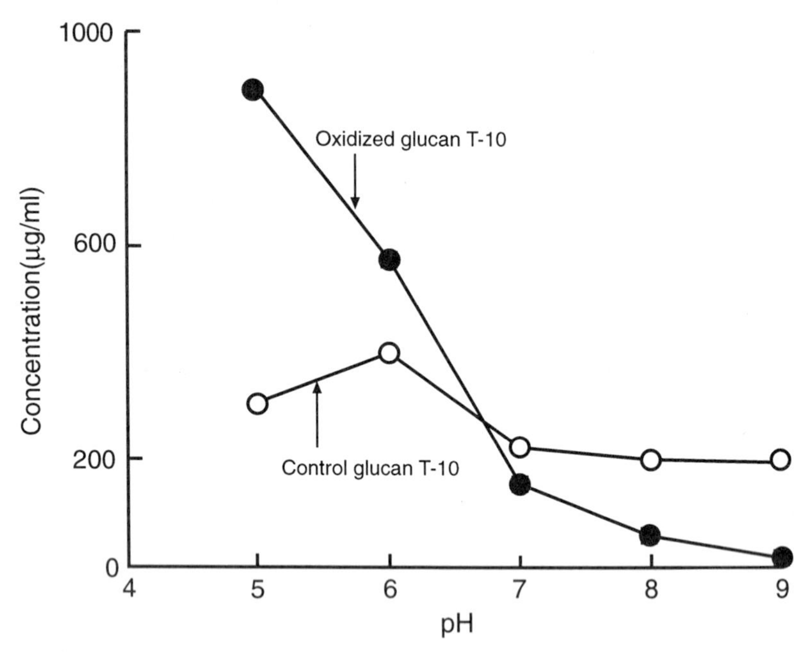

Figure 8. Oxidized glucan T-10 and glucan T-10 inhibit GBL activity of *S. sobrinus* at different hydrogen ion concentrations.

Table 3. Mean caries scores in experimental rats

Group	X±SD
Sucrose only	20.7±4.4
T-10 plus sucrose	13.6±2.0
OX-T-10 plus sucrose	9.4±1.4

The caries score of the sucrose group is significantly higher than the T-10 and oxidized T-10 groups ($p<0.05$). The caries score of glucan T-10 group is significantly higher than oxidized T-10 group ($p<0.05$). Rats were purchased from the Charles River Laboratory, 20 days of age. The rats were weighed and randomly divided into 3 groups, 10 to each group. All animals were infected with a swab saturated with an overnight culture of *S. sobrinus* 6715 each day for 10 days. The bacteria were collected from individual animals with swabs at the 11th day and cultured on mitis-salivarius agar. Single colonies were selected to culture in Trypticase soy broth medium. Colonies were assayed for GBL with glucan T-2000-induced aggregation. Three groups of rats were fed individual diets (Michalek *et al.*, 1975) but containing: group 1: 50% sucrose (Sigma Chemical Co) and 10% oxidized glucan T-10; group 2: 50% sucrose and 10% glucan T-10 (Sigma Chemical); group 3:50% sucrose. The experiment was continued for 45 days at which time all the animals were weighed, and blood was withdrawn from the heart. All the rats were sacrificed with the aid of rodent anaesthetic. The heads were autoclaved 5 minutes to loosen soft tissue. Teeth of the three groups were obtained and stained with 4 mg/ml murexide in 70% (v/v) alcohol. Caries were scored by the Keyes (1958) technique. After the serum was obtained, blood glucose was measured by use of a blood glucose diagnostics kit (Sigma Company). When the rats were sacrificed, the blood of every rat was collected into a micro-tube. The tube was centrifuged at 3000 x g for 3 minutes, then read for the volume of red blood cell hematocrit. The data were evaluated by a statistical unpaired T-test.

supplemented to 50% (w/w), and when used, glucans were supplemented to final 10% (w/w) concentrations. During the first 10 days of the experiment, the animals were subjected to *S. sobrinus* 6715 via cotton swabs and in drinking water. At the end of 45 days the animals were sacrified and the teeth prepared for analyses of caries. The caries scoring was done according to Keyes (1958) using both frequency and size of the carious lesion. The results (Table 3) clearly show the efficacy of oxidized glucan in reducing the caries scores. The results were statistically significant at the 0.05 level. Based on these results, the oxidized glucans are promising anti-caries agents. Each of the rats employed in the caries studies were weighed prior to, and following, the experiment. The weights were obtained in order to assess any effects of the glucan(s) on growth of the rats. The results show that the controls (no glucan, but sucrose-supplemented) were identical to the glucan T-10 plus sucrose, with respect to average animal weight post-experiment. The results contrast with those fed oxidized glucan-T10. The oxidized glucan group weighed only 78% of the controls, suggesting toxicity. Blood glucose levels of all groups were the same, based on a statistical analysis of the results. Similarly, hematocrit values were also statistically identical. Even though hematocrit and blood glucose measurement do not suggest toxicity, the weights of the animals revealed that the Ox-T10 group was developmentally retarded.

CONCLUSIONS

The research was concerned with the use of oxidized α-1,6 glucan as inhibitors of streptococcal GBLs and as potential anti-caries agents. The results can be summarized as: i) Oxidized glucan T-10, a low molecular weight α-1,6 glucan, was superior to the unoxidized or oxidized-reduced glucan in inhibiting the GBL of *S. sobrinus;* ii) Glucan T-2000 was superior to oxidized glucan T-2000 in promoting aggregation of the bacteria; iii) Inhibition of GBL by oxidized α-1,6 glucans was only slowly reversible and became irreversible when mixed with the reducing agent borohydride; iv) Inhibition by oxidized glucan T-10 was more pronounced at pH 8-9 than at pH 5-6; v) Oxidized glucan T-10 retained its stability in distilled water for a period > 14 days; vi) Oxidized glucan T-10 was a highly effective anti-caries agent when rats were given a diet rich in sucrose; vii) Animals fed oxidized glucan T-10 gained less weight, compared to animals fed unoxidized glucan T-10, over a several week time period; viii) Oxidized α-1,6 glucan was a better inhibitor than the unoxidized glucan of sucrose-dependent adhesion of the streptococci to glass surfaces.

Based on the foregoing results, it is suggested that oxidized α-1,6 glucans have multiple effects on the surface proteins of *S. sobrinus*.

Earlier, Ono *et al.* (1981) showed that oxidized dextran, but not oxidized starches, inactivated the dextransucrases of oral streptococci. It is known that the dextransucrases possess glucan-binding domains. It is therefore likely that the oxidized glucans acted on the glucan-binding domains of the enzyme(s). In the present studies, it was shown that the oxidized α-1,6 glucans were potent inhibitors of aggregation of *S. sobrinus* by high molecular weight α-1,6 glucans. Because aggregation of oral streptococci is lectin-dependent, and independent of dextransucrases or glucosyltransferases, it is likely the oxidized glucan acted on the GBL. Thus, oxidized glucans probably act at two sites to modify sucrose-dependent adhesion.

No efforts were made to isolate GBLs and study amino acid side chain modification (Singh *et al.*, 1993) following glucan-cell interactions. Based however, on the known chemistry of dialdehyde-protein interactions, it is suggested the lectin is inactivated via Schiff's base formation. This mechanism requires that the aldehydo- function condense with an amino function, yielding an unsaturated intermediate. The unsaturated intermediate (Schiff's base) is reversible, but when reduced with borohydride, becomes irreversible.

When borohydride was added to mixtures of the aldehyde-cells, the GBL activity could not be restored by washing away the derivatized dextran.

The fact that the oxidized dextran retains its inhibitory powers upon prolonged aqueous incubation suggests that the dextran could be used in water-based formulations. This is of course assuming the derivative is efficacious in preventing caries and is non-toxic. Results clearly showed that oxidized glucan was an anti-caries agent, even when the diet was 50% (w/w) sucrose. It is interesting that fully hydroxylated glucan T-10 was not a highly effective anti-caries agent. It may be possible that very low amounts of aldehydoglucan may be anti-cariogenic when animals are fed "normal" levels of sucrose. Yet these same levels of aldehydo-dextrans may have no toxicities at all. As far as it is known, the aldehydo-dextrans used in this study are superior to any other carbohydrate as an anti-caries agent when experimental rats are fed sucrose.

ACKNOWLEDGMENTS

This research was supported in part by NIH-DE-7199 and NIH-DE-5102.

REFERENCES

Adams, B., and L. Lerner. 1992. Observation of hydroxyl protons of sucrose in aqueous solution: no evidence for persistent intra-molecular hydrogen bonds. J. Am. Chem. Soc. 114:4827-4829.

Banas, J.A., R.R.B. Russell, and J.J. Ferretti. 1990. Sequence analysis of the glucan-binding protein of *Streptococcus mutans* Ingbritt. Infect. Immun. 58:667-673.

Bauer, P.D., C. Trapp, D. Drake, K.G. Taylor, and R.J. Doyle. 1993. Acquisition of manganous ions by mutans group streptococci. J. Bacteriol. 175:819-825.

Bock, K., and R.U. Lemieux. 1982. The conformational properties of sucrose in aqueous solution: intramolecular hydrogen-bonding. Carbohydr. Res. 100:63-74.

Brown, G.M., and H.A. Levy. 1973. Further refinement of the structure of sucrose based on neutron-diffraction data. Acta Cryst. B29:790-797.

Cheetham, N.W.H., M.E. Slodki, and G.J. Walker. 1991. Structure of the linear, low molecular weight dextran synthesized by a D-glucosyltransferase (GTF-S3) of *Streptococcus sobrinus*. Carbohydr. Polym. 16:341-353.

Ciardi, J. 1983. Purification and properties of glucosyl-transferases of *Streptococcus mutans*: a review, p. 51-64. In: Glucosyltransferases, Glucans, Sucrose and Dental Caries. Doyle, R.J. and J.E. Ciardi (eds.). IRL Press, Arlington, VA

Cox, S.D., M.O. Lassiter, K.G.Taylor, and R.J. Doyle. 1994. Fluoride inhibits the glucan-binding lectin of *Streptococcus sobrinus*. FEMS Microbiol. Letts. 123:331-334

Davis, H.M., H.B. Hines, and J.R. Edwards. 1986. Structure elucidation of a water-insoluble glucan produced by a cariogenic oral *Streptococcus*. Carbohydr. Res. 156:69-77.

Doyle, R.J., and K.G. Taylor. 1994. Sucrose, glucan-binding proteins and oral streptococcal adhesion. Cells Mater. 4:91-100.

Drake, D., K.G. Taylor, A.S. Bleiweis, and R.J. Doyle. 1988a. Specificity of the glucan-binding lectin of *Streptococcus cricetus*. Infect. Immun. 56:1864-1872.

Drake, D., K.G. Taylor, and R.J. Doyle. 1988b. Expression of the glucan-binding lectin of *Streptococcus cricetus* requires manganous ions. Infect. Immun. 56:2205-2207.

Germaine, G.R., C.F. Schachtele, and A.M. Chludzinski. 1974. Rapid filter paper assay for the dextransucrase activity from *Streptococcus mutans*. J. Dent. Res. 53:13555-1360.

Gibbons, R.J., and R.J. Fitzgerald. 1969. Dextran-induced agglutination of *Streptococcus mutans*, and its potential role in the formation of microbial plaques. J. Bacteriol. 98:341-346.

Inoue M., and E.E. Smith. 1980. Specific inhibition of glucosyl-transferase of *Streptococcus mutans*. Carbohydr. Res. 80:163-177.

Keyes, P.H. 1958. Dental caries in the molar teeth of rats. J. Dent. Res. 37:1088-1099.

Landale, E.C., and M.M. McCabe. 1987. Characterization by affinity electrophoresis of an α-1,6-glucan-binding protein from *Streptococcus sobrinus*. Infect. Immun. 55:3011-3016.

Liang, L., D. Drake, and R.J. Doyle. 1989. Stability of the glucan-binding lectin of oral streptococci. J. Dent. Res. 68:1677.

Loesche, W.J. 1986. Role of *Streptococcus mutans* in human dental decay. Microbiol. Rev. 50:353-380.

Lü-Lü, J.S. Singh, M. Galperin, D. Drake, K.G. Taylor, and R.J. Doyle. 1992. Chelating agents inhibit activity of and prevent expression of streptococcal glucan-binding lectins. Infect. Immun. 60:3807-3813.

Ma, Y., M.O. Lassiter, J.A. Banas, M. Yu Galperin, K.G. Taylor, and R.J. Doyle. 1996. Multiple glucan-binding proteins of *Streptococcus sobrinus*. J. Bacteriol. 178:1572-1577.

Ma, Y., M.Y. Galperin, K.G. Taylor, and R.J. Doyle. 1994. Glucan-binding proteins of *Streptococcus sobrinus,* p. 275-286. Lectins: biology, biochemistry, clinical biochemistry, vol. 10. In: van Driessche, E., J. Fisher, S. Beeckmans, and T.G. Bϕg-Hansen (eds.), Textop, Hellerup, Denmark.

McAlister, D., S. Singh, K.G. Taylor, and R.J. Doyle. 1986. Inhibition of the glucosyltransferases of *Streptococcus mutans* by 6-aldehydosucrose and 6-aldehydo-methyl-α-D-glucopyranoside, p. 413-419. In: Molecular Microbiology and Immunology of *Streptococcus mutans*. Hamada, S, S.M. Michalek, H. Kiyono, L. Menaker, and J.R. McGhee (eds.). Elsevier, Amsterdam.

McAlister, D., S. Nambiar, K.G. Taylor, and R.J. Doyle. 1989a. Glucosyltransferases of *Streptococcus sobrinus* C211 are both stimulated and inhibited by hydrogen peroxide. Oral Microbiol. Immunol. 4:146-152.

McAlister, D., R.J. Doyle, and K.G. Taylor. 1989b. Inhibition by maltose, isomaltose and nigerose of the synthesis of high-molecular-weight D-glucans by the D-glucosyltransferases of *Streptococcus sobrinus.* Carbohydr. Res. 187:131-138.

McCabe, M.M., and R.M. Hamelik. 1978. Multiple forms of dextran-binding proteins from *Streptococcus mutans.* Adv. Exp. Med. Biol. 107:749-760.

McCabe, M.M., R.M. Hamelik, and E.E. Smith. 1977. Purification of dextran-binding protein from cariogenic *Streptococcus mutans.* Biochem. Biophys. Res. Comm. 78:273-278.

McCabe, M.M., and E.E. Smith. 1975. Relationship between cell-bound dextransucrase and the agglutination of *Streptococcus mutans*. Infect. Immun. 12:512-520.

Michalek, S.M., J.R. McGhee, and J.M. Navia. 1975. Virulence of *Streptococcus mutans*: a sensitive method for evaluating cariogenicity in young gnotobiotic rats. Infect. Immun. 12:69-75.

Mooser, G., and C. Wong. 1988. Isolation of a glucan-binding domain of glucosyltransferase (1,6-α-glucan synthase) from *Streptococcus sobrinus*. Infect. Immun. 56:880-884.

Mooser, G. 1992. Glycosidases and glycosyltransferases. The Enzymes 20:187-233.

Nesbitt, W.E., R.J. Doyle, K.G. Taylor, R.H. Staat, and R.A. Arnold. 1982. Positive cooperativity in the binding of *Streptococcus sanguis* to hydroxylapatite. Infect. Immun. 35:157-165.

Ono, K., D.W. Nuessle, and E.E. Smith. 1981. Oxidized saccharides as inhibitors of α-glucan synthesis by *Streptococcus mutans* glucosyltransferase. Carbohydr. Res. 88:119-134.

Robyt, J.F. 1983. Synthesis of glucans from the reducing end by insertion mechanisms, p. 123-133. In: Glucosyltransferases, Glucans, and Dental Caries. Doyle, R.J. and J.E. Ciardi (eds.). IRL Press, Arlington VA.

Robyt, J.F. 1995. Mechanisms in the glucansucrase synthesis of polysaccharides and oligosaccharides from sucrose. Adv. Carbohydr. Chem. and Biochem. 51:133-168.

Schilling, K.M., and W.H. Bowen. 1992. Glucans synthesized *in situ* in experimental salivary pellicle function as specific binding sites for *Streptococcus mutans.* Infect. Immun. 60:284-295.

Schilling, K., and R.J. Doyle. 1995. Bacterial adhesion to hydroxylapatite. Meth. Enzymol. 253:536-542.

Seymour, F.R., R.D. Knapp, and B.L. Lamberts. 1980. Structural analysis of soluble D-glucans from strains of *Streptococcus mutans* by ^{13}C-nuclear magnetic resonance spectrometry. Carbohydr. Res. 84:187-185.

Shimamura, A. 1989. Use of ^{13}C-N.M.R. spectroscopy for the quantitative estimation of 3-O-and 3,6-di-O-substituted D-glucopyranosyl residues in α-glucans formed by the D-glucosyltransferases of *Streptococcus sobrinus*. Carbohydr. Res. 185:239-248.

Singh, J.S., K.G. Taylor, and R.J. Doyle. 1993. Essential amino acids involved in glucan-dependent aggregation of *Streptococcus sobrinus*. Carbohydr. Res. 244:137-147.

Singh, S., C.M. Maynard, F.T. Lin, K.G. Taylor, and R.J. Doyle. 1983. Solution structure and chemical modification of sucrose. p.155-160. In: Glucosyltransferases, Glucans, Sucrose and Dental Caries. Doyle, R.J. and J.E. Ciardi (eds.). IRL Press, Arlington, VA.

Smith, D.J., H. Akita, W.F. King, and M.A. Taubman. 1994. Purification and antigenicity of a novel glucan-binding protein of *Streptococcus mutans.* Infect. Immun. 62:2545-2552.

Thaniyavarn, S., S. Singh, K.G. Taylor, and R.J. Doyle. 1983. Kinetic analysis for the inhibition of dextransucrase by aminosugars, p. 161-170. In: Glucosyltransferases, Glucans, Sucrose and Dental Caries. Doyle, R.J. and J.E. Ciardi (eds.). IRL Press, Arlington, VA.

Thaniyavarn, S., S. Singh, K.G. Taylor, and R.J. Doyle. 1982. Pyridine analogs inhibit the glucosyltransferase of *Streptococcus mutans*. Infect. Immun. 37:1101-1111.

Thaniyavarn, S., S. Singh, C.M. Maynard, K.G. Taylor, and R.J. Doyle. 1981. Amino sugars: a new class of inhibitors of dextransucrase. Carbohydr. Res. 96:134-137.

Wang, J., S. Singh, K.G. Taylor, and R.J. Doyle. 1995. Streptococcal glucan-binding lectins do not recognize methylated α-1,6 glucans. Glycoconjugate J. 12:109-112.

Wu, Q., Q. Wang, K.G. Taylor, and R.J. Doyle. 1995. Subinhibitory concentrations of antibiotics affect cell surface properties of *Streptococcus sobrinus*. J. Bacteriol. 177:1399-1401.

Wu-Yuan, C.D., and R.E. Gill. 1992. An 87-kilodalton glucan-binding protein of *Streptococcus sobrinus* B13. Infect. Immun. 60:5291-5293.

ABSTRACTS

ENTEROPATHOGENIC *E. COLI* TRIGGERS EPITHELIAL SIGNALS TO FORM A FUNCTIONAL RECEPTOR THAT MEDIATES BACTERIAL ATTACHMENT AND ACTIN REARRANGEMENT

Ilan Rosenshine,[1,2] Sharon Ruschkowski,[2] Markus Stein,[2] Dieter J. Reinscheid,[2] and B. Brett Finlay[2]

[1]Department of Biotechnology and Molecular Genetics, The Hebrew University, Faculty of Medicine, POB 12272, Jerusalem 91120, Israel, and [2]Biotechnology Laboratory, University of British Columbia, Vancouver, B.C., Canada, V6T-1Z3.

Enteropathogenic *E. coli* (EPEC) belongs to a group of bacterial pathogens that induce actin accumulation beneath adherent bacteria. We found that EPEC adherence on epithelial cells mediates the formation of finger-like pseudopods (up to 10μm) beneath bacteria. These actin-rich structures also contain tyrosine phosphorylated host proteins concentrated at the pseudopod tip beneath adherent EPEC. Intimate bacterial adherence (and pseudpods formation) occurred only after prior bacterial induction of tyrosine phosphorylation of an epithelial membrane protein, Hp90, which then associates either directly or indirectly with an EPEC adhesin, intimin. These interactions lead to cytoskeletal nucleation and pseudopod formation. This is the first example of a bacterial pathogen that triggers signals in epithelial cells to form a suitable receptor which mediates binding of a specofoc bacterial ligand and subsequent cytoskeletal rearrangement.

REFERENCES

Rosenshine I., Ruschkowski, S and Finlay B B. (1996). Expresion of attacking/effacing activity by enteropathogenic *Escherichia coli* depends on growth phase, temperature, and protein synthesis upon contact with the host cell. Infect Immun. In press

Rosenshine I., Ruschkowski S., Stein M., Reinsceid D., and Finlay B. B. (1996). A pathogenic bacterium triggers epithelial signals to form a function bacterial receptor that mediates pseudopod formation. EMBO J. In press.

HEMADSORPTION OF *MYCOPLASMA GALLISEPTICUM* INVOLVES SWITCHING OF VARIABLE SURFACE MEMBRANE PROTEINS

A. Athamna,[1] R. Rosengarten,[1] S. Levisohn,[2] I. Kahane,[1] and David Yogev[1]

[1]Department Membrane and Ultrastructure Research, The Hebrew University-Hadassah Medical School, Jerusalem 91120, Israel. [2]Div. of Avian Diseases, Kimron Veterinary Institute, Bet Dagan 50250, Israel.

Hemadsorbing (HA+) and spontaneously derived nonhemadsorbing (HA-) mutants of *Mycoplasma gallisepticum* (MG) were characterized with respect to their specific protein profiles and the ability to restore the hemadsorbing capability. MG A5969 strain, in which less than 0.01% of the colonies adsorb erythrocytes (RBC), enable us to study the molecular events involved in the transition between the two phenotypes (HA+ HA-). Our Study shows that HA of MG involved several membrane surface proteins undergoing high-frequency phase variation. The most interesting observation was the identification of colonies exhibiting, within a colony, sectorial regions of hemadsorbing RBC. Antiserum prepared against an hemadsorbing clone (HA+) was used in Western immunoblot analysis to identify antigenic differences between HA+ and HA- phenotypes. Several proteins (P80, P50, P48, and P16) were found in the HA+, but not in HA-, clones. This HA+ antigenic profile was further monitored throughout sequentially subcloned HA population of MG (HA+ HA- HA+). The appearance of these additional proteins was consistent with the HA+ phenotype. Experiments in which HA+ proteins were bound to RBC revealed that P80 and P50 were selectively bound. The role of these phase-variable proteins in hemadsorption is suggested. These proteins were defined by detergent-phase fractionation with Triton X-114 as integral membrane proteins. The selective susceptibility of these proteins to trypsin treatment of intact organisms and the identification of positive, negative, or sectored colonies in the colony immunoblot analysis with P50/P80 antiserum demonstrated that these proteins were exposed on the cell surface and subject to high-frequency phenotypic switching. The molecular characterization of these proteins is under way.

ADHERENCE OF *MYCOPLASMA PNEUMONIAE* TO HUMAN ALVEOLAR MACROPHAGES

Abed Athamna,[a] Mordechai R. Kramer,[b] and Itzhak Kahane[a]

[a] Dept of Membrane and Ultrastructure Research, The Hebrew University-Hadassah Medical School, Jerusalem, Israel, and [b] Pulmonary Institute, Hadassah University Hospital, Jerusalem 91120, Israel.

Alveolar macrophages are the predominant phagocytic cell in the lung that combat invading microorganisms. Their antimicrobial functions require receptor-ligand interactions between the surfaces of the phagocytic cells and the microorganisms. Opsonins and lectins are mediators of such interaction of phagocytes with bacteria and, among them, the mycoplasmas. Studies of the latter were primarily conducted employing rodent macrophages. No data is available on human alveolar macrophages (HAØ). This is the aim of the present study. HAØs were obtained from bronchoalveolar lavage performed routinely in patients. The human pathogen, *Mycoplasma pneumoniae* (MP) causes primary atypical-cold agglutinin-positive pneumonia. Alveolar macrophages attach more of the glass-adhering subpopulation of a given MP culture. The attachment is dose-responsive and enhanced by human serum. It is inhibited by dextran-sulfate, but not by dextran or several monosaccharides, showing that sulfated components on the macrophage surface act as receptors for MP binding. In addition, sialylated compounds, such as fetuin and 1-acid glycoprotein, were found to be potent inhibitors of the attachment, also indicating the role of sialic acid residue in recognition and attachment of MP to human alveolar macrophages.

ADHESINS OF *HELICOBACTER PYLORI* THAT BIND H TYPE 2 AND LEWIS BLOOD GROUP ANTIGENS: AN EXPLANATION OF INCREASED SUSCEPTIBILITY OF BLOOD GROUP O AND NON-SECRETORS TO PEPTIC ULCERS

A. H. Alkout, C.C. Blackwell, D.M. Weir, W. Luman, and K. Palmer

Department of Medical Microbiology, University of Edinburgh, and Department of Gastorenterology, Western General Hospital, Edinburgh, Scotland.

Before *Helicobacter pylori* was identified, it was observed that individuals of blood group O or non-secretors of the ABO blood group antigens were over-represented among patients with peptic ulcers (Mourant *et al.*, 1978). Adhesins on H. pylori that bind fucose have been identified, and evidence from the studies of Boren *et al.* (1993) implied that Lewis^b was the only fucose containing blood group antigen to which these bacteria bound. Since non-secretors are genetically incapable of producing Lewis^b antigen, this could not explain their increased susceptibility. We proposed that the H type 2 antigen found on almost all individuals except the extremely rare Bombay phenotype is a key receptor for *H. pylori* and that the higher levels of this antigen expressed on cells of blood group O would contribute to density of colonization by the bacteria. For the enhanced susceptibility of non-secretors there were two explanations. The Lewis^a antigen present in non-secretors could be a key receptor for the bacteria. While non-secretors can produce only Lewis^a, secretors also express this antigen but in highly variable amounts (Saadi *et al.*, 1993). If the terminal fucose of H type 2 is an important epitope in the receptor, the terminal fucose of Lewis^b in the mucus of secretors might block the adhesins and reduce density of binding; its absence in non-secretors might result in enhanced binding compared with secretors.

There were 3 objectives for this study: 1) to determine if H type 2 and the Lewis blood group antigens were expressed on gastric mucosa cells and the Kato III cell line and buccal epithelial cells to be used in the model for bacterial binding: 2) to test the hypothesis that in addition to Lewis^b, H type 2 and Lewis^a are receptors for *H. pylori*; 3) to isolate surface components of the bacteria that bind H type 2 or the Lewis antigens by affinity purification with synthetic blood group antigens as in our previous studies on *Staphylococcus aureus* (Saadi *et al.,* 1994).

Binding of *H. pylori* (NCTC 11637) and binding of monoclonal antibodies to H type 2 (Serotec), Lewis^a (Scottish Antibody Production Unit, SAPU) and Lewis^b (SAPU) to epithelial cells was assessed by flow cytometry. Gastric mucosal cells and buccal epithelial cells were obtained from patients undergoing gastroscopy from whom informed consent had been obtained. To assess the effects of smoking and fasting on bacterial binding to cells of healthy donors, buccal epithelial cells were obtained from 30 male Muslim postgraduate students during RAMADAN when they were fasting 12-16 hours per day and two weeks after the end of the fast. Synsorb beads to which synthetic H type 2, Lewis^a or Lewis^b were conjugated were used in affinity purification of adhesins from outer membrane proteins. Biotinylated H type 2, Lewis^a and Lewis^b (Syntesome, GmbH, Munchen, Germany) were used in dot blot assays with whole cells and outer membrane proteins of NCTC 11637 and in a spectrophotometric assay to examine binding of these antigens whole cells of *H. pylori*, NCTC 11637 and 51 local patient isolates.

Monoclonal antibodies bound to each of the cell types tested indicating that the antigens were expressed on the gastric mucosa and on the cells to be used for assessment of bacterial binding. Cells from patients of blood group O bound significantly more monoclonal anti-H type 2 than cells of non-O-blood groups (P 0.05). Because the patients had fasted overnight before gastroscopy, we assessed the effect of fasting on expression of blood group antigens on cells of healthy donors. The level of binding of anti-H type 2 was significantly reduced during fasting, but not binding of the anti-Lewis antibodies.

Compared with cells obtained from healthy non-smokers those from smokers bound significantly lower levels of antibodies to H type 2 (P 0.001), Lewis^b (P 0.005) and Lewis^a (P 0.002).

Binding of bacteria to cells of healthy donors was significantly associated with binding of antibodies to H type 2 (p 0.005) and Lewis^b (P 0.001), but not Lewis^a. Binding of bacteria to cells from smokers was significantly lower than those of non-smokers (P 0.001). For binding studies with cells from patients, we used cells from non-smokers. Binding of *H. pylori* to cells of group O secretor patients (n=7) was significantly higher than for patients who were non-O secretors (n=7) (P 0.05). Binding of *H. pylori* to cells of O secretors (N=7) was significantly higher compared with cells of O non-secretors (N=7) (P 0.01).

Binding of *H. pylori* to buccal epithelial cells from patients and to Kato III cells was inhibited by pre -treatment of the cells with monoclonal antibodies to H type 2, Lewis^a or Lewis^b. Although the Kato III cells had the highest binding levels for anti-Lewis^a, the greatest inhibition of bacterial binding was observed when the cells were treated with anti-H type 2.

A 61 kDa protein was isolated by affinity purification with synthetic H type 2, Lewis^b or Lewis^a. Larger amounts of the protein were obtained in the eluates from H type 2, followed by that from Lewis^b and the least from Lewis^a. Pre-treatment of Kato III cells with the proteins reduced binding of *H. pylori* compared with binding of the bacteria to untreated cells. The percentage inhibition reflected the amount of protein present in the eluate. In dot immunoblot assays with whole cells or outer membrane proteins, intensity of the blots to which biotinylated H type 2 were added gave the darkest color, Lewis^b an intermediate color and the palest was observed with Lewis^a.

Each of the 52 strains tested in the whole cell assays for binding of biotnylated blood group antigens bound H type 2, Lewis^b and Lewis^a. For the majority of strains (72 %), the optical densities obtained with H type 2 were greater than those for Lewis^b and Lewis^a the lowest. This pattern was observed for strains obtained from patients with gastric disease and for patients with no signs of ulcers or gastritis. There was no association between patterns of binding of the three antigens and ABO blood group or secretor status of the patient.

Our studies indicate that smoking can affect expression of the blood group antigens studied and that this needs to be controlled for in any assays in which bacterial binding of *H. pylori* to human epithelial cells is being examined . The results also indicate that while smoking is a risk factor for ulcers, if similar interaction occur on the mucosa of the stomach, smoking is not a factor that contributes to density of colonization by these bacteria. H type 2, Lewis^b and Lewis^a are receptors for *H. pylori*. The terminal fucose molecules of H type 2 and Lewis^b appear to be important epitopes in the host receptors as binding of monoclonal antibodies to these two antigens is significantly correlated with binding of the bacteria, more of the 61 kDa protein was eluted from these two antigens and there was greater binding of the biotinylated H type 2 and Lewis^b to whole cells and purified proteins of *H. pylori* . If density of colonization is important in initiation of these disease process leading to ulcers, our studies help explain the epidemiological associations between peptic ulcer disease, blood group O and secretor status. We are currently assessing antibodies to this adhesin in serum and secretions to determine if they are effective in reduction of bacterial binding.

REFERENCES

Boren T, Falk P, Roth KA, Larson F and Normark S. Attachment of *Helicobacter pylori* to human gastric epithelium mediated by blood group antigens. Science 1993; 262: 1892-1895.

Mourant AE, Kopec AC, Domaniewska-Sobczak K. Blood groups and diseases. Oxford: Oxford University press, 1978.

Saadi AT, Blackwell CC, Raza MW, James VS, Stewart J, Elton RA and Weir DM. Factors enhancing adherence of toxigenic *Staphylococcus aureus* to epithelial cells and their possible role in sudden infant death syndrome. Epidemiol. Infect. 1993; 110: 507-517.

Saadi AT, Weir DM, Poxton IR, Stewart J. Essery SD, Blackwell CC, Raza MW and Busuttil A. Isolation of an adhesin from *Staphylococcus aureus* that bind Lewis^a blood group antigen and its relevance to sudden infant death syndrome. FEMS Immunol. Med. Microbiol. 1994; 8: 315-320.

PURIFICATION OF THE GAL1-4 GAL-BINDING ADHESIN FROM *STREPTOCOCCUS SUIS*

Kaarina Tikkanen,[a] Jukka Hytönen,[b] and Jukka Finne[b]

[a]Department of Biochemistry and Biotechnology, A. I. Virtanen Institute, University of Kuopio, FIN-70211 Kuopio, Finland, and [b]Department of Medical Biochemistry, University of Turku, FIN-20520 Turku, Finland.

Two variant adhesion activities, P_N and P_O, have been discovered in the Gal 1-4 Gal binding, gram-positive meningitis-associated bacterium *Streptococcus suis*. The Gal 1-4 Gal binding adhesin was purified from *S. suis* type P_N. Purification was monitored by blot binding with a specific inhibitor of the adhesins, pigeon ovomucoid, a glycoprotein with blood group P_1 determinants. The purified adhesin had a molecular weight of 18 kDa and an isoelectric point of 6.4. When the adhesin was bound to latex- particles, a strong agglutination of erythrocytes was achieved. The agglutination was inhibited by the same inhibitors (galactose, N-acetylgalactosamine and pigeon ovomucoid) as agglutination with whole bacterial cells. The N-terminal amino acid sequence of the adhesin was the same in the examined two other trains of type P_N and two strains of type P_O . The adhesion was detected in all 24 *S. suis* strains examined representing different serotypes (1, 2, 4, 5, 7 and 8) by Western blot analysis using polyclonal antibodies. The adhesin was highly immunogenic and showed opsonizing activity. As a potential virulence factor present in different serotypes the *S. suis* the adhesin thus represents a potential vaccine against *S. suis* infections.

REFERENCES

Tikkanen, K., Haataja, S., François-Gerard, C. & Finne, J., (1995): Purification of a galactosyl-1-4galactose binding adhesin from the gram-positive meningitis-associated bacterium *Streptococcus suis*. J.Biol. Chem. 270, 28874-28878

PROTEIN F2, A NOVEL FIBRONECTIN BINDING PROTEIN FROM *STREPTOCOCCUS PYOGENES*, POSSESSES TWO BINDING DOMAINS

Joseph Jaffe, Shira Natanson-Yaron, and Emanuel Hanski

Department of Clinical Microbiology, The Hebrew University-Hadassah Medical School, Jerusalem 91010, Israel.

Binding of group A streptococcus (**GAS**) to respiratory epithelium is mediated by the fibronectin (**Fn**) binding adhesin, protein F. Previous studies have suggested that certain **GAS** strains express **Fn** binding proteins different than protein F. In this study, we have cloned, sequenced and characterized a gene (*prtF2*) from **GAS** strain 100076 coding for novel **Fn** binding protein, termed protein F2. Insertional inactivation of *prtF2* in strain 100076 dramatically reduces binding to both soluble **Fn** and extracellular matrix, indicating that protein F2 is accountable for the **Fn** binding activity in this strain. *prtF2*-related genes exist in most of **GAS** strains that bind **Fn** but lack *prtF1* (previously denoted as *prtF*). The existence of *prtF2*-related genes was demonstrated in eight **M49**, three **M53**, two **M70**, and three **M71** strains that bind **Fn** but lack *prtF1*. These observations suggest that protein F2 is a major **Fn** binding protein. Protein F2 is highly homologous to **Fn** binding proteins from *Streptococcus dysgalactiae* and *Streptococcus equisimilis*, particularly at its carboxy-terminal portion. The **Fn** binding activity of protein F2 is localized to two domains. One domain is composed of three consecutive repeats of which two are of 37 amino acids each and the third partial repeat is of 21 amino acids. This domain that was termed **FBRD** (Fn binding repeat domain), is located near the carboxy-terminal region of the protein. The other domain, which was denoted **UFBD** (upstream Fn binding domain), is located 100 amino acids amino-terminal of the first domain and it resides on a stretch of approximately 100 amino acids. Expression of **FBRD** and **UFBD** as separated proteins demonstrated that each of them is capable to inhibit **Fn** binding by streptococcal cells. The expression of protein **F2** is regulated in response to alterations in the atmospherical O_2.

SURFACTANT PROTEIN A ENHANCES PHAGOCYTOSIS OF *KLEBSIELLA PNEUMONIAE* K21 A SEROTYPE BY INCREASING THE ACTIVITY OF MANNOSE RECEPTOR OF ALVEOLAR MACROPHAGES AND BY OPSONIZING THE BACTERIA

K. Kabha[1], Y. Keisari[1], H. Parolis[2], J. Schlepper-Schafer[3], J. Schmegner[3], I. Ofek[1]

[1]Tel-Aviv Univ., Tel-Aviv, Israel, [2]Rhodes Univ., Grahamstown, S. Africa, and [3]Univ. of Konstanz, Konstanz, Germany.

Surfactant protein A (SP-A) is a major protein surfactant of the lung and has been previously shown to posses mannose specific lectin-like activity and to enhance the phagocytosis of bacteria by macrophages. In the present study, we sought to determine the ability of SP-A to interact with and opsonize *Klebsiella pneumoniae* (KP). For this purpose we employed a K21a serotype and purified capsular polysaccharides (CPS) which contains Man-α3-Man sequences and a K2 serotype and its CPS which lacks such sequences. Binding to and killing of the bacteria by macrophages were estimated microscopically after mixing fluorescein-labeled bacteria with macrophages and by viable counts, respectively, in the presence or absence of SP-A and other agents. Binding of SP-A to the bacteria and its CPS was estimated by ELISA using immobilized SP-A, electron microscopy with SP-A coated gold particles and by bacterial agglutination tests. Pretreatment of the macrophages or the bacteria resulted in enhanced attachment to and killing by the macrophages of K21a serotype and to a much less extent of K2 serotype. K21a serotype and its CPS bound to SP-A while serotype did not. The binding of K21a was Ca^{++} dependent and was specifically inhibited by K21a CPS, mannan and mannosylated BSA but not by other polysaccharides including K2 CPS. Binding of yeast mannan (or K21a serotype) and anti-mannose receptors by macrophages pretreated with SP-A was increased and not affected, respectively, as compared to non treated macrophages. The results suggest that SP-A enhances phagocytosis of *K. pneumoniae* K21a serotype by alveolar macrophages via two mechanisms one of which involves increased activity of the macrophage mannose receptors and the other involves binding of SP-A to the K21a capsular polysaccharides and specific receptors on the macrophages resulting in opsonization of the bacteria.

SYNTHETIC AVENUES TO CLUSTER-GLYCOSIDES SUITABLE FOR THE INHIBITION OF CARBOHYDRATE-DEPENDENT MICROBIAL ADHESION

Thisbe K. Lindhorst and Christoffer Kieburg

University of Hamburg, Institut für Organische Chemie, Martin-Luther-King-Platz 6, D-20164 Hamburg, Germany; email tklind@chemie.uni-hamburg.de.

The interaction between a microbial lectin and a specific carbohydrate epitope on the surface of the host cell is often essential for microbial adhesion and the resulting infection.

However, an oligovalent interaction is required for effective adhesion. Clustered glycosides can mimic the natural receptor structures and thus act as powerful inhibitors of protein-carbohydrate interactions, a finding which has been termed the clustering effect.

For the success of this concept feasible synthetic avenues to potent cluster glycosides are required. Several synthetic strategies are presented that use simple non-carbohydrate and dendritic cores which are equipped with saccharides on the periphery. Tri-, tetra-, and hexavalent clusters were obtained by using glycosidic bonds as well as thio urea and peptide linkages. Mainly clustered -mannosides were prepared for the interaction with type-I-fimbriae from *Escherichia coli* but also other monosaccharides as well as disaccharides were successfully used. the NMR-analytic of the products will be discussed regarding the conformational characteristics of the molecules.

VITRONECTIN BINDING BY STAPHYLOCOCCI MAY ENHANCE ADHESION TO AND COMPLEMENT ACTIVATION ON BIOMATERIAL SURFACES

Fredrik Lundberg, Jianlin Yu, Maria Stollenwerk, and Åsa Ljungh

Department of Medical Microbiology, University of Lund, Sölvegatan 23, S-22362 Lund, Sweden.

Infections constitute an important complication to the use of materials implanted temporarily or permanently in different parts of the body. Microbes from the normal skin microflora, like coagulase-negative staphylococci, CNS, are the most common etiologic agents. Once established, these infections are impossible to cure and in general lead to extirpation of the material.

Immediately after insertion, host proteins adsorb to the biomaterial surface. Of these, fibronectin, Fn, and also fibrinogen have been proposed to mediate adhesion of *S. aureus*. The way proteins adsorb determines if microbial binding, platelet, complement factor and/or eukaryotic cell binding domains are exposed. Heparinization has been shown to provide an antithrombogenic surface with low complement activation. Many host proteins have heparin binding domains and adsorb differently to heparinized versus unheparinized surfaces.

We have studied adhesion of CNS strains isolated from biomaterial-associated infections in cerebrospinal fluid, CSF, and CNS strains from uninfected biomaterials to native polyvinylchloride, PVC, and to end-point heparinized PVC, hep-PVC, in a perfusion model with and without preperfusion of the system with CSF. Eleven of 19 CNS strains adhered at higher numbers to unheparinized than to hep-PVC. Preperfusion of surfaces with CSF reduced bacterial adhesion of all but 3 strains, and 5 strains adhered at higher numbers to hep-PVC than to PVC. Twelve of 16 strains from infections expressed binding of vitronectin, Vn, 14/16 binding of Fn and heparin. The three control strains bound heparin and two of them Vn and Fn. Preincubation of bacterial cells with Vn significantly reduced adhesion of 13/27 CNS strains to polystyrene, and preincubation with Fn reduced adhesion of 8/27 CNS strains. When the PVC, perfused with CSF, was preincubated with antibodies to Vn, adhesion of cells of a Vn binding *S. epidermidis* strain was reduced whereas antibodies to Fn did not block binding of either Fn or Vn binding strains.

Soluble Vn is well known to bind the terminal complement complex C5-9 and inhibit insertion in the membrane to give lysis. We compared complement activation as measured by lysis of rabbit erythrocytes by different CNS strains on heparinized and unheparinized polystyrene. CNS strains varied in their ability to induce complement activation. CNS strains expressing Vn binding enhanced complement activation on heparinized as well as on unheparinized surfaces in the presence of whole but not in Vn-depleted serum.

Multiple factors, including Vn and Fn binding, may mediate adhesion of staphylococci to biomaterials. Expression of Vn binding was shown to enhance complement activation. These data indicate that blocking of Vn binding may decrease the rate of complication to the use of biomaterials particularly in the blood stream.

EXPRESSION OF TYPE 1 FIMBRIAE AND CAPSULE IN *KLEBSIELLA PNEUMONIAE* K21A

R. Matatov[1], I. Schechter[2], R. Perry[1], K. Kabha[1], H. Sahly[3], R. Podschun[3], I. Ofek[1] and J. Goldhar[1]

[1]Department of Human Microbiology, Tel-Aviv University, [2] Central Laboratories, Ministry of Health, Jerusalem, Israel, [3] Department of Medical Microbiology and Virology, University of Kiel, Germany.

Expression of virulent factors by pathogenic bacteria is considered to be coordinated in order to provide maximum advantage to the pathogen to eventually cause disease. In the present study we examined the expression of type 1 fimbriae in relation to capsule expression by *Klebsiella pneumoniae* K21a. Capsule formation was monitored by immunoelectron-microscopy using K21a capsule antiserum and expression of type 1 fimbriae was monitored by mannose-sensitive yeast aggregation. The parent *K.pneumoniae* K21a strain was sensitive to an ph/k21a phage, resistant to ph8/30 phage and lacked fimbriae activity. We have obtained three derivatives of the parental strain, all of which were devoid of visible capsule and resistant to the ph/K21a but sensitive to the ph8/30. Two of the nonencapsulated derivatives were spontaneous mutants, selected by either picking of random agar-grown colonies (K21a/8/30) or by picking ph/K21a resistant colony (K21a/PhR). They exhibited type 1 fimbriae activity. In contrast, type 1 fimbriae activity was not detected in the third non-encapsulated derivative (K21a/521) obtained by genetic manipulation of the capsule genes (Kabha *etal.* Infect. Immun. 63:847-51), nor in a phenotype, in which the capsule formation was suppressed by growing the parental K21a strain in salicilate supplemented broth. We speculate that the expression of type 1 fimbriae by the spontaneous non-encapsulated derivatives is due to the selection of clones harboring a putative locus, which negatively co-regulates the expression of these two virulent factors in *K.pneumoniae* K21a.

THE *S. AUREUS* COLLAGEN-BINDING MSCRAMM CONTAINS A CRYPTIC FIBRONECTIN BINDING EPITOPE

Victor Nikolaevi[1*], Cecilia Ryden[1], Amy Schneider[2], Joseph M. Patti[2], and Magnus Hook[2]

[1]Department of Medical and Physiological Chemistry, Biomedical Center, University of Uppsala, Sweden; [2] Center for Extracellular Matrix Biology, Institute of Bioscience and Technology, Texas A&M University, Houston, Texas.

Staphylococcus aureus express cell surface adhesins that mediate microbial adhesion to extracellular matrix components (ECM) within host tissues. We have used the acronym MSCRAMMs (Microbial Surface Components Recognizing Adhesive Matrix Molecules) to describe this family of microbial adhesins. A collagen-binding MSCRAMM (CNA) and fibronectin-binding MSCRAMMs (FnBPA and B) have been purified and the genes encoding these proteins cloned and sequenced. In addition, the ligand binding sites of the individual have also been identified. Recent binding assays with recombinant versions of CNA suggested the presence of a putative fibronectin binding epitope. Polypeptides M17 (CBD 151-297), M19 (CBD 151-318), M31 (CBD 61-343), and M55 CBD (30-529) representing different lengths of CNA were overexpressed in *E. coli* as HIS6-fusions and purified by metal chelating affinity chromatography. M19, M31, and M55 exhibited significant collagen binding activity, whereas M17 did not bind collagen. Interestingly, M17 and M19 also bound intact fibronectin in a dose-dependent manner using an ELISA based assay. Binding of fibronectin to M17 was inhibited with intact fibronectin and proteolytic fragments of 120 kD and 40kD, but not with the amino terminal 29 kD fragment. The fibronectin-binding activity seems to be expressed in staphylococcal CNA since (i) CNA-positive *S. aureus* strain Phillips adhered significantly better to fibronectin-coated polystyrene than the CNA-negative mutant of the same strain and (ii)M19 inhibited the binding of *S. aureus* strain Phillips to fibronectin-coated polystyrene.

---

* Permanent address: Mechnikov's Research Institute for Vaccines and Sera, Moscow, RUSSIA.

ADHESION OF *CANDIDA ALBICANS* MUTANT STRAINS TO EXFOLIATED EPITHELIAL CELLS AND A TISSUE LINE

Z. Roth[1], Z. Altboum[2], H. Sandovsky-Losica[1], and E. Segal[1]

[1]Department of Human Microbiology, Sackler School of Medicine, Tel-Aviv University
[2]Department of Microbiology, Israel Institute for Biological Research, Ness-Ziona, Israel.

Adhesion of *Candida albicans* to host cells is considered a significant step in the development of candidiasis and a fungal virulence factor. *C. albicans* strains may differ in adherence ability, a feature believed to be correlated to their pathogenicity. We, therefore, initiated a study to investigate whether mutant-strains differ in adherence potential from their parent wild-type. We assessed the adherence *in vitro* of

C. albicans CBS562, a strain extensively studied in our laboratory, and two mutant-strains derived from this wild-type. We studied a histidine auxotrophic mutant, SAG5, which was isolated following mutagenesis of the wild-type CBS562 with N nitrosoguanidine (NTG). The second mutant, SAR1, is a respiratory deficient strain - a petite mutant, which was isolated from SAG5 after mutagenesis with NTG.

The adhesion was tested *in vitro* using two systems: 1) adhesion to exfoliated epithelial cells (human buccal cells) that were obtained from healthy volunteers. Adhesion was evaluated microscopically; and 2) adhesion to a tissue line (HaCat Cells), evaluated by a direct ELISA technique. The results indicated a reduced adhesion of the mutant-strains.

EPITOPE MAPPING OF YAD A, THE YERSINIA ENTEROCOLITICA ADHESIN

Yasmin El Tahir and Mikael Skurnik

Turku Centre for Biotechnology, PO Box 123, 20521 Turku, Finland.

Yersinia enterocolotica (*Y.e.*) is an enteric pathogen of animals and humans. YadA is a fibrillar outer membrane protein encoded by the 70-kb virulence plasmid of *Y.e.* and has been associated with important biological functions such as autoagglutination, serum resistance, cell adherence, and binding to extracellular matrix proteins like collagen, fibronectin, and laminin. The YadA fibrils are formed as trimers of identical subunits. Amino acid sequence of YadA reveals hydophobic and hydrophilic domains suggesting that YadA is an amphipathic protein. A molecular weight of 44,439 is predicted from the amino acid sequence of YadA of *Y.e.* serotype 0:3 (Ye03).

To map the functional epitoped of $YadA_{Ye03}$, 96 overlapping peptides (16 residues long) biotinylated to their N-terminus were synthesized (Chiron Mimotopes, Australia). The peptides cover the entire $YadA_{Ye03}$ sequence except the carboxyterminal membrane-anchoring region. The ability of the peptides to bind different target molecules was tested by immobilizing the peptides on strepavidin coated plates. Three monoclonal antibodies (3G12, 2G12, and 2A9) against YadA were tested and a linear immunodominant epitope for all three antibodies was mapped to two overlapping peptides covering $YadA_{Ye03}$ residues 154-173. A polyclonal antiserum from rabbits immunized with live virulent Ye03 bacteria specifically reacted with the most aminoterminal peptide of $YadA_{Ye03}$, but not at all or very weakly with the peptides reacting with the monoclonal antibodies.

FLAGELLAR DISPLAY FOR BACTERIAL ADHESIVE EPITOPES

B. Westerlund-Wikström, J. Tanskanen, R. Virkola, M. Lindberg,[1] J. Hacker,[2] and T. K. Korhonen

Department of Biosciences, Division of General Microbiology, University of Helsinki, Helsinki, Finland; Department of Microbiology, Swedish University of Agricultural Sciences, Uppsala, Sweden[1]; Institute für Molekulare Infektionsbiologie, Universität Würzburg, Würzburg, Germany[2].

In *Escherichia coli* more than 40 genes are involved in the biosynthesis and function of flagella. The flagellar filament is built up of thousands of identical flagellin (FliC) subunits helically arranged around a hollow central channel (1). The filament is terminated by the FliD protein and connected to the hook and basal body by the hook-associated proteins FlgK and FlgL.

We have previously studied adhesive proteins of pathogenic bacteria (2) and in order to determine the binding epitopes in these proteins we developed a flagellar display technique. The method takes advantage of the fact that parts of the FliC variable domain can be deleted and replaced without loss of flagellar polymerization. As a model, chimeric flagella carrying one, two or three fibronectin-binding repeats (D1, D2, D3) of the FnBPA protein of *Staphylococcus aureus* (3) were created and tested for binding characteristics. the chimeric flagella bound soluble as well as immobilized fibronectin and the binding was most efficient with the flagella containing three D repeats of FnBPA. The flagella bound to frozen sections of human kidney and to cultured human fibroblasts in a pattern similar to that shown by a monoclonal anti-fibronectin antibody, indicating that they recognize tissue- and cell-bound fibronectin also.

Chimeric flagellar filaments expressing hundreds of intimately-associated adhesive epitopes offer efficient tools to analyze adhesin-receptor interactions of pathogenic bacteria.

REFERENCES

1. Namba, K., Yamashita, I. and Vonderviszt, F. 1989. Nature 342;648-654.
2. Westerlund, B. and Korhonen, T.K. 1993. Mol. Microbiol. 9;687-694
3. Signäs, C., Raucci, G., Jönsson, K., Limdgren, P. E., Anantharamaiah, G.M., Höök, M. 1989. Proc. Natl. Acad. Sci. USA 86:699-703

DIFFERENTIAL RECOGNITION OF GLYCOLIPID ADHESION MOTIFS WITHIN HUMAN SKIN BY APATHOGENIC AND PATHOGENIC STAPHYLOCOCCI

J. Bünger, G. Sauermann, U. Hoppe, and F. Wolf

Department of Experimental Dermatology, Paul Gerson Unna Skin Research Centre, Beiersdorf AG, Unnastrasse 48, D-20245 Hamburg, Germany.

Healthy human skin is extensively colonized by various apathogenic or putatively pathogenic microorganisms. The healthy cutaneous microflora is mainly composed of members of the micrococcaceae, coryneform bacteria and *P. ovale*, the latter restricted to the scalp. In addition, exposure to nosocomial pathogens poses an additional threat for the skin. Endogenous and/or exogenous disturbances of the meticulously maintained equilibrium, e.g. by medical treatment, radio- or chemotherapy or immunosuppresion induced by various virus infection or intense UV irradiation, result in sometimes dramatic alterations of the cutaneous microflora. In addition, a number of cutaneous diseases are accompanied by bacterial or fungal superinfections of the skin, e.g. seborrhoic eczema and pityriasis versicolor are accompanied by superinfections with *P. ovale*, and in atopic eczema/neurodermatitis approximately 50% of the patients are severely superinfected with the pathogen *S. aureus*. The role of *S. aureus* in the etiology of atopic eczema is an up to now open question; however, as *S. aureus* produces a number of endo- and exotoxins resulting in profound stimulation of the cutaneous immune system, it is obvious that at least aggravation of the disease by *S. aureus* superinfections will be the case. It could be demonstrated that eradication of the pathogen by antibiotic therapy resulted in significant improval of the overall disease. However, due to lacking specificity of antimicrobial agents, the resident microflora is eradicated as well as the pathogen. In order to overcome these limitations, adhesion mechanisms of micrococcaceae belonging to the resident microflora and of pathogenic were investigated with special emphasis on glycolipid adhesion motifs. It was found that *S. epidermidis* and *S. aureus* utilize structurally different sets of adhesion motifs for receptor-driven interaction with human skin. Whereas *S. epidermidis* specifically recognized glycolipids containing the GalNAc-Gal-Gluc core structure, *S. aureus* selectively bound to cholesterol and cholesterol-containing lipids. Based on these findings a number of active ingredients suitable for anti-adhesive therapy of cutaneous infections were developed, circumventing the increasing risk of microbial multiresistance to classical antimicrobial principles. In addition, novel species-specific anti-adhesive agents were discovered directed against *S. aureus*, thus facilitating selective removal of pathogenic staphylococci from superinfected skin without altering the resident microflora of patients. It is proposed that by using anti-adhesive agents targeted against pathogenic staphylococci a significant improval of patients suffering from atopic eczema will occur, combining the benefits of neglectable risk of creating multiresistance and avoiding damages to the natural flora of man. In addition, this strategy opens the way for selective "anti"microbial therapy of infections with nosocomial pathogens increasingly gaining importance in clinical medicine, thus decreasing the risk of nosocomial superinfection and improving survival chances of severely wounded patients in intensive care units.

32

CONCLUDING REMARKS

Itzhak Kahane and Itzhak Ofek

Several reasons have stimulated the need to discuss the topics of this seminar. The former include the fast progress in our understanding of the processes involved in microbial adhesion, the onset of the technology for production of the carbohydrate inhibitors of adhesion and the emergence of the multi-drug resistant strains of microorganisms urging the need for an alternative therapy.

Data presented in the seminar and reported here, have indicated that indeed it is prime time for our discussions, because they support the notion that the design of anti-adhesion approaches and compounds against diverse microbial species is promising. Although *E. coli* continues to be a target for such therapy, the anti-adhesion approach is now evaluated for many microorganisms including fastidious bacteria, mycoplasmas, viruses, fungi and protozoa.

Moreover, several important issues have been discussed to improve the efficacy of the treatment and have passed like a golden thread in many presentations throughout the seminar. For example, the fact that many microorganisms produce multiple sorts of adhesins, each specific for a distinct receptor, must be considered and may require a mixture of adhesins' inhibitors. We may improve anti-adhesion therapy by designing an approach or agents which influence the balance of the regulatory systems -e.g. stimulons, modulons and regulons, that sense different environmental factors including oxidative stress, acid stress and starvation which alter gene expression including those required for the production of adhesins. The possibility to switch the adhesins to On/Off position may add another aspect in the design of the strategy to combat the disease. It may be better to choose the right conditions to keep the organisms in the Off conditions.

Other factors that should be considered in the design of new inhibitors relate to the consequences of adhesion of microorganisms to host cells. For example, several studies indicate that adhesion influences microbial growth and provides the necessary access to host cell receptors that induce cytokines production leading to the inflammatory response .

An extremely interesting aspect, that will hopefully unravel in future studies, is the sequence data indicating resemblance of some adhesins to that of cytoplasmic enzymes though that they have membrane leading anchor.

One can see, therefore, that the weaknesses in our knowledge will deserve broad and basic studies in order to design new agents for the suggested new strategies to combat microbial infectious diseases.

INDEX